Ecology of Fragmented Landscapes

Ecology of Fragmented Landscapes

SHARON K. COLLINGE

Foreword by Richard T. T. Forman

The Johns Hopkins University Press
Baltimore

© 2009 The Johns Hopkins University Press
All rights reserved. Published 2009
Printed in the United States of America on acid-free paper
2   4   6   8   9   7   5   3   1

The Johns Hopkins University Press
2715 North Charles Street
Baltimore, Maryland 21218-4363
www.press.jhu.edu

Library of Congress Cataloging-in-Publication Data
Collinge, Sharon K.
Ecology of fragmented landscapes / Sharon K. Collinge ; foreword by
Richard T. T. Forman.
p. cm.
Includes bibliographical references and index.
ISBN-13: 978-0-8018-9138-0 (hardcover : alk. paper)
ISBN-10: 0-8018-9138-8 (hardcover : alk. paper)
1. Fragmented landscapes.   2. Landscape ecology.   I. Title.
QH541.15.F73C65   2009
577.27—dc22   2008021040

A catalog record for this book is available from the British Library.

*Special discounts are available for bulk purchases of this book. For more
information, please contact Special Sales at 410-516-6936 or
specialsales@press.jhu.edu.*

The Johns Hopkins University Press uses environmentally friendly
book materials, including recycled text paper that is composed of at
least 30 percent post-consumer waste, whenever possible. All of our
book papers are acid-free, and our jackets and covers are printed on
paper with recycled content.

Contents

Foreword

Imagine getting a beautiful multicolor, with-everything, chopped-up pizza for a special person. Suddenly, while crossing the campus, it slips out of your grasp—and you are left pondering the pieces. Though devastated, you happen to notice the overall pattern of fragments, plus their many sizes, shapes, color combinations, and spatial arrangements. That pizza and the pattern on the ground provide clues to the pages ahead. The big picture, like the view from a plane or a satellite, is the land around us and how to understand it, and perhaps even how to mold its face for the future.

Given its complex topic, this book is surprisingly quite readable. The author takes us through the key ecological subjects as if we were sharing a conversation. Ideas and theories are delightfully clear, from intellectual history to today's understanding to glimpses of tomorrow. The ecology of fragmented landscapes comes alive.

Now back to the pizza, a whole one with no disaster this time. If you eat the left half, the loss, like eliminating the western half of a forest, leaves a large unfragmented piece. With land, habitat loss, degradation, and fragmentation are major processes producing different results. Habitat loss is the giant, with the degradation of remaining habitat coming right behind, yet fragmentation in landscapes is conspicuous and important worldwide.

Consider the two big fragmentation processes, roads and development. A wooded landscape crisscrossed by logging roads, or "rides" in an English wood, is fragmented for many tiny animals but not for most birds, depending on whether the corridor slices create major barriers to movement. If farmland or a residential development extends along the roads, the former forest is now fragmented for most animals, which often remain as smaller populations with demographic and genetic problems. Cities can spread outward in concentric zones, or around satellite cities, or along transportation corridors, or in dispersed sites (sprawl). The first two patterns involve no fragmentation, the third explicitly fragments the land, and the fourth does so when the development sites coalesce. New highways also split human communities. In addition, the rapid expansion of development and roads, arguably a more urgent problem than climate change, threatens already fragmented land, even a valuable emerald network of large, connected, natural patches. So, how can we mold the land so both nature and people thrive long term?

Unfragmented is better than fragmented in most cases, including for beautiful pizzas, but sometimes the opposite is true. Fragmented habitat often slows the spread of many pests and diseases, yet it may accelerate this process for others. Small woods contain few rare species, but they have many common species packed together, to the delight of families observing wildlife in a park.

Natural populations and communities are at the forefront of this book. However, fragmentation also strongly affects the network structure and function of streams and rivers, soil erosion, groundwater quantity and quality, and accumulations of nutrients and chemicals. Curiously, we haven't yet figured out what the predominant widespread fragmentation patterns are, and we still can identify only a very small number of better and worse ones. Overall, good patterns include a few large natural patches, many small patches or corridors surrounding a large patch, and an elongate cluster of small patches between two large ones. A pattern with just a few small natural patches remaining is bad, but what about all the other fragmentation patterns in landscapes?

Visualize a natural area next to a built area. Movements and flows between them can have four effects: the built area either negatively or positively affects the natural area, or vice versa. Which has the greatest and which the least effect? The negative impact of a built area on a natural area is greatest, due, for example, to movements of domestic animals, stormwater pollutants, traffic noise, and even people themselves. The positive effect of a built area on a natural area is the least evident, and overall both the positive and negative effects of nature on built areas are intermediate. In short, fragmenting nature with built areas also degrades the natural area.

These interactions between adjacent land uses also highlight the value of the book's landscape- rather than patch-centric perspective on fragmentation. Landscapes are a highly heterogeneous mosaic of many habitat types and land uses. Indeed, when ecologists "climbed from the sea onto the land," the model of island biogeography was gradually replaced by the patch-corridor-matrix model, in part because a patch is surrounded by diverse habitats, with each being a source of effects on the patch, a species source, and differentially suitable for movement between patches. Also, the patch edge-to-interior ratio and the roles of generalist edge species and specialist interior species are important mechanisms controlling species richness on patches of different sizes and, especially, shapes.

Do patch sizes, shapes, and their arrangement help predict the future? Large natural patches seem more likely to be perforated, dissected, or shrunk, whereas small ones could easily disappear. Convoluted patches are vulnerable to fragmentation, dissection, and shrinkage, while it is probable that square ones will persist unchanged. A group of small patches is more likely to disappear if they are clustered rather than dispersed. Thus natural populations and communities are strongly affected by such pervasive, changing land patterns.

Still, we know surprisingly little about the diverse sequences of patterns in landscapes that are being fragmented. What are the ecologically optimum sequences when one starts with different patterns, and which are optimal, for example, for residential development versus land restoration? Indeed, knowing an optimum sequence, we could usefully serve society by highlighting the best and worst locations for the next local change, be it a mall or a park.

The principles arrayed in the pages of this book are directly usable in bioconservation, transportation, agriculture, recreation, forestry, range management, wildlife management, and water resources—all fields dealing with the land. What is perhaps of greatest significance is their linkage with urban land planning, where the focus has been on providing jobs, housing, transportation, and economic development. A promising future depends on natural systems and their human uses at a landscape or regional scale taking center stage, alongside these traditional needs of society.

The treasure in your hand will reveal rich food for thought and frontiers for research. I especially valued the entrées for diverse types of experiments, species movement patterns, the roles of species interactions, restoration approaches, and ecological land planning. Its solid grounding in the evolution of ideas and theories will make the book useful for a long time as a reference on my shelf. Unraveling good science of direct use to society is a joy to see. So—read, absorb, and enjoy.

Richard T. T. Forman
PAES Professor of Landscape Ecology
Harvard University

Preface

Dismantle is defined as to "disassemble" or "take apart." When I originally conceived this book, I thought of the phrase "dismantling nature" because it seemed to aptly describe a fundamental process occurring around us. In our zest for transforming lands for human use, we have created fragmented landscapes in which many natural communities and ecosystems are effectively being taken apart. The structural pieces—such as populations and species of plants, insects, birds, mammals, reptiles, and fish—are declining worldwide. The functional aspects of our ecosystems are also being lost, including pollination, seed dispersal, the decomposition of dead organic matter, carbon sequestration, and water filtration. But perhaps this process of disassembly is not irreversible. My main motivation for writing this book was to summarize the abundant scientific literature on the ecology of fragmented landscapes, so that students and practitioners could apply this knowledge toward efforts to protect and re-assemble natural ecosystems, thereby maintaining both species and ecosystem services.

Spatial variation in landscape structure, also called *habitat patchiness*, arises from naturally occurring environmental heterogeneity (for example, variation in soil chemistry or soil moisture), as well as from anthropogenic causes (such as forest clear-cutting or agricultural land conversion). The study of habitat loss and fragmentation provides an important synthesis for ecology, because it unites fundamental ecological theories and concepts with an extensive body of empirical literature. It also provides important links to environmental applications, including conservation, restoration, and planning. There are literally thousands of papers that have been published in the past 40 years on the ecology of fragmented landscapes. An exhaustive review of this body of work would make a book that is much too long to read, so my intention was to synthesize the key areas of research and emphasize what we know and how we got here. I have undoubtedly omitted some important works in the process of selecting papers for review, but I have made every attempt to discuss the most relevant concepts and findings.

Several excellent papers on the topic of habitat loss and fragmentation have been published in recent years, including those by McGarigal and Cushman (2002), Fahrig (2003), and Ewers and Didham (2006). A recent book-length review by Lindenmayer and Fischer (2006) provides an outstanding summary of scientific research in this field. Three volumes focused specifically on connectivity and conservation appeared in 2006—A. Anderson and Jenkins, Hilty et al., and Crooks and Sanjayan—suggesting that a synthesis on this topic was indeed due.

This book differs from previously published works in that it not only provides a synthetic review of theory and empirical research in ecological fragmentation, but also highlights the application of research findings to the conservation of biological diversity, ecological restoration, and ecological planning. These topics reflect my own research interests and experiences over the past 20 years. My graduate training included emphases on population and community ecology, landscape ecology, and landscape architecture and planning, and I have continued to integrate those perspectives in my research and teaching.

I am grateful to the many students who have taken courses in conservation biology, landscape ecology, and conservation planning at Harvard University, the University of California–Davis, and the University of Colorado at Boulder. They have helped clarify my thinking about these topics and urged refinement of my explanations of concepts into an understandable form. I intend for this book to be read and used by upper-level undergraduates and graduate students in the fields of conservation biology, ecology, environmental studies/environmental science, and landscape architecture and planning. The book could also be used in courses such as conservation biology, conservation planning, landscape ecology, and population and community ecology, as well as graduate seminars or discussion groups focused on ecology, conservation, restoration, or planning.

I am indebted to many people for their support, assistance, and good humor as I prepared and completed a rather enormous project. I have had the good fortune of remarkable teachers and mentors in my professional career, and I especially thank Svaťa Louda and Richard Forman for their brilliance and their advice, encouragement, and confidence in me. I, in turn, have had the pleasure of being a teacher and mentor to an exceptional group of undergraduate and graduate students, each of whom has brought a unique and meaningful perspective to my own learning process—these people include Abby Benson, Jory Brinkerhoff, Dave Conlin, Sara Jo Dickens, Fritz Gerhardt, Whit Johnson, Kimberly Kosmenko, Laura Makar, Amy Markeson, Jaymee Marty, Katherine McClure, Jenny Ramp, Becky Rawlinson, and Sue Rodriguez-Pastor.

Staff members of Boulder's land management agencies have provided a supportive and stimulating research environment for my work in Boulder grass-lands, and so I thank Mark Brennan, Mark Gershman, Brian Pritchett, Lynn Riedel, and Heather Swanson for their collaboration. I've also benefited over the years from friends and colleagues in Boulder, as well as more distant places, who have helped to shape my thinking about ecology, conservation, and restoration, including Susan Beatty, Mary Cadenasso, Jonathan Chase, Tom Crist, Elizabeth Crone, Kevin Crooks, Brent Danielson, Dan Doak, Lenore Fahrig, Nick Haddad, Steven Handel, Susan Harrison, Karen Holl, Gary Huxel, Margaret Palmer, Todd Palmer, Kevin Rice, Mark Schwartz, Dave Theobald, and Kim With.

During the writing of this book, I have enjoyed the company, encouragement, and helpful advice of several people. Pete and Chris Coppolillo facilitated my in-credible journey to Tanzania for my sabbatical leave in fall 2005, where the be-ginnings of this book were written near the banks of the Great Ruaha River. Chris Ray has been a close colleague and friend for many years, and I have learned vol-umes from her about ecology, modeling, and conservation. Several colleagues and friends reviewed chapters or provided inspiration, supporting material, and good ideas, including Dave Armstrong, Carl Bock, Emilio Bruna, Ellen Dam-schen, Kendi Davies, Marcel Holyoak, Martha Hoopes, Pieter Johnson, Erik Jules, Claire Kremen, Andy Martin, Valerie McKenzie, and Brett Melbourne. My editor at the Johns Hopkins University Press, Vince Burke, had the confidence to invite me to embark on this project, and has patiently, kindly, and persistently guided me through the process.

Close friends joined with me to enjoy good food and laugh often, and so my thanks to the "gals," Kimman, Connie, and Jane, for putting it all into perspec-tive. My biological family—my Mom and Dad, Lola Jane and Irwin Collinge, and my siblings, Mike, Linda, and Judy, and their spouses and children—provided love, laughter, and lots of encouragement. I imagine that my Dad would have said "Way to go!" when this book was finally finished. And my two biggest rounds of applause go to Jake, who I hope will continue to live in a world of wonder and bi-ological wealth; and to my best fan, Joan Mary Laubacher, who inspires me the most and has been here for the whole thing.

Ecology of Fragmented Landscapes

CHAPTER ONE

Introduction

Framing the Issues

Habitat fragmentation is widely regarded as a—if not *the*—
central issue in conservation biology.

—*John A. Wiens (1996)*

My home town of Emporia, Kansas, sits almost at the geographic center of the contiguous United States and is also smack in the middle of the tallgrass prairie ecosystem of the North American Great Plains. When I was growing up, it seemed as though I was surrounded by acres and acres of grassy pastures, the flowering stalks of the grasses reaching well above my head in particularly wet years. Endless grass that stretched for miles, or so it seemed. As a teenager, I often longed for the more dramatic Rocky Mountains to the west, but little did I know that I was living in a very special place. Not only was this the largest extent of tallgrass prairie that remained in Kansas, and in the United States, but it was the greatest area of tallgrass prairie left in the world! Only when I got to graduate school and began learning about conservation biology did I understand the unique value of the place where I was born and raised. That's when I was startled to see the squiggly blotch on conservation maps that depicted the outline of the Kansas Flint Hills and to find out that less than 1% of the original tallgrass prairie ecosystem remained intact. Most of it was converted to cornfields in the 19th and 20th centuries to support the burgeoning agricultural industry, resulting in the corn belt within the states of Illinois, Iowa, Indiana, and Ohio, as well as parts of South Dakota, Nebraska, Kansas, Minnesota, Wisconsin, Michigan, Missouri, and Kentucky.

The midwestern United States is certainly not the only landscape that has been transformed over the past 200 years. Globally, about 50% of tropical dry forests were converted to other uses by 1950, and another 10% was lost between 1950 and 1990. Nearly 70% of the native cover of temperate grasslands disappeared by 1950, and an additional 15% has vanished since then. In the last two decades, 35% of all mangroves have been lost; in the last several decades, 20% of known coral reefs have been destroyed and another 20% degraded. Most of the world's biomes have experienced a 20%–50% conversion to human use. More land was turned into cropland in the 30 years after 1950 than in the 150 years between 1700 and 1850. These figures—from the most comprehensive global analysis of the status of the world's natural lands and biological diversity (Millenium Ecosystem Assessment 2005)—show that rates of land conversion exceed 50% for half of the world's biomes, and projections indicate further losses in the next 50 years. Regional analyses can be even more grim, depending on how land cover types are categorized. For example, Hoekstra et al. (2005) defined *critically endangered* ecoregions as those where habitat conversion exceeded 50% and where the ratio of habitat converted to habitat protected (the *conservation risk index*) was greater than 25%. They identified 64 such ecoregions, including (1) the Atlantic coastal forests of Brazil, Argentina, and Paraguay, (2) Mediterranean forests, woodlands, and scrub, and (3) the Pacific temperate rainforests of western North America.

To most readers, it is no surprise that these extensive and intensive land use changes have associated environmental costs. Habitat destruction and degradation are the leading causes of declines in biological diversity worldwide (Wilcove et al. 1998). Further, the loss and fragmentation of natural landscapes diminishes their capacity to provide vital ecosystem services (Daily and Ellison 2002; Kremen et al. 2002; Millenium Ecosystem Assessment 2005). As we transform the planet to fill our needs, we are leaving behind a legacy of scattered remnants of native habitat surrounded by agricultural fields, shopping centers with vast parking lots, and clear-cut forests. The scope of this problem is inherently local, in that land use change happens every day, based on decisions made by individual people going about their livelihoods. But these collective decisions mean that everywhere we look across the planet, we see some form of human impact. So these issues extend beyond our own towns, forests, and pastures to the entire Earth.

Habitat loss and fragmentation are not the only environmental challenges, of course. Many other such issues loom large, including climate change, invasive species, disease, and overexploitation. And habitat loss and fragmentation can interact synergistically with these other factors to produce harmful effects on species and ecosystems (e.g., Ewers and Didham 2006). For example, overexploitation may be strongly influenced by landscape change, since hunters are likely to have easier access to fragmented habitats or habitats bisected by roads (Peres 2001; Tabarelli, Cardosa da Silva, and Gascon 2004). Likewise, invasive species may be more likely to succeed along the edges of fragmented habitats, causing further negative impacts on native species (Hansen and Clevenger 2005; Lockwood, Hoopes, and Marchetti 2007). Habitat loss and fragmentation are also likely to limit the migrations of species in response to climate change (Schwartz, Iverson, and Prasad 2001; J. Hill et al. 2006). This book will primarily focus on loss and fragmentation effects, but I discuss these sorts of synergisms where and when they are particularly relevant. I would argue that each of these environmental challenges interact with and are ultimately affected by patterns of land use change, so understanding the ecological aspects of fragmented landscapes is also crucial to resolving these other impacts.

This book is designed to help the reader appreciate the collective body of research that clarifies the consequences of these sorts of landscape changes on populations and communities. It presents a rich font of scientific information being used in many hopeful activities—such as ecological restoration and conservation planning—that are helping to reverse declines in biodiversity. My challenge for the reader is to critically evaluate the accumulated knowledge presented here and use it to move forward to devise creative solutions that will help to alleviate the threats of habitat loss and fragmentation for biological diversity and ecosystem services.

LOSS VERSUS FRAGMENTATION

The term habitat fragmentation has now been used so broadly in conservation biology that the meaning intended by its original usage has become somewhat ambiguous (e.g., Haila 2002). It is worth clarifying the differences between habitat loss and fragmentation to encourage more accurate knowledge and use of these terms. *Habitat loss* occurs any time a piece of land is converted from its current state to some other land use or land cover type, such as when a stand of old-growth forest is cleared to make way for a housing development. *Loss* in this sense refers to the fact that the overall area of native habitat has been reduced. So, if we started with 100 ha of forest and now have only 10 ha, we have lost 90% of that habitat. *Habitat fragmentation*, in contrast, denotes a particular spatial process of land conversion. In the strict sense of the word, *fragmentation* refers to breaking a whole into smaller pieces while controlling for changes in the amount

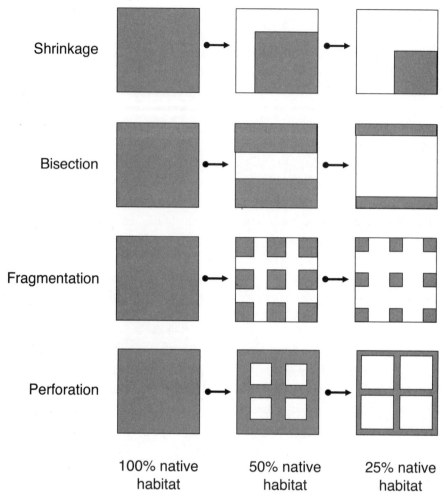

Shrinkage

Bisection

Fragmentation

Perforation

100% native
habitat

50% native
habitat

25% native
habitat

Figure 1.1. Four different spatial processes of landscape change. In each scenario, a block
of continuous habitat is reduced to 50% and then 25% of its original area over time.
The top row represents *shrinkage*, where there is reduction in habitat area but no
fragmentation or subdivision. The second row is *bisection*, where the original habitat
is initially divided into two equal-sized areas that shrink in size. The third row is
fragmentation, where the initial habitat is divided into nine equal-sized patches that
shrink in size. The bottom row is *perforation*, where the native habitat is perforated by
the transformed habitat. In each scenario, the dark gray areas are native habitat and the
white areas are transformed habitat. Modified from Forman (1995) and redrawn from
Collinge and Forman (1998).

of habitat (Forman 1995; Collinge and Forman 1998; Fahrig 2003; Ewers and Didham 2006). In one conceptualization of different land conversion sequences (fig. 1.1), all of the sequences have the same amount of habitat in each stage, but the bisection and fragmentation sequences are fragmented in the strict sense, while the shrinkage and perforation sequences are not, because they continue to maintain contiguous areas no matter how much habitat remains. But even the bisection and fragmentation sequences, like those for shrinkage and perforation, have undergone habitat loss. In actual landscapes undergoing real processes of land conversion, it is impossible to have habitat fragmentation without habitat loss, because the creation of fragments requires the conversion of part of the original land area. But it *is* possible to have loss without fragmentation (shrinkage and perforation in fig. 1.1; Forman 1995; Fahrig 2003).

Why is it necessary to be so concerned about this distinction between habitat loss and fragmentation? It's not just because scientists are picky people. There are at least two reasons why it is vital to distinguish between their effects. It becomes meaningful when we want to understand the specific mechanisms responsible for shifts in biological diversity in relation to landscape change, and it is also critical when we consider the most effective strategies for reversing the trends of biodiversity losses. One particular species may decline primarily due to the loss of its preferred habitat, so as that habitat shrinks from 100% to 50% of its original area, the abundance of that species may correspondingly diminish. But another species may be quite sensitive to fragmentation, perhaps due to an enlargement of the perimeter-to-area ratio of small versus large habitat patches and its accompanying increase in the amount of edges. This second species may be more responsive than the first to the spatial configuration of habitat patches, while the first species may be more responsive to the overall amount of habitat available. Knowing the difference is crucial in devising appropriate conservation and management strategies to prevent declines of these species. Because land conversion patterns vary in their spatial configurations, they may also differ substantially in their influence on important ecological processes (Franklin and Forman 1987; Wiens et al. 1993; S. Harrison and Fahrig 1995; Kareiva and Wennergren 1995; Fahrig 1997, 2003; S. Harrison and Bruna 1999; McGarigal and Cushman 2002; Lindenmayer and Fischer 2006).

This issue of habitat loss versus fragmentation was extensively examined by Lenore Fahrig (2003), and her synthesis goes a long way toward clarifying the confusion that exists in designing and interpreting studies of landscape change. Fahrig reviewed the vast quantity of literature on habitat fragmentation (in 2002 she counted over 1600 articles that contained the phrase "habitat fragmentation";

there were nearly 5000 at my last count) and made two key observations. First, she noted that many research efforts involve measurements of fragmentation at the patch scale, not the landscape scale. Second, a considerable number of studies are designed in such a way that it is not possible to distinguish between habitat loss and fragmentation per se. As an example of the initial problem, quite a few of these efforts measure species' responses to changes in patch size but do not take into account the overall amount of habitat available to them in the broader landscape. But landscapes that have larger patches may also be ones in which there is a greater amount of habitat in general. If the researcher fails to recognize the relationship between local patch size and the quantity of available habitat, then conclusions based solely on patch size may actually reflect species' responses to the total amount of habitat.

To address the second problem, Fahrig provided some guidance for "unconfounding" the confounded. She suggested that one way to disentangle the effects of habitat loss and fragmentation would be to experimentally vary habitat spatial configuration while maintaining the amount of habitat, as some researchers have done in fine-scale field studies. Another way would be to statistically control for the quantity of habitat when measuring landscape fragmentation. Thus, Fahrig argued, to measure the effect of fragmentation per se, researchers must control for the effect of habitat loss either experimentally or statistically (p. 499). Ideally, the most statistically robust way to study fragmentation in real landscapes would be to select a number of fragmented landscapes (say, 5–10) and pair them with the same number of intact landscapes that were similar in every possible physical and biological parameter except for the spatial arrangement of habitat. In that way, response variables—such as species richness, population abundance, species interactions, or genetic diversity—could be measured in these two types of landscapes and compared. But this approach is virtually logistically impossible for a study that takes place outdoors, so this tactic has been used largely in microcosm experiments in the laboratory (see chapter 4) and in simulation models (see chapter 9). Because most field studies have been unable to examine the whole, they tend to study the parts—for example, looking at the effects of particular components of landscape change and including such factors as variation in patch size, isolation, edge effects, and landscape context.

To summarize, *anthropogenic landscape change* (the conversion of native landscapes to other land uses or land cover types) involves both habitat loss and habitat fragmentation. But in most field-based research, it has been difficult to disentangle the separate effects of habitat loss versus habitat fragmentation. At the end of her 2003 paper, Fahrig suggested that two generalizations can be made from studies of habitat fragmentation: first, habitat loss has large, negative effects on biodiversity; and second, habitat fragmentation, in the strict sense, has rather weak effects on biodiversity, and these effects can be both positive and negative.

Even though most studies of habitat fragmentation published in ecological journals are actually studies of habitat loss, making it difficult to infer the impacts of fragmentation per se, their results do have implications for species' responses to landscape fragmentation. For example, later chapters will discuss a famous, long-term, experimental study of forest fragments in the Amazon rainforest (W. Laurance et al. 2002). This work involved the creation of different-sized forest fragments surrounded by areas cleared for cattle grazing. The researchers comprehensively studied the changes in species richness for many taxonomic groups in response to changes in forest area, comparing data from a continuous forest to that from small, medium, and large fragments. The results generally show that patch area had strong positive effects on species richness, and patch edges had strong negative effects on species' abundance patterns. Because the amount of edge habitat increases in fragmented landscapes, the findings of this comprehensive study suggest that edge-sensitive species will likely be detrimentally affected by forest fragmentation, which is an important contribution toward understanding the impacts of habitat loss and fragmentation. But, as Fahrig (2003) indicates, the Amazon-forest-fragments study is technically a study of forest loss, not fragmentation.

Throughout this book, I've tried to maintain a strict usage of the term fragmentation. Terms such as *landscape change, habitat loss and isolation,* or *habitat loss and fragmentation* refer to the patterns and processes of anthropogenic changes in land cover and land use. The phrase *fragmented landscapes* generally pertains to areas that have experienced both habitat loss and isolation. *Fragmentation* by itself is employed where specific mention is made of experimental studies where habitats have been subdivided (i.e., fragmented), but where their overall area was kept constant. These terms are used as consistently as possible, so that this discussion will be enlightening rather than confusing.

NATURALLY OCCURRING PATCHINESS VERSUS HUMAN-CAUSED FRAGMENTATION
Origins of Patchiness

If you take a close look, you'll notice that landscapes are patchy. Palm oases are splashes of green in desert landscapes, occurring where water rests close to the surface. Light gaps reveal breaks in tropical forest canopies where trees have blown down during wind storms. Spatial and temporal variation in the distribution and abundance of vital resources, as well as in geological and ecological processes, result in landscape heterogeneity, often called *habitat patchiness* (e.g., in Wiens 1997). In many terrestrial systems, patchiness involves spatial variation in bedrock, soils, nutrients, or water that affects the distribution of plant species, which, in turn, at least partially determines animal distributions. In marine or freshwater systems, patchiness in species' distributions may be the result of such factors as differences in substrate type, water depth, or period of inundation. Natural disturbances, including fires, floods, disease outbreaks, wind storms, and wave action, also create patchiness by altering the structure of populations, communities, and ecosystems and by causing changes in resource availability or the physical environment (Levin and Paine 1974; Delcourt, Delcourt, and Webb 1983; Pickett and White 1985). In addition, species may generate their own patches, independent of any underlying environmental heterogeneity, by their clumped dispersal patterns (e.g., *spatial pattern formation* in Tilman and Kareiva 1997).

Ecologists have a long history of observing and studying the underlying causes of patchiness (Watt 1947; Levin and Paine 1974; Wiens 1976). Concepts and methods from landscape ecology, which focus specifically on the causes and consequences of spatial heterogeneity for ecological processes, including disturbance (Forman and Godron 1986; Urban, O'Neill, and Shugart 1987; M. Turner 1987, 1989; Wiens et al. 2006), have increasingly been mainstreamed in ecology in general. The bulk of this book features studies of the patterns and outcomes of human-caused landscape heterogeneity, rather than naturally occurring patterns and processes. This dichotomous categorization is a convenient way to distinguish types of disturbances, but it is important to recognize that in most—if not all—landscapes, natural patchiness and human activities interact to create the spatial patterns that we see (e.g., Foster, Fluet, and Boose 1999). As a basis for understanding the impacts of anthropogenic landscape change on natural systems, however, it is useful to briefly review here what is known about the ecological characteristics of naturally patchy systems.

Differences between Patches

Given that nature is inherently patchy, some may wonder why anthropogenic landscape changes that increase habitat patchiness are necessarily a bad thing.

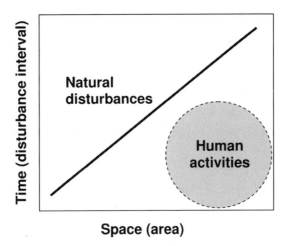

Figure 1.2. Conceptualization of the relationship between spatial extent and occurrence interval for natural disturbances and human activities. Natural disturbances (represented as falling along the solid line)—such as fires, pathogen outbreaks, or floods—that extend over small areas tend to occur at frequent intervals (lower left end of solid line), whereas those that cover large areas tend to occur infrequently (upper right end of solid line). Human activities, such as clear-cut forestry or industrial agriculture, tend to occur over broad spatial scales relatively frequently (light gray circle in lower right part of graph). Redrawn based on figures and text in Delcourt, Delcourt, and Webb (1983) and Urban, O'Neill, and Shugart (1987).

For example, does it matter if industrial forestry produces landscapes filled with small patches of forest surrounded by clear-cuts, if we know that fires in these same ecosystems can result in small patches of forest surrounded by burned areas? To answer that question, it would be helpful to know something about the structural and functional qualities of patches and landscapes that are naturally patchy versus those that are human modified. Unfortunately, specific differences in species' responses to naturally patchy versus human-fragmented landscapes have rarely been compared directly, so it is difficult to draw robust conclusions. However, the spatial and temporal scales of natural disturbances versus human activities have been explored in some detail. For example, Delcourt, Delcourt, and Webb (1983) presented a hierarchical framework to describe what they called the space-time domains of factors that influence vegetation dynamics. The key concept from this framework was that fine-scale events that shape native vegetation communities tend to occur with relatively high frequency, whereas events that occur over relatively broad spatial scales occur rather infrequently (fig. 1.2). The authors pointed out, however, that some human activities tend to be exceptional in that they occur over broad areas with relatively high frequency, which places them outside of the typical space-time domain.

This notion of hierarchical space-time patterns in terrestrial landscapes was elaborated by Urban, O'Neill, and Shugart (1987), particularly with regard to the impacts of human activities. These authors argued that anthropogenic activities influence landscapes via four particular mechanisms, each of which may pose particular challenges for native species. First, human activities re-scale landscape patterns in space and time. For instance, fire suppression alters fire regimes so that when fires do eventually burn, they tend to be larger and hotter then they would have been in the absence of suppression. Because the characteristics of these fires differ from those experienced in the historical context of a community, the authors argue that tree mortality may be higher, and regeneration dynamics may be slower, than under more natural conditions. Second, human activities re-scale natural regions by imposing barriers—such as roads, pipelines, buildings, and dams—that retard the flow of species, disturbances, nutrients, or materials across landscapes. These newly bounded habitat fragments may therefore be too small to encompass natural disturbance regimes or dispersal patterns, thereby reducing habitat suitability for some species. Third, human activities may disrupt ecosystems in ways that fall outside the range of space-time scales experienced under natural disturbance regimes. For example, cultivation associated with industrial agriculture tends to occur over large areas with high frequency, which violates the typical natural regimen of small events happening with high frequency and large events with low frequency. Fourth, human activities may remove the rather intricate internal patch structure of natural habitats. As the authors point out, naturally occurring forests tend to have tree-fall gaps, fallen logs, and vertical stratification of vegetation, whereas human-created forest fragments are often much more homogeneous.

In addition to these differences in the spatial and temporal scales of natural disturbances and human activities, we can make some guesses as to the likely distinctions in natural versus human-modified landscapes. Structurally, naturally occurring and anthropogenic fragments may vary in their shapes, their degree of contrast with their surroundings, and the magnitude of their edge effects. For example, human-created fragments tend to have straighter boundaries, whereas natural disturbances create more sinuous boundaries. The contrast is often greater between human-created patches and their surroundings than it is for naturally occurring patches (Forman 1995), which may result in less pervasive edge effects in the latter (see chapter 5). As fragmentation proceeds, the size distribution of the fragments becomes increasingly dominated by a large number of small patches and fewer and fewer large patches (e.g., see Harris 1984).

These differences in landscape structure are likely to alter the functional qualities of human-fragmented landscapes. In general, human-induced loss and fragmentation result in a reduction of habitat area and connectivity, and species may not readily adapt to these changes, especially if they have evolved in the context of large, continuous habitats. In most cases, there is probably insufficient time for species to adapt behaviorally or physiologically in order to successfully cope with these novel environmental conditions. However, because the study of species' responses to broad-scale landscape change is relatively recent, it is not yet clear whether species will be able to evolve and overcome problems associated with these altered landscapes in the long term, or if extinction is inevitable for them (Ewers and Didham 2006).

LESSONS FROM PATCHY HABITATS

Despite the differences in naturally patchy versus human-modified landscapes, research on population and community dynamics from a range of naturally patchy landscapes offers some clues for understanding species' responses to human-caused landscape changes. Studies of organisms that occupy spatially discrete habitats include the classic work of MacArthur and Wilson (1967) on oceanic islands (see chapter 2), as well as books, essays, and papers on pikas on talus patches in western North America (Peacock and Smith 1997; Peacock and Ray 2001); plants and invertebrates in vernal pools in Mediterranean climates (Holland and Jain 1981; J. King, Simovich, and Brusca 1996; Grillas et al. 2004; Gerhardt and Collinge 2007; pictured in fig. 1.3); invertebrates in rock pools (Romanuk and Kolasa 2004), pitcher plants (Kneitel and Miller 2003), and tree holes (Kitching 2001); multiple plant and animal species in limestone glades in Missouri (Van Zandt et al. 2005); and comprehensive long-term research on populations of the Glanville fritillary and its natural enemies in dry meadows of the Åland Islands of Finland (Hanski 1999).

To illustrate the relevance of research in naturally patchy systems for human-fragmented landscapes, there is a noteworthy collection of papers by Susan Harrison and colleagues that have focused on plant and animal populations and communities that inhabit patchily distributed serpentine soils in northern California (fig. 1.4). This body of research has convincingly shown that habitat patchiness influences butterfly population dynamics, patterns of plant species richness and composition, plant-pollinator interactions, and patterns of invasions by exotic

Figure 1.3. An example of a naturally patchy habitat, an ephemeral (temporary) wetland during the aquatic phase, near the Tour du Valat Research Station, 30 km south of Arles in southern France. Photo by S. K. Collinge.

species (S. Harrison, Murphy, and Ehrlich 1988; S. Harrison 1989, 1997, 1999; S. Harrison, Rice, and Maron 2001; Wolf and Harrison 2001; S. Harrison et al. 2006).

Serpentine soils in California are the surface expressions of complex geologic processes involving the ancient collision of tectonic plates, but the relevant point here is that these rocky outcrops tend to be patchily distributed across the landscape, have unusual soils, and support a relatively specialized flora and fauna (Kruckeberg 2006). Harrison's dissertation research on the population dynamics of the Bay checkerspot butterfly (*Euphydryas editha bayensis*), which occurs on serpentine patches, is often cited as a textbook case of metapopulation dynamics (S. Harrison, Murphy, and Ehrlich 1988; S. Harrison 1989; see also chapter 2). In her study of butterfly populations inhabiting scattered serpentine outcrops, one large serpentine patch supported thousands of butterflies, whereas several small serpentine patches located near the large patch supported tens to hundreds of butterflies. Small patches even farther from the large one were not even occupied by the butterflies, despite containing apparently suitable habitat for them. This

Figure 1.4. A non-serpentine meadow (foreground, lighter gray) and a serpentine out-crop (small hilltop in background, darker gray) in Lake County, California. Photo by S. K. Collinge.

pattern of patch occupancy appeared to reflect the limited dispersal ability of but-terflies (S. Harrison 1989). Patches far from the large serpentine outcrop were not easily colonized by the butterflies and so were unoccupied, but nearby patches were easily accessible to them. Harrison described these spatially struc-tured, local populations as a *mainland-island metapopulation*, by analogy to sys-tems where small island populations in a lake or ocean exist near a mainland (S. Harrison 1991; S. Harrison and Taylor 1997; see also chapter 2). In collections of populations that vary substantially in size, small local groups may occasionally disappear, but the entire collection does not become extinct, due to the persis-tence of the large mainland population. This finding contributed greatly to a clarification of the dynamics of such populations, including those that become subdivided or fragmented due to human activities.

Spatially structured populations occur within the context of species assem-blages, and the question of how landscape spatial structure influences species di-versity is fundamental to ecology and conservation biology (S. Harrison 1997, 1999; S. Harrison et al. 2006; see also chapter 3). In a study of plant communities on serpentine outcrops, both woody and herbaceous plants were sampled in 24 small serpentine patches and in 24 similarly sized and spaced sites in large, continuous serpentine patches to assess patterns of local and regional plant diversity in relation to habitat patchiness (S. Harrison 1997, 1999). By comparing the occurrences of plant species in these different settings, Harrison found that for serpentine endemic woody and herbaceous species, local diversity (the number of species observed per sampling site) was higher on continuous serpentine habitat than on small patches. But patchiness led to a higher differentiation in community composition among the sites, such that the overall regional diversity was higher on patchy versus continuous serpentine habitat. Additionally, small patches tended to be more frequently invaded by alien species or species that were not normally found on serpentine soils than were large, continuous patches. These combined results have useful implications for conservation: large serpentine patches were more likely to support a higher diversity of endemic species and were invaded less frequently by exotic species. But the observation that small patches differed more from one another than from equal-sized sampling sites within continuous areas suggests that collections of small patches may also substantially contribute to capturing plant diversity in this system (S. Harrison 1999).

To further explore the observation that alien species tended to invade small patches more frequently than large ones, S. Harrison, Rice, and Maron (2001) studied the occurrence patterns of two common alien grasses, *Avena fatua* and *Bromus hordeaceus*, that were both frequently observed on small patches in the previous study (S. Harrison 1999). *Bromus* was more abundant at the edges of large serpentine patches than in the interiors of large patches, but the amount of *Avena* did not vary significantly with the distance from patch edge to interior. Soil characteristics of large serpentine patch edges also did not differ significantly from those of patch interior sites, so the higher invasion by *Bromus* at the patch edges was probably due to the influx of seeds from the matrix surrounding the patches, rather than to the seedlings' inability to germinate and survive in patch interiors. Seeds appeared to easily invade patch edges but were unable to disperse over the long distances to the interior of large patches. These results suggest that the surrounding matrix has strong effects on native habitat patches, an important theme in studies of habitat loss and fragmentation that will be explored in more detail in subsequent chapters.

To assess spatial factors affecting the local and regional diversity of serpentine plant assemblages, Harrison ambitiously extended the scope of her research to serpentine patches throughout the entire state of California (S. Harrison et al.

2006). She and her colleagues sampled plant species richness in 109 study sites within the four broad geographic areas in which serpentine soils occur in California. They also measured a suite of spatial characteristics (patch area, isolation, and shape) and environmental variables (soils, rainfall, temperature, and productivity), which they then used as explanatory variables for local and regional plant species richness. The results showed strong correlations of environmental variables (especially soils and rock cover) with local plant species richness, but little effect of spatial attributes of the patches. Regional richness was positively correlated with the overall amount of serpentine habitat within the region, but it was not associated with other measures of spatial structure. The authors concluded that the lack of strong association between richness and spatial structure at the local scale was perhaps due (1) to the wide range of patch sizes examined (in contrast to the very small and very large patches sampled in S. Harrison 1997, 1999), and (2) to the fact that these serpentine patches were not necessarily true islands, in that there is an exchange of many species between the surrounding matrix and the serpentine patches. Local environmental conditions, therefore, appeared to swamp variation in species richness due to patch spatial characteristics. The observed positive association between the overall amount of serpentine habitat in the region and regional species richness suggests that the amount of habitat, but not necessarily its spatial configuration, influenced species richness. This is consistent with Fahrig's (2003) review described above, where most studies of fragmented habitats find strong effects of habitat loss, but not necessarily of habitat fragmentation per se.

Whether habitat patchiness affects the interactions among species is also a critical question in conservation (Soulé et al. 2003; see also chapter 7). Pollinator behavior may be disrupted in fragmented landscapes, with serious implications for plant reproductive output. Studies of the reproductive biology of the serpentine morning glory (*Calystegia collina*) in this same system of serpentine outcrops revealed that flower and fruit production were higher on large (>300 ha) serpentine patches than on small (<5 ha) patches (Wolf 2001; Wolf and Harrison 2001). There were no detectable differences, however, in the number of bees that visited flowers in small versus large patches. A closer examination of plant reproduction showed that the higher fruit production on large patches compared to small patches was probably because bees were more effective pollinators on large patches. Plant patches occurred in clusters on large serpentine patches but not on small ones, so the former areas were where bees were more likely to carry compatible pollen. On small patches, there were greater possibilities for bees to carry pollen from other species, thereby increasing the proportion of incompatible pollen in their pollen loads. These results imply that large serpentine patches were more likely to support the high plant densities and clustered spatial configurations necessary for successful plant reproduction, whereas small patches failed to do so, despite equivalent abundances of pollinators among the patches.

Collectively, these studies of naturally patchy serpentine outcrops in California provide useful models for understanding the potential effects of human-caused fragmentation of natural systems on ecological processes. Serpentine outcrops come in different sizes, shapes, and spatial configurations and support many rare species that are of conservation concern. In addition, they are invaded at their edges by alien species, similar to what happens with human-created fragments. Lastly, the size, isolation, and internal spatial heterogeneity of serpentine patches affect species interactions that ultimately influence plant performance and persistence. Many of the lessons learned from studies of naturally patchy habitats have been fruitful in guiding and interpreting research on habitat loss and fragmentation.

SYNOPSIS

This book is intended to provide a succinct synthesis and review of the ecology of fragmented landscapes. It is obviously written from my own perspective of the field, based on my research over the past two decades. My hope is that it is sufficiently concise, yet comprehensive enough to provide necessary information while saving the reader from having to review the hundreds (really, thousands) of papers that have been published on this topic. The book is geared toward advanced undergraduates, beginning graduate students who want to enter this field, nature enthusiasts, conservation scientists and practitioners, and professional ecologists in other research circles who want to know the state of our knowledge about habitat loss and fragmentation. Because spatial structure in ecology has been examined in multiple ways, the various chapters blend approaches from population, community, and landscape ecology. They emphasize key ecological theories and concepts, review a rich collection of empirical literature, and forge vital links to environmental applications, such as the conservation of biological diversity, ecological restoration, and conservation planning. There are some brief historical perspectives for each topic, highlighting the major conceptual foundations of and developments in the field in the past 40 or so years. The chapters attempt to provide enough detail from featured studies to give readers a sense of the diverse methods used by researchers in their work, since it will be useful for students or novices in this field to learn different observational, experimental, and modeling approaches.

This book often just scratches the surface of particular concepts or processes, but it does contain sufficient information to encourage further exploration (for example, the SLOSS debate in chapter 3 and metacommunity theory in chapters 2 and 7). The examples in it are mostly from terrestrial systems, reflecting my own professional bias, but some freshwater and marine examples are included where appropriate. There is not much direct discussion here of population genetics, although it is mentioned occasionally. Allendorf and Luikart (2007) have a comprehensive and readable text on conservation genetics. I strongly encourage readers to dig deeper into topics of interest, and I offer an extensive reference list to guide those explorations. Also, because many concepts are linked, they may initially appear in one chapter and then reappear in another. There are also several reviews of the study of habitat loss and fragmentation that provide good summaries of some of the key issues (Saunders, Hobbs, and Margules 1991; Debinski and Holt 2000; Haila 2002; McGarigal and Cushman 2002; Fahrig 2003; Ewers and Didham 2006). Finally, I urge students of this subject not to neglect the classics. Most chapters refer to these fundamental articles and books, and in many cases re-reading papers published decades ago can offer fresh new insights.

This book is organized logically to move from the building blocks of theory and observation to more complex experiments and interactions. It has a recurring theme throughout: habitat area and connectivity are crucial to population and community processes. Habitat patches must be both large enough to support viable populations and connected enough to allow a sufficient movement of individuals, thus preventing genetic and demographic erosion. The book begins with a summary of the major theories and conceptual frameworks from population, community, and landscape ecology that have contributed to the study of habitat loss and fragmentation. The next few chapters critically survey the vast quantity of literature on the effects of patch size and isolation (chapter 3) and context (chapter 5) on species and communities, with separate treatments of observational (chapter 3) and experimental (chapter 4) studies. The movement of organisms may explain species' responses to habitat loss and fragmentation, and chapter 6 explores this subject. Explicit considerations of species interactions are examined in chapter 7, with a separate chapter devoted to interactions involving hosts, vectors, parasites, and pathogens (chapter 8). Modeling studies have built upon and informed empirical studies, and several basic modeling approaches are reviewed in chapter 9. Chapters 10 and 11 are meant to be hopeful, focusing on restoration (chapter 10) and conservation planning (chapter 11), two sets of approaches that may help to connect the pieces of fragmented landscapes into a functioning whole. The final chapter summarizes what we've learned and where our scientific knowledge has been successfully applied in conservation settings (and where it hasn't yet) and then offers some suggestions for promising research directions.

This is a rich and rewarding field of study, with exciting opportunities to link rigorous ecological theory with practical applications to the conservation of biological diversity and ecosystem services. Increasingly and fortunately, the barrier between basic and applied ecology is fading. I am heartened by the ever-enlarging batch of scientists and practitioners who seek to learn more about critical challenges in studies of the environment and are committed to using their intellectual power to find creative solutions to pressing issues. I am also encouraged by the brilliant and thoughtful work of my colleagues in this field, who have markedly propelled our knowledge and techniques for conducting research forward since I first started reading about this topic in the late 1980s. We have amassed an enormous body of scientific information, even though we often despair that we don't yet know enough about certain topics or systems. It is essential that we continue to communicate what we do know to the public and the media and to engage with non-traditional partners in industry, organizations, and government agencies to enhance the well-being of the collective biological wealth of this planet Earth.

CHAPTER TWO

Conceptual Frameworks

A good theory points to possible factors and relationships in
the real world that would otherwise remain hidden and thus
stimulates new forms of empirical research.
—*Robert H. MacArthur and Edward O. Wilson (1967)*

Now that the motivations for studying the ecological consequences of habitat loss
and fragmentation have been explained, it's important to understand the conceptual frameworks that underlie this field of study. Several key ecological theories contribute to our understanding of the consequences of loss and fragmentation for individuals, populations, and communities. Chapter 1 noted that natural
systems become patchy for a variety of reasons, and that this book focuses primarily on human-induced habitat loss and fragmentation. But some of the theories reviewed in the present chapter were developed for entirely different reasons.
Only subsequently have they been applied to studies of habitat loss and fragmentation in order to explain and hopefully predict ecological responses to landscape change.

This chapter describes and reviews the original theories, later modifications to
the theories, and the perspectives all of them offer for examining ecological fragmentation. It starts with the basic idea of species-area curves, then moves on
to the two main bodies of ecological theory—island biogeography theory and
metapopulation theory—that relate population and community dynamics to
patch size and isolation. Both theories were developed in the late 1960s and have

undergone modifications since then that are particularly relevant to discussions of fragmented landscapes. There are several additional conceptual frameworks that provide important insights into changing landscape patterns and species' responses, including metacommunities, meta-ecosystems and metalandscapes: percolation theory, hierarchy theory, patch dynamics and shifting mosaics, and fractal analysis.

SPECIES-AREA CURVES

One of the few truisms in ecology is that larger areas support a greater number of species. Although ecologists often lament the fact that many of the patterns we observe in nature are species specific or habitat specific, and thus not easy to generalize, the species-area relationship is surely one of the most satisfying patterns, due to its predictability and consistency. That the number of species in an area increases with its size was observed by early naturalists such as Darwin and Wallace, and this was formalized as an ecological principle in the early part of the 20th century (Arrhenius 1921). Ecologists have repeatedly noted a positive relationship between the area of an island (or a chunk of habitat) and the number of species present on it (fig. 2.1). The islands of the West Indies are a classic example, since various ecologists have studied relationships between island area and several different taxa on these islands. In all cases there is a distinct positive relationship between area and species richness that can be described by the power law:

$$S = cA^z$$

where S is the number of species on the island, A is the area of the island, and c and z are constants fitted to the data (MacArthur and Wilson 1967; B. Wilcox 1980; Rosenzweig 1995; Gotelli 1995). By taking the logarithm of each side of the equation, the relationship becomes a straight line, with c as the y-intercept of the line and z as the slope of the line. The z-value can be described in words as the extent to which species richness changes with island area. The logarithmic transformation allows for easy comparisons of the parameters of this linear equation among taxa and study systems. In the West Indies example, a comparison of the slopes (z-values) for different species groups reveals that species richness increases at a higher rate with island area for non-flying mammals and reptiles and amphibians ($z = 0.48$ and 0.38, respectively) than for birds and bats ($z = 0.24$ and 0.24, respectively) (cited in B. Wilcox 1980). These differences in slopes suggest that the less mobile species are more confined in their distribution; thus new

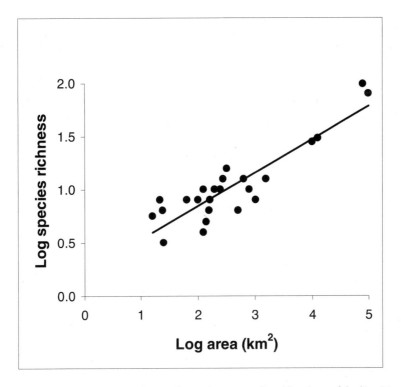

Figure 2.1. A species-area curve for reptiles in the West Indies. The slope of the line (z) = 0.30. Redrawn based on data from Wright (1981).

species of these less-mobile animals are encountered more rapidly as area increases than what is observed for the more mobile species, birds and bats.

Three explanations have consistently been posited for the species-area relationship: the passive sampling hypothesis, the habitat heterogeneity hypothesis, and island biogeography theory. The *passive sampling hypothesis* (Arrhenius 1921; Connor and McCoy 1979; Williamson 1981) says that simply by increasing the size of an area sampled, you are more likely to encounter more individuals. And as you increase the number of individuals that you sample, you are also more likely to encounter new species. So the increase in species richness with greater area merely reflects increased sampling effort. The *habitat heterogeneity hypothesis* (Williamson 1981; Boecklen 1986; Rosenzweig 1995) states that larger areas are more likely to contain more types of habitats. And as the number of habitat types increases, so, too, will the number of species encountered, since many species are habitat specialists. *Island biogeography theory* (MacArthur and Wilson 1967), discussed in detail below, asserts that species richness increases with island area due to the dynamic equilibrium between the rate of species colonization of an island and the rate of species extinctions on it. Rosenzweig (1995) provides an excellent, detailed discussion of species-area curves and their explanations.

The relative importance of each of these three explanations of the species-area relationship has been widely discussed, and no clear consensus has yet emerged on the primary mechanism responsible for it (Boecklen 1986; Herkert 1994; Holt et al. 1999; Ricklefs and Lovette 1999; Harte, Blackburn, and Ostling 2001; Cam et al. 2002; Chittaro 2002; V. Smith et al. 2005; Ouin et al. 2006). As with many ecological questions, the answer is probably that some combination of these factors influences patterns of species richness in relation to area. For the purposes of this discussion, the main take-home message is the following: there are multiple hypotheses for the species-area pattern, and although island biogeography theory is perhaps the most familiar explanation to most readers, it is not the only conceptual model that has been advanced to explain this relationship.

ISLAND BIOGEOGRAPHY THEORY

Robert MacArthur and E. O. Wilson worked together in the 1960s to develop a general theory that could explain the richness of species on oceanic islands. Their observations of birds and ants on tropical islands had revealed strong patterns of increasing species richness with area, consistent with the general species-area relationship described above. Together they constructed a conceptual framework to explain these patterns (MacArthur and Wilson 1967). Island biogeography theory (IBT) asserted that species richness increases with island area, due to the balance between colonization and extinction rates on an island (fig. 2.2a). Since it should be relatively easy for individual plants or animals to reach islands that are close

Figure 2.2. (*opposite*) Conceptual representations of the equilibrium theory of island biogeography (IBT): (a) as originally conceived by MacArthur and Wilson (1967), (b) with the rescue effect of J.H. Brown and Kodric-Brown (1977), and (c) with the predicted effect of habitat corridors of A. Bennett (1990). Rates of immigration are the top four (a) or two (b and c) lines shown on the left side of each graph, and rates of extinction are the bottom four (a and b) or two (c) lines on the left side of each graph. Species richness for a particular island distance-size combination is determined by dropping a perpendicular line to the x-axis from where the two lines depicting immigration and extinction rates cross each other. N = near islands, F = far islands, L = large islands, S = small islands.

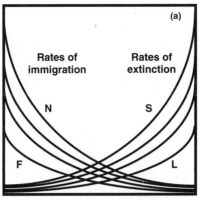

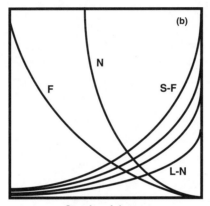

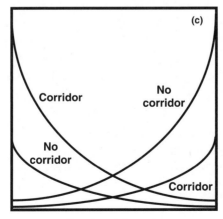

to a mainland, colonization rates would be primarily influenced by an island's distance from a mainland source of colonists. So near islands would have higher colonization rates than far islands. Extinction rates would be primarily influenced by the size of the island, since smaller islands would only be able to support small populations of a given species, compared to what larger islands could maintain, and small populations are more vulnerable to extinction than large ones. Thus the equilibrium number of species that could occur on an island of a particular size and degree of isolation would be determined by the point at which the rate of colonization met the rate of extinction. Large islands close to a mainland would support the highest number of species, whereas small islands distant from a mainland would harbor the fewest number of species.

At the same time that MacArthur and Wilson were constructing this theory based on oceanic islands, they recognized the analogy of oceanic islands with the terrestrial "islands" that remained as landscapes were converted from forests to agricultural fields. In fact, in the first chapter of their classic monograph, titled "The Importance of Islands," they describe insularity as a "universal feature of biogeography" and almost immediately (in the second paragraph) present the analogy of oceanic islands with habitat fragments: "The same principles [of insularity] apply, and will apply to an accelerating extent in the future, to formerly continuous natural habitats now being broken up by the encroachment of civilization" (p. 4).

MODIFICATIONS TO IBT
The Rescue Effect

Recall that the key prediction from IBT regarding habitat isolation was that immigration rates should vary inversely with the distance of a fragment from a source of colonists. Islands that were farther from a mainland would therefore have fewer species, since it would be more difficult for colonists to reach these islands compared to those close to the mainland. For terrestrial habitat fragments, a continuous expanse of native habitat could be considered a mainland colonist source. J.H. Brown and Kodric-Brown (1977) suggested a modification to IBT by claiming that island distance may not only influence immigration, but may also significantly influence extinction rates (fig. 2.2b). Specifically, in insular habitats or closely spaced resource patches that are easily accessible, immigration rates may be very high (and relatively higher than extinction rates), shifting the balance of these two rates in favor of immigration. These high rates of immigration may effectively decrease the probability of extinction by rescuing populations before they reach precariously small sizes, which J.H. Brown and Kodric-Brown referred to as the *rescue effect*. The main contribution of the rescue effect concept was that *both* island size and distance (not just size, as in the original IBT) may influence the extinction rate of species on islands and, consequently, patterns of species richness.

Corridors

Additional conceptual developments regarding the effects of island or habitat isolation on species richness suggested that any mechanism that reduced isolation among habitat patches should also enhance immigration rates to those patches. If a patch is less isolated, organisms should be able to disperse to that patch more easily, which ultimately would increase the colonization rate. *Habitat corridors*, which are linear strips of vegetation that link otherwise isolated fragments, were proposed as a means to reduce isolation among habitat patches. For this very reason, the use of habitat corridors appeared as one of Diamond's (1975) suggested principles of reserve design. If habitat corridors successfully reduce the effective distance between habitat patches, then they should enhance colonization rates and facilitate the rescue effect (fig. 2.2c). It logically follows that habitat fragments connected by corridors should support larger populations, and perhaps a higher number of species, than completely isolated fragments of equal size (e.g., Simberloff and Cox 1987; A. Bennett 1990; Saunders and Hobbs 1991).

Landscape Effects

Because island biogeography theory was based largely on terrestrial species inhabiting oceanic islands, there was really no need to invoke the ocean's characteristics as an unsuitable source of or site for colonization to explain species richness on islands. In applying this theory to terrestrial fragments, however, many authors have argued that these landscapes do not have a similarly inhospitable matrix, so the theory requires some modification in its application to fragmented landscapes (Burgess and Sharpe 1981; Wilcove, McLellan, and Dobson 1986; Forman 1995; Wiens 1996; H. Murphy and Lovett-Doust 2004; Hilty, Lidicker, and Merenlender 2006). Forman (1995) suggested that, although species richness on oceanic islands could be described as a function of island size and isolation, the species richness of patches in a landscape mosaic would be influenced by additional factors. He devised the following equation to encapsulate those additional factors:

$$S = [f(\text{habitat diversity } (+), \text{ disturbance } (- \text{ or } +), \text{ area of patch interior } (+),$$
$$\text{age } (+ \text{ or } -), \text{ matrix heterogeneity } (+), \text{ isolation } (-)]$$

where S equals the species richness of a patch in a landscape mosaic, and the plus or minus sign following each patch characteristic indicates its proposed relationship with species richness. For example, habitat diversity, area of patch interior, and matrix heterogeneity are all expected to positively influence species richness in a patch, and patch isolation to affect it negatively. Two variables, disturbance and patch age, may have either negative or positive effects on species richness.

In summary, island biogeography theory and its modifications made predictions about how island size and isolation (as well as additional factors in the case of terrestrial habitat fragments) should influence species richness via their effects on the rates of immigration and extinction. The theory implicitly included population processes, since colonization and extinction occur at the level of populations. It also assumed that small populations would be more vulnerable to extinction than large ones. But the overall goal of IBT and its modifications was to make explicit predictions about species richness—a community property—on islands and habitat fragments.

METAPOPULATION THEORY

Island biogeography theory's emphasis on rates of colonization and extinction to explain patterns of species richness corresponds closely with metapopulation theory (Levins 1969), another major ecological framework constructed to explain the dynamics of patchy systems. The formulation of metapopulation theory was published soon after MacArthur and Wilson's (1967) IBT, but emerged from studies of a very different ecological problem. Levins was tasked by applied entomologists to work out the best way to introduce persistent populations of insect biological control agents (predators or parasitoids) into agricultural fields to effectively control pest insects. Levins recognized that pests attack crops over a broad geographic area, so that biological control interventions would likely require multiple, possibly interacting populations of biological control agents. His mathematical formulation included local population extinctions followed, after a relatively short lag period, by migration and recolonization from other populations. He used the term *metapopulation* to describe the collection of local populations that periodically experienced extinction but were linked by colonization (immigration).

As originally conceived by Levins, metapopulation theory emphasized that local populations of organisms undergo periodic colonization and extinction, but the metapopulation can persist indefinitely if rates of extinction are balanced by rates of colonization. However, failed colonization of empty patches may occur if patches are extremely isolated (Hansson 1991; S. Harrison 1991; Fahrig and Merriam 1994). Elaborations of Levins' original metapopulation theory explored how variation in patch characteristics among local populations may lead to very different metapopulation dynamics (e.g., S. Harrison, Murphy, and Ehrlich 1988; Hanski and Gilpin 1991; S. Harrison 1991; S. Harrison and Taylor 1997; Hanski 1999; Hanski and Gaggiotti 2004). For example, S. Harrison (1991) and S. Harrison and Taylor (1997) noted at least four types of spatially-structured populations, each with quite different dynamics (fig. 2.3). Superficially, these spatially structured populations may look the same—they are collections of spatially separated patches that occur across a broad landscape. But functionally, these populations differ substantially. The *classic* metapopulation of Levins (1969) (fig. 2.3a) consists of (1) habitat patches that are invariant in size, isolation, and quality, with each patch having an equal probability of extinction and recolonization, and (2) asynchronous patch dynamics, that is, the dynamics on each habitat patch are independent of one another. One variation on this theme is the *mainland-island system* (fig. 2.3b) described by S. Harrison, Murphy, and Ehrlich (1988) for checkerspot butterflies on serpentine patches (see chapter 1). Here the probability of extinction is unequal among the patches, due to variation in patch or population size: populations on small patches may periodically go extinct, but the large patch persists indefinitely. Mainland-island metapopulations are analogous to *source-sink* metapopulations, except the designation of a population as a source or sink depends upon habitat quality, not necessarily patch size. *Sources* are defined as populations with positive growth rates that export surplus individuals (dispersers) to adjacent populations, whereas *sink* populations are those with negative growth rates that are sustained only via immigration from sources (Pulliam 1988; Pulliam and Danielson 1991). Another alternative is the *patchy* population (fig. 2.3c), which is not really a metapopulation at all, but simply a single large population that happens to occupy several habitat patches. Think of the daily foraging movements of squirrels among trees in several suburban backyards, where the backyards are not discrete populations but merely patchily distributed resources used by a single squirrel population. A third variant is the *non-equilibrium* metapopulation (fig. 2.3d), which is the most troubling for conservation biologists.

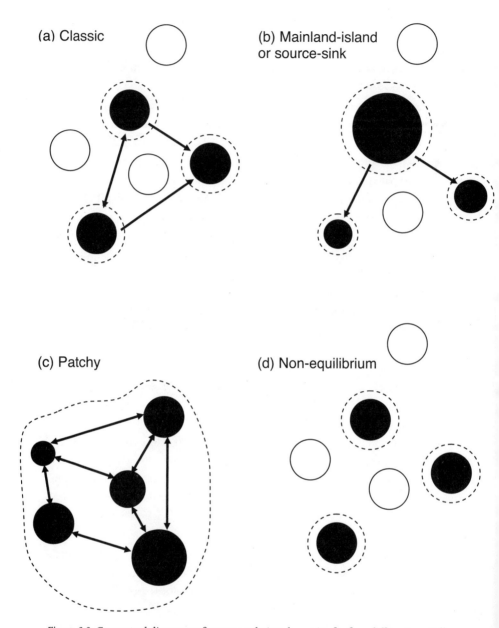

Figure 2.3. Conceptual diagrams of metapopulation dynamics for four different spatially structured populations: (a) classic metapopulation, (b) mainland-island or source-sink metapopulation, (c) patchy population, and (d) non-equilibrium metapopulation. Redrawn from S. Harrison and Taylor (1997, figs. 1a, b, c, and d). Filled circles represent occupied habitat patches; empty circles, vacant habitat patches; dotted lines, boundaries of local populations; arrows, dispersal between patches.

This is because the populations are so isolated from one another that migration among populations is nonexistent; thus neither rescue from extinction nor recolonization following extinction can occur. So if a local population goes extinct, presumably there is no hope for the re-establishment of a population, except perhaps with human intervention. Finally, a metapopulation may functionally exhibit more than one of these types of dynamics in different parts of the metapopulation (S. Harrison 1991; S. Harrison and Taylor 1997, fig. 1e).

These distinctions between types of spatially structured populations are of more than just academic interest. Because the dynamics of these populations differ substantially, management interventions must also vary accordingly. And management actions based on a superficial inspection of populations could be misguided. For example, if a set of patchily distributed populations functions as a mainland-island system, then conservation and management actions should prioritize the long-term protection of the large mainland population, since it plays an essential role in the persistence of the overall metapopulation. But if it were assumed, in this case, that all populations were equally prone to extinction (the classic metapopulation), then conservation efforts would not necessarily focus on the mainland population. In some instances, a species may exhibit functionally variant metapopulations across its geographic range, so management efforts must be similarly fluid. For example, Stith et al. (1996) conducted detailed population studies of the threatened Florida scrub jay (*Aphelocoma coerulescens*) and observed mainland-island, non-equilibrium, and classic metapopulation dynamics in different parts of the species' range in peninsular Florida.

As was the case for island biogeography theory, metapopulation theory and its modifications have been applied widely to the population dynamics of species in human-induced habitat fragments (Dunning et al. 1995; S. Harrison and Fahrig 1995; McCullough 1996). Because failed colonization of empty patches may occur due to patch isolation (Hansson 1991; S. Harrison 1991; Fahrig and Merriam 1994), the presence of either stepping stones or corridors, or the management of the matrix between fragments, should increase the regional (metapopulation) persistence of native species by reducing isolation effects and increasing colonization probability. The purported benefit of connecting otherwise isolated patches has been similarly advocated, based on the predicted positive relationships between colonization and species richness in IBT. However, the role of corridors, in particular, has been debated fervently (Simberloff et al. 1992; Hanski and Simberloff 1997), as is discussed in chapters 3 and 6. With regard to metapopulation theory, in the extreme case that corridors promote so much movement among patches that populations become synchronized, then the persistence of the overall metapopulation may be threatened by random variability (*stochasticity*) in the environment (the patchy populations of S. Harrison and Taylor 1997; see also Koelle and Vandermeer 2005; chapter 6 contains a further discussion of movement).

BUT FRAGMENTS ARE NOT ISLANDS

The similarities between IBT and metapopulation theory have been evident to many ecologists and conservation biologists. Both focus on immigration and extinction in patchy landscapes (Hanski and Gilpin 1991; Hanski and Simberloff 1997; Hanski 1999), and so both should be relevant to understanding populations and communities in fragmented landscapes. IBT emphasizes species richness on islands, while metapopulation theory focuses on the persistence of single-species populations. They differ in that in IBT there is a mainland that serves as a source of colonists, while in metapopulation theory, this may or may not be the case (it exists only for mainland-island metapopulations). In classic metapopulation theory (Levins 1969), patches do not differ in isolation, either, as they do in IBT. The theories are similar in that both IBT and metapopulation theory are equally nebulous about variation in habitat quality among patches or islands, as well as about characteristics of the matrix that may affect rates of colonization and extinction (e.g., Forman 1995; Wiens 1997). Both consider the matrix as an essentially nondescript background.

In terms of their applicability to conservation, both theories have been extremely useful in generating hypotheses and stimulating empirical research. And because of their limitations, they have also helped the field move forward by motivating researchers to refine these simple, elegant models into messier but more accurate representations of the world. Island biogeography theory was effectively eclipsed in the 1990s by metapopulation theory, with the criticism that IBT had too many limitations for its applicability to fragmented landscapes. Hanski (1999) pointed out that one of the possible reasons for this shift, aside from issues of the inhospitable matrix of IBT described above, was that the metapopulation paradigm posits a key role for the conservation of small habitat patches. In terms of conservation planning, IBT projects that small patches will contain few species, so they may not be a high priority for conservation actions. But metapopulation theory does not similarly undervalue small patches, since they contribute to overall metapopulation persistence. I would argue that these two theories have been and continue to be equally useful as guiding frameworks for conservation, but they are perhaps equally limited in their *literal* applicability for conservation action. Both, however, have had major impacts on how we think about fragmented populations and communities and on what actions we might take to ameliorate the negative consequences of isolation and extinction.

METACOMMUNITY THEORY

Although the concept of a metacommunity has only recently been developed in depth (Holyoak, Leibold, and Holt 2005), early metapopulation researchers appreciated the existence of metacommunities (Hanski and Gilpin 1991). The dynamics of host-parasitoid and predator-prey interactions in patchy landscapes lent themselves particularly well to the theoretical formulations and empirical tests of metacommunity theory (Murdoch, Chesson, and Chesson 1985; A. Taylor 1990). Recent elaborations of metacommunity ecology are particularly relevant to understanding species interactions in fragmented landscapes. Metacommunity thinking is essentially an extension of metapopulation theory—but instead of populations of a single species blinking in and out of scattered habitat patches, we now have multiple species that occur and interact across patchy landscapes. A *metacommunity* is broadly defined as a collection of communities connected by dispersal (Hanski and Gilpin 1991; Leibold et al. 2004; Chase 2005; Holyoak, Leibold, and Holt 2005), while a *community* is a collection of individuals that interact either directly or indirectly by partitioning the resources within a patch of shared habitat (Hubbell 2001; Holyoak, Leibold, and Hold 2005). In community ecology, conventional theory has typically focused on local, deterministic interactions—such as predation and competition—to explain the structure of local communities. Metacommunity theory, on the other hand, extends the ideas of metapopulation theory to multiple species and suggests that regional dynamics such as dispersal, disturbance, and extinction can help to explain the broader distribution and coexistence of species. The strength and influence of these regional processes, such as colonization or extinction, are likely to be influenced by the spatial arrangement of local communities, a recurrent theme throughout this book. Recent works have called for the unification of these local and regional approaches, so that we can refine our understanding of the suite of factors involved in structuring communities and use this knowledge to predict outcomes of anthropogenic change, such as habitat loss and fragmentation, species invasions, and climate change (Hubbell 2001; Chase 2005; Holyoak, Leibold, and Holt 2005).

Neutral Theory versus Niche Theory

Hubbell (2001) proposed the neutral theory of biodiversity, where patterns of species richness and composition are related to the size of the metacommunity, the rates of dispersal within the metacommunity, and the rates of formation of new species. Hubbell's work derived from his observations of highly diverse tropical tree communities, where all species compete for the same resources (nutrients, water, light) and obtain resources in similar ways, yet hundreds to thousands of different tree species can coexist in these forests. The *neutral theory* thus assumes that species do not differ in their traits. Chase (2005) argued that Hubbell's neutral theory has a limited use in making predictions about the likely responses of species to fragmentation. Neutral theory assumes that all species have the same traits, yet we know that fragmentation is likely to affect species differently, depending on traits such as body size, home range size, trophic level, or competitive ability. Chase noted that "as the neutral theory disregards differences in species traits, it can say little about how fragmentation will differentially alter the composition of species, as well as their interactions with other species in the community" (p. 184).

Niche theory, on the other hand, assumes that species coexistence is based on differences among species in traits that are related to resource acquisition or response to the environment. Modern niche theory expands upon earlier descriptions of the niche and argues that additional factors influence species coexistence, including other species, such as enemies or mutualists, and temporal and spatial variation in resource availability (Chase 2005). The emphasis on spatial variation in both resource availability and species traits may be especially relevant for studies in fragmented landscapes.

The subject of mechanisms of species coexistence is obviously relevant to discussions of persistence in fragmented habitats, but it is well beyond the scope of this book. For an excellent review of the theory on how spatial heterogeneity affects two-species interactions, see Hoopes, Holt, and Holyoak (2005), and to catch a glimpse of future developments in metacommunity ecology, see Holt, Holyoak, and Leibold (2005). What is important to emphasize here is the fact that species interactions vary in response to habitat spatial structure, which has been well documented for many types of interactions and is explored in detail in chapter 7. Metacommunity theory adds to the formal conceptual framework for considering how species interactions might respond to variation in habitat spatial structure, and thus can provide a context for predicting and interpreting the ecological consequences of habitat loss and fragmentation.

META-ECOSYSTEMS AND METALANDSCAPES

But it doesn't stop with metapopulations and metacommunities. There are also meta-ecosystems and metalandscapes. Loreau, Mouquet, and Holt (2003) extended the idea of metapopulations and metacommunities to *meta-ecosystems*, which they defined as "a set of ecosystems connected by spatial flows of energy, materials and organisms across ecosystem boundaries" (p. 674). For readers familiar with landscape ecology, this sounds a lot like the definition of a landscape: "a mosaic where a cluster of local ecosystems is repeated in similar form over a kilometers-wide area" (Forman 1995, p. 39). Landscape ecology also emphasizes the flow of energy, materials, and organisms across spatially heterogeneous areas. But Loreau, Mouquet, and Holt (2003) claimed that meta-ecosystems were different from landscapes, particularly because (1) landscapes are continuous, but meta-ecosystems may be discontinuous—in this way, the meta-ecosystem concept directly translates the notion of a collection of spatially disjunct populations and communities to spatially disjunct ecosystems; (2) a landscape has a characteristic spatial scale (but this is debatable, since some advocate that landscapes are defined by the perceptions of different organisms; see Wiens 1989), but a meta-ecosystem can be at any spatial scale; and (3) the meta-ecosystem concept is not just about examining multiple scales, but is a "new tool to understand the emergent constraints and properties that arise from spatial coupling of local ecosystems" (p. 675). An example of a meta-ecosystem could be a series of spatially disjunct wetland ecosystems that exchange nutrients via sequential visitation by migrating waterfowl. Birds visit the wetlands for a few days to forage, and perhaps they leave behind some nutrients and possibly a few feathers. While the wetlands are spatially discontinuous, they are connected by flows of nutrients and materials mediated by the migrating birds.

This theme of spatially separate but functionally connected populations, communities, and ecosystems has also been expanded to landscapes. Based on their observations of the population dynamics of migrating songbirds in the midwestern United States, With, Schrott, and King (2006) surmised that some bird species may exhibit regional source-sink population dynamics. In other words, populations in a continuous, unfragmented landscape may rescue populations in a distant, highly fragmented landscape via immigration, or metalandscape connectivity. Simulation models varied the levels of habitat loss and fragmentation, as well as immigration, to explore the conditions under which metalandscape connectivity could effectively rescue regional populations from extinction. Thus what these authors proposed is essentially a broad-scale extension of metapopulation theory: in the same way that connectivity is critical to the persistence of a metapopulation, metalandscape connectivity may be critical to regional population persistence.

PERCOLATION THEORY

Percolation theory derives from studies of porous materials and has been used in landscape ecological research to examine the patterns and consequences of landscape structure (e.g., Gardner and O'Neill 1991; Gardner, O'Neill, and Turner 1993). The use of percolation theory in spatial ecology involves the construction of two-dimensional computerized maps and the evaluation of ecological processes that percolate through the map, such as the spatial spread of a natural disturbance like fire (Turner, Romme, and Gardner 1994) or the spread of an invasive species (With 2004). *Neutral landscape models*, in particular, are maps that are generated via random processes or spatially correlated processes, but they are not based on landscape patterns derived from known ecological processes. These neutral, or *null*, models are thus analogous to null models in other areas of ecology (e.g., Connor and Simberloff 1979). With regard to studies of habitat loss and fragmentation, the most relevant neutral models are those that simulate the consequences of habitat spatial configuration for populations and communities (With 1997). These neutral models are comprised of a rectangular lattice in which cells are designated as either habitat or non-habitat (fig. 2.4). A *percolating cluster* is defined as a collection of connected habitat cells that extends from one side of the lattice to the other (With 1997). A special feature of neutral models derived from percolation theory is that, when cells in the lattice are randomly assigned to one category (let's say suitable habitat in this case), there exists a *critical threshold* where the lattice suddenly transitions from being unconnected to forming a percolating cluster (Gardner, O'Neill, and Turner 1993; With 1997). In this example, critical threshold refers to the proportion of cells that are designated as habitat, and the threshold is reached when about 59% of the cells have been randomly assigned to the habitat category. In other words, if you were to pop inside a randomly simulated landscape with 59% habitat, it would be possible to walk across the entire lattice without stepping outside the habitat. As the proportion of cells assigned as habitat approaches 100%, the path across the lattice becomes straighter.

This critical threshold has a useful application to studies of habitat loss and fragmentation, because it defines the landscape as connected. But the critical threshold depends upon how the modeler defines the rules of the game. Con-

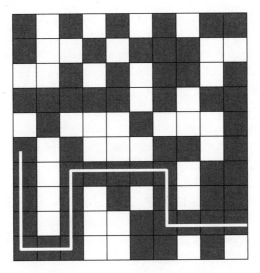

Figure 2.4. A rectangular lattice of cells generated randomly as a neutral model of landscape structure. According to percolation theory, these randomly placed cells provide a connected network when 59% of the cells are selected, as shown, using the 4-cell rule. The gray cells represent habitat, the white cells represent non-habitat, and the white line shows the continuous path of a percolating cluster through the habitat cells.

nectivity within a neutral model may be defined using different neighbor rules, which include 4-, 8-, or 12-cell neighbor rules. The *4-cell rule* is also called the nearest neighbor rule, and it goes like this: in order for a single cell that is considered suitable habitat in the lattice to be considered connected to another cell of suitable habitat (i.e., part of the same patch), it must share one side in common with an adjacent suitable habitat cell in one of the four cardinal directions. In this rule, cells of similar type that touch only along the diagonal do not count as connected. In the *8-cell rule*, or next-nearest neighbor rule, a cell is connected if it shares sides along either a cardinal or a diagonal direction, so there are eight cells that surround a single cell that could potentially be connected. The *12-cell rule*, or third-nearest neighbor rule, extends to the four cells in the four cardinal directions that are two cells away from the single focal cell. While this is a brief overview, applications of percolation theory to ecological modeling studies will be reviewed in detail in chapter 9.

SPATIAL PATCHINESS AND SCALING

The previous discussions of habitat patchiness and of moving from metapopulations to metalandscapes both raise the issue of scaling. The basic idea here is that the ecological patterns and processes that we observe at one scale may shift in character as we move to another scale of observation (Allen and Starr 1982; O'Neill et al. 1986; Wiens 1989; Levin 1992). Spatial patterns obvious at one scale may not be discernible at another scale. For instance, if you were an ant within a 1000 ha forest fragment, you probably would not notice that the forest is isolated from other forests, but if you were an eagle flying 1000 m above the Earth, you would perceive the forest as a patch. So the ideas of pattern and scale are intimately linked. Another example concerns spatial distribution, where individuals or species may be aggregated at local scales but uniformly distributed at broader scales. Moreover, the dynamics at one level of biological organization, such as the community, emerge as the collective behavior of components at the next-lower level of biological organization, in this case, the population. The notion of scale was also discussed briefly in chapter 1, with the description of the space-time domains of scale (Delcourt, Delcourt, and Webb 1983; Urban, O'Neill, and Shugart 1987). This chapter succinctly reviews several conceptual frameworks that relate explicitly to scale and are relevant to understanding habitat loss and fragmentation—including hierarchy theory, patch dynamics, and fractal geometry.

Hierarchy theory (Allen and Starr 1982; O'Neill et al. 1986; Allen and Hoekstra 1992) posits that processes in nature are inherently multi-scaled. Spatial patchiness occurs at multiple scales, as well, and we may understand a particular landscape as a hierarchy of patches within patches. The fact that landscapes are patchy is a recurrent theme throughout this book, and concepts from hierarchy theory suggest that organisms will respond to patchiness at different scales. So a patchy landscape may be viewed as fragmented for one species, but as continuous for another species (e.g., Haila 1990). This concept is critical when designing ecological studies of fragmentation, as well as for interpreting species' responses to fragmented landscapes. For example, landscape structure can be quantified at different spatial scales and then related to a particular ecological variable, such as population abundance or species richness. Significant associations between the ecological response variable (say, population abundance) and the landscape pattern at a particular spatial scale can inform researchers about the scale at which populations respond to landscape pattern (e.g., Collinge, Prudic, and Oliver 2003).

The related ideas of patch dynamics and a shifting mosaic largely differ from each other in the scale at which patterns and processes are considered. Patch dynamics was formalized by Pickett and White (1985), based on observations that ecological systems are inherently variable over space and time. *Patch dynamics* focuses on the event or agent causing a patch and the changes in species in the patch over time. A *shifting mosaic*, in contrast, is a large area in equilibrium that contains many patches in various successional stages. Thus in patch dynamics, the focus is on changes within patches, whereas in a shifting mosaic, the focus is on the whole landscape (Forman 1995). For example, a neotropical forest typically contains light gaps, caused by trees being blown over during windstorms. Within each patch, species composition shifts over time from light-loving species that thrive immediately after the gap is created to shade-loving species that eventually dominate the gap as the canopy closes. If we look at the forest at a broader spatial scale, at any one time there are likely to be many light gaps of various ages and in various stages of succession from light-loving to shade-loving species (Denslow 1987).

Fractal geometry can be used to quantify spatial patterns in many natural phenomena, including soil aggregates, root systems, and complex landscapes (Milne 1991). The classic application used to describe fractal geometry is in measuring a coastline. The measurement of the length of a coastline depends upon the scale of measurement used. If you used a 1 km long ruler to measure the coastline of California, you would say it was shorter than someone who measured the same coastline using a 1 cm long ruler. Fractal analysis has been used in a couple of key ways in understanding fragmented landscapes. First, it has been employed to analyze the shape of a patch or pathway—a two-dimensional line through the landscape. For example, this tool has been used to measure the pathway taken by an organism during movement and ask whether different organisms respond differently to landscape structure (With 1994). Second, fractal analysis has been used to assess the complexity of patch shapes, which may have implications for species persistence in a landscape. A fractal dimension of 1.0 is a straight line or a perfectly linear path, whereas a fractal dimension approaching 2.0 represents a highly complex, convoluted pathway or patch boundary. Fractal dimension has been used as one of a few simple indices to describe landscape pattern, which will be elaborated further in chapter 9.

SYNTHESIS

Ecological theories derived from population and community ecology, including island biogeography theory and metapopulation theory, predict that any mechanism that reduces isolation among habitat patches should also expedite the movement of organisms between patches, thereby reducing the rates of species loss and enhancing the probability of fragment recolonization. These theories provide a mantra that will be examined and reiterated throughout this book—that habitat loss and fragmentation influence the rates of species colonization and extinction, based on whether patches are large enough to support viable populations or connected enough to facilitate the exchange of individuals among patches. Metacommunity, meta-ecosystem, and metalandscape frameworks all extend the notions of metapopulation theory to higher levels in the ecological hierarchy. These concepts, in turn, emphasize the importance of spatial structure for species interactions, the movement of organisms and materials among ecosystems, and the movement and viability of organisms among landscapes. The relevance of multiple scales of ecological patterns and processes is highlighted in conceptual frameworks such as percolation theory and hierarchy theory. In relation to studies of habitat loss and fragmentation, these concepts provide a formal means for observing and interpreting responses of different species under various ecological conditions.

CHAPTER THREE

Fragment Size and Isolation

One of the most profound developments in the application
of ecology to biological conservation has been the recognition that virtually all natural habitats or reserves are
destined to resemble islands, in that they will eventually
become small isolated fragments of formerly much larger
continuous natural habitat.

—*Bruce A. Wilcox (1980)*

Larger areas of land or water support more species than smaller areas, a point
that has been made clear with repeated illustrations of species-area curves and is
fundamental to our ecological understanding of the natural world. Yet despite the
fact that this is old news, ecologists are still talking about this relationship and papers are still being published on this topic. Why? Probably because it is one of the
most ubiquitous patterns in nature, and because ecologists have still not fully explained why this pattern is so. There are also profound implications of the
species-area relationship for the conservation and restoration of biodiversity.
Along with patch area, patch isolation has been widely investigated as an explanation for patterns of species composition in fragmented habitats. Patch isolation may substantially alter ecological processes within and between habitat fragments, including the rates of animal and plant dispersal among fragments and
the persistence of populations and communities.

The strong emphasis on investigations of patch size and isolation in the literature on ecology and conservation biology undoubtedly derives from the influence of MacArthur and Wilson's (1967) theory of island biogeography. Their explanation of species richness on oceanic islands, and their extension of these ideas to habitat fragments, provided an invitation to ecologists to explore the significance of habitat fragment size and isolation for populations and communities. In the ensuing 40 years, a plethora of papers has been published that are devoted to these topics. During the same time period, strong links were forged between island biogeography theory (IBT) and conservation biology, which has prompted much discussion and controversy regarding options for reserve designs that would maximize the conservation of biological diversity. In particular, questions arose regarding the utility of habitat corridors (are they "vital linkages" or "vastly expensive failures"?), and researchers argued about whether a single large reserve or several small reserves (the SLOSS debate) would be optimal for biodiversity protection. The paradigm of "fragments as islands" derived from IBT is pervasive in the literature on fragmented landscapes. This has been fruitful, but it has also perhaps limited our advancement. We have continued to think of fragments as islands, but terrestrial landscapes are much more heterogeneous and complex.

This chapter attempts to distill the vast literature on species richness and composition in relation to patch size and isolation, as well as studies of the mechanisms responsible for these patterns. This subject is clearly fundamental to understanding the effects of fragmentation, so many of the topics mentioned in this chapter will appear in greater detail in other chapters. And the fundamental issues explored in detail in this chapter will provide the foundation for concepts discussed in later chapters. To provide a brief road map, this chapter will emphasize fragments of anthropogenic origin, in contrast to studies of naturally occurring patchy habitats. It will cover observational studies of patch size and isolation, focus primarily on population and community patterns in relation to habitat isolation, and consider whether habitat corridors that link otherwise isolated fragments serve to promote species persistence.

FRAGMENT SIZE AND SPECIES RICHNESS

Because island size clearly influences species richness—recall the ubiquitous species-area curves—and terrestrial habitat fragments often resemble islands, much of the early work concerning habitat loss and fragmentation focused on the relationship between fragment size and species richness (e.g., Saunders et al. 1987). There are literally dozens of papers on this topic, representing many groups of organisms and many habitat types (table 3.1); the most well studied are the birds of temperate and tropical forests. In general, research on birds demonstrates that the number of species within isolated forest fragments decreases as fragment area decreases (Forman, Galli, and Leck 1976; Whitcomb et al. 1981; Verner, Morrison, and Ralph 1986; Blake and Karr 1987; Lynch 1987; Newmark 1991; Beier, Van Drielen, and Kankam 2002; see fig. 3.1), with the commonly noted interpretation that the probability of local extinction increases as fragment size decreases. For example, a study of grassland bird communities in the midwestern United States showed that approximately 79% of the grassland bird species were present in a 1000 ha grassland fragment, but only 31% of the bird species occurred in 10 ha grassland fragments (Herkert 1994). This habitat-fragment species-area pattern has been documented for birds in other habitats (e.g., Soulé et al. 1988; Crooks et al. 2001; Crooks, Suzrea, and Bolger 2004), as well as for mammals (Picton 1979; Newmark 1986; A. Bennett 1990; Verboom and van Apeldoorn 1990; Cutler 1991; Lindenmayer et al. 2000), insects (Webb and Hopkins 1984; Ås 1993; Bruhl, Eltz, and Linsenmair 2003; Feer and Hingrat 2005; Cane et al. 2006), and herbaceous plants (Simberloff and Gotelli 1984; Webb and Vermaat 1990; Leach and Givnish 1996; Krauss et al. 2004; Petit et al. 2004; Guirado, Pino, and Roda 2006). These positive correlations between fragment size and species richness are commonly observed and have often been used as the basis for decisions regarding biodiversity protection.

FRAGMENT SIZE AND POPULATION ABUNDANCE

These consistent, positive relationships between species richness and habitat area ultimately represent the collective responses by populations of different species to shifts in habitat area. In other words, larger fragments may sustain a greater number of species, because the habitat is extensive enough to support viable populations of more species than would be possible in smaller fragments. Recall the right-hand half of the IBT diagram (in chapter 2), which predicts that larger fragments have more species because extinction rates of populations are higher in smaller fragments. To understand the components of overall species-area patterns, many investigators have examined the responses of individual species to changes in fragment size.

Some species may be more tolerant of reductions in habitat fragment size than others, based on their particular life-history attributes (Terborgh 1986; W. Laurance 1991; Webb and Thomas 1994; MacNally and Bennett 1997; Kolozsvary and Swihart 1999; Deng and Zheng 2004; Ewers and Didham 2006; but see

Table 3.1

Selection of 35 observational studies of species richness in relation to fragment area. Studies were chosen to represent a wide range of organisms, habitats, and geographic locations and are listed in chronological order.

Reference	Organism(s)	Habitat type	Location	Fragment size range
Forman, Galli, and Leck 1976	Birds	Temperate broadleaf forest	United States	0.01–40 ha
Picton 1979	Large mammals	Alpine	United States	30–11,700 km^2
Blake and Karr 1987	Birds	Temperate broadleaf forest	United States	1.8–600 ha
Soulé et al. 1988	Birds	Chaparral	United States	0.4–104 ha
Fonseca and Robinson 1990	Small mammals	Tropical rainforest	Brazil	60–36,000 ha
Webb and Vermaat 1990	Plants	Heathland	England	5–200 ha
Estrada, Coates-Estrada, and Meritt 1993	Bats	Tropical rainforest	Mexico	1–2000 ha
Herkert 1994	Birds	Grasslands	United States	0.5–650 ha
Bellamy, Hinsley, and Newton 1996	Birds	Temperate deciduous forest	England	0.02–30 ha
Fukamachi, Iida, and Nakashizuka 1996	Plants	Temperate mixed broadleaf/conifer forest	Japan	1.8–261 ha
Leach and Givnish 1996	Plants	Grasslands	United States	0.2–6.0 ha
Grashof-Bokdam 1997	Plants	Temperate broadleaf forest	The Netherlands	0.04–770 ha
Kruys and Jonsson 1997	Lichens	Boreal forest	New Zealand	0.4–15.9 ha
Hinsley et al. 1998	Birds	Temperate broadleaf forest	The Netherlands, United Kingdom, Denmark, Norway	≤0.5–15 ha
Shanker and Sukumar 1998	Small mammals	Montane forest	India	0.2–60 ha
Grant and Berkey 1999	Birds	Aspen forests	United States	<1–250 ha
Lindenmayer et al. 2000	Mammals	Temperate broadleaf forest	Australia	0.2–125 ha
Borgella, Snow, and Gavin 2001	Hummingbirds	Tropical rainforest	Costa Rica	0.3–226 ha

Reference	Taxon	Habitat	Location	Area range
Shochat, Abramsky, and Pinshow 2001	Birds	Scrubland and planted conifer forest	Israel	2.5–3000 ha
Beier, Van Drielen, and Kankam 2002	Birds	Semi-deciduous forest	Ghana, West Africa	3–33,000 ha
Gibb and Hochuli 2002	Arthropods	Heathland and woodland	Australia	0.04–164.4 km²
Bruhl, Eltz, and Linsenmair 2003	Ants	Tropical rainforest	Malaysia	1.46–438 km²
Ochoa-Gaona et al. 2004	Woody plants	Tropical rainforest	Mexico	10–265 ha
Summerville and Crist 2004	Moths	Temperate broadleaf forest	United States	1.7–289 ha
Watson, Whittaker, and Dawson 2004	Birds	Littoral forest	Madagascar	0.30–464 ha
Zhu et al. 2004	Plants	Tropical rainforest	China	3–75,000 ha
Feer and Hingrat 2005	Dung beetles	Tropical rainforest	French Guiana	1.1–38 ha
Murakami, Maenaka, and Morimoto 2005	Ferns and fern allies	Temperate forest	Japan	0.11–60.3 ha
Gignac and Dale 2005	Lichens and bryophytes	Boreal forest	Canada	0.002–17 ha
Harcourt and Doherty 2005	Primates	Tropical forest	Africa, Asia, Madagascar, South America	0.01–100 km²
Martínez-Morales 2005	Birds	Tropical cloudforest	Mexico	0.6–16,289 ha
Michalski and Peres 2005	Primates and carnivores	Tropical forest	Brazil	0.47–13,551 ha
Guirado, Pino, and Roda 2006	Understory plants	Forests	Spain	8–18,000 ha
Lovei et al. 2006	Ground beetles	Temperate broadleaf forest	Hungary and Ukraine	41–3995 ha
Peak and Thompson 2006	Birds	Riparian forests	United States	Width: 55–530 m

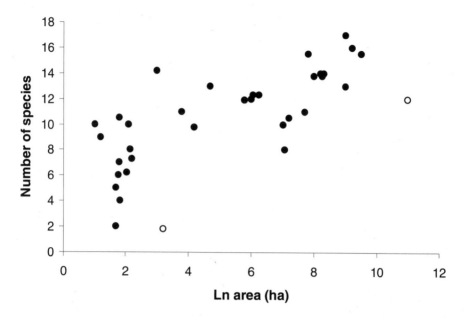

Figure 3.1. The number of forest bird species per transect in 35 forest fragments in Ghana, West Africa, in relation to forest fragment area. The two open circles represent outliers: the lower left outlier has fewer species than expected for a fragment of its size, but this was the most isolated forest fragment; and the upper right outlier also had fewer species than expected for its size, but this fragment was dominated by an exotic tree species. Redrawn from Beier, Van Drielen, and Kankam (2002).

also MacNally, Bennett, and Horrocks 2000). For example, insect species with poor dispersal abilities persisted in large heathland fragments but not in small ones (Hopkins and Webb 1984), presumably because the small fragments were insufficient to maintain viable populations, and these insects were unable to disperse to more suitable habitat. Similarly, the relatively sedentary Florida scrub lizard (*Sceloporus woodi*) is a poor disperser and appears to be constrained to large habitat patches in central Florida (Hokit, Stith, and Branch 1999). Rare species, which tend to be more specialized in their feeding habits than common species, have been shown to be particularly sensitive to decreases in habitat fragment size (Terborgh and Winter 1980; W. Laurance 1990, 1991). For example, Cabot's tragopan (*Tragopan caboti*), a pheasant endemic to southeastern China, is a habitat specialist, has poor dispersal abilities, is non-migratory, and was absent from small, isolated forest patches (Deng and Zheng 2004). Ecologists on other continents have made similar observations for relatives of these comparatively large, chickenlike birds. In particular, for hazel grouse (*Bonasa bonasia*) in Sweden (Åberg, Swenson, and Andrén 2000) and greater prairie chickens (*Tympanuchus cupido*) in the midwestern United States (Winter and Faaborg 1999), the probabilities of occurrence were greater in large than in small habitat fragments.

The studies just described noted the presence/absence patterns of species in relation to fragment size. Alternatively, species may be present in small fragments, but persist at either lower (e.g., Crooks et al. 2001; Crooks, Suarez, and Bolger 2004) or higher (MacNally, Bennett, and Horrocks 2000) population sizes than in larger fragments. The magnitude of the effects of patch size on population abundance may be influenced by characteristics of the species themselves (as noted above) or of the landscapes in which they occur. Bender, Contreras, and Fahrig (1998) performed a meta-analysis of 25 different studies that reported patch-size effects on populations of 134 different species of birds, mammals, and insects. They related the magnitude of the patch-size effect to the general features of the species and the landscapes in which they occurred, including such factors as habitat affinities of the species, whether the species were migratory or resident, taxonomic group, trophic status, geographic location, and percent cover of habitat in the surrounding landscape. They concluded that habitat association explained most of the variation among species in their response to changes in patch size. Specifically, the population density of edge species (those species primarily associated with habitat edges; see chapter 5) was negatively associated with patch size, but the population density of interior species (those species primarily associated with core habitats) was positively related to patch size. For generalist species (those with no clear association with edge or interior habitats), there was no discernible effect of patch size and population density. The authors also noted that resident species were more strongly affected by reductions in patch size than were migratory species, and that western hemisphere species (North and South America) showed greater responses to changes in patch size than did eastern hemisphere species (Europe, Africa). The latter result is intriguing, because it suggests that there may be fewer area-sensitive birds remaining in the Eastern Hemisphere, since it has been inhabited by humans for a greater length of time than the Western Hemisphere (Bender, Contreras, and Fahrig 1998).

If the population sizes or densities of particular species in smaller fragments are indeed lower than those in larger fragments, what is the cause? Four basic processes are responsible for changes in population size—birth, death, immigration, and emigration—and these demographic factors are likely to vary under different conditions. Much of what we know about demographic shifts in relation to patch size comes from experimental fragmentation studies, which will be discussed in the following chapter. But several observational studies—such as the recent meta-analysis of demographic responses by birds to forest fragmentation by Lampila, Monkkonen, and Desrochers 2005—point to specific demographic processes that explain changes in population abundance with fragment size. For example, the reproductive output by ovenbirds (*Seiurus aurocapillus*) in deciduous forests in Pennsylvania was 20 times lower in small forest fragments than in large forested areas (Porneluzi et al. 1993). Large habitat patches occupied by New England cottontails (*Sylvilagus transitionalis*) in the northeastern United States were more resource-rich than small habitat patches, resulting in a nutrient limitation for rabbits on small patches and, ultimately, higher mortality rates on small as opposed to large patches (Villafuerte, Litvaitis, and Smith 1997). In Oregon, recruitment of the forest understory herb *Trillium ovatum* was reduced significantly for plants growing near clear-cut edges than for those in interior habitats in the forest (Jules 1998). Because smaller fragments have a proportionately greater edge habitat, a logical inference from this study is that smaller fragments should have lower *Trillium* recruitment than larger fragments. Each of the studies mentioned above demonstrates a change in a particular demographic parameter in relation to habitat fragment size or edge effects.

Shifts in demographic parameters in patches of different sizes may ultimately be caused by altered interspecific interactions. Further research on the understory herb *Trillium ovatum* revealed that the lower recruitment of plants near forest edges was most likely due to decreased seed production, which in turn was caused by shifts in pollination and increased seed predation by rodents (Jules and Rathcke 1999). Similar findings of reduced pollination in fragmented habitats have been found for a meadow herb in Sweden (Jennersten 1988) and for several plant species that inhabit the Argentinean Chaco (Aizen and Feinsinger 1994). Higher seed predation on *T. ovatum* by deer mice (*Peromyscus maniculatus*) in forest edge habitats bolstered mouse survival and dispersal, resulting in three- to four-fold differences in mouse densities in forest fragments as compared to continuous forest (Tallmon et al. 2003). Avian nest predation and brood parasitism, which have been studied in great detail, tend to be higher in small forest fragments than in large fragments or a continuous forest, and have caused the increased mortality and decline of songbird populations (Brittingham and Temple 1983; Wilcove 1985; J. Gibbs and Faaborg 1990). These intriguing domino effects of landscape change involve multiple species and their interactions.

FRAGMENT ISOLATION

By now it should be quite obvious that there is almost always a positive relationship between the size of habitat remnants and species richness within those fragments. For some species, but not all, there is also a positive relationship between patch size and population density or abundance. But the effects of patch isolation, to which we now turn, are not so ubiquitous and consistent. Many studies that examine both patch size and isolation in relation to species richness or population density find a primary effect of patch size and a secondary or negligible effect of patch isolation (e.g., Bellamy, Hinsley, and Newton 1996; Bruun 2000; Brotons and Herrando 2001; Fernández-Juricic 2004; Krauss et al. 2004; Watson, Whittaker, and Dawson 2004; Harcourt and Doherty 2005), while others show effects of isolation that are equal to or greater than the effects of patch size (Estrada, Coates-Estrada, and Meritt 1998; Kehler and Bondrup-Nielsen 1999; Kolozsvary and Swihart 1999; Deng and Zheng 2004; Ficetola and De Bernardi 2004; Piessens et al. 2004; Parris 2006). For example, Feer and Hingrat (2005) noted that the species richness and abundance of dung beetles in tropical forest fragments in French Guiana were positively related to fragment size, but not significantly related to measures of fragment isolation. In contrast, Piessens et al. (2004) studied plant species richness in remnant heathland patches in Belgium and found that species richness increased with area, but was even more strongly influenced by measures of patch isolation.

Part of the difficulty in discerning consistent isolation effects is that patch isolation is rather tricky to measure. Recall that island biogeography theory described isolation in terms of the distance of an island from a mainland source of colonists, which is quite sensible for terrestrial species that cannot possibly inhabit the marine environment. But for habitat patches in fragmented landscapes, isolation may be measured in various reasonable ways, depending on landscape characteristics and species' sensitivities to habitat types or to land uses surrounding the habitat remnants. This section will briefly review the ecological theories underlying investigations of patch isolation, explore how isolation has typically been measured in the field and in computer simulations, cite evidence for species' responses to fragment isolation, and discuss some factors that influence the various responses of species to isolation.

Theory Related to Isolation

Two main bodies of ecological theory, island biogeography theory and metapopulation theory, relate population and community dynamics to patch isolation. Each of these theories provides testable predictions for understanding organisms' responses to isolation. Recall from chapter 2 that the key prediction from MacArthur and Wilson's (1967) island biogeography theory regarding habitat isolation was that closer islands should have higher immigration rates than more distant islands. J.H. Brown and Kodric-Brown (1977) contributed the idea of the rescue effect, suggesting that both island size and distance may influence the extinction rate of species on islands and, consequently, patterns of species richness. Later developments suggested that habitat corridors may facilitate movement among fragments and effectively serve to reduce habitat isolation (E. Wilson and Willis 1975; Simberloff and Cox 1987; A. Bennett 1990).

Metapopulation theory describes the population dynamics of single species in patchy or fragmented habitats. Recent advances in metapopulation theory also suggest that habitat corridors can modify species persistence by enhancing the rates of colonization among metapopulations (e.g., McCullough 1996). Too much movement among patches, however, may cause the population dynamics in patches to become synchronized, so that the persistence of the overall metapopulation may be threatened.

Measuring Isolation

Patch isolation may meaningfully influence the number of species that can persist in a habitat fragment, as well as the abundance of a particular species within a fragment. But as mentioned above, measuring isolation is more complicated than it may appear at first glance. It is tempting to claim here that patch isolation is far more difficult to measure than patch size. That is probably true, although defining just what constitutes a habitat patch for a particular organism may prove annoyingly elusive. In most published studies where patch size has been reported, the measurements are for clearly defined habitat fragments, such as forest remnants in agricultural landscapes or chaparral fragments within urbanized landscapes. Moreover, as discussed at length in chapter 1, Fahrig (2003) pointed out that many studies measure species' responses to changes in patch size at the local scale, but do not measure the overall amount of habitat in the broader landscape. Landscapes that have larger patches usually also have a greater overall amount of habitat. Thus conclusions based solely on local patch size may actually reflect species' responses to the amount of habitat in the broader landscape.

To return to the related issue of isolation, many measures of patch isolation, unfortunately, have been confounded with measures of the amount of habitat in the landscape (Fahrig 2003). Specifically, patch isolation, as described in most studies, is actually a measure of the amount of habitat in the broader landscape. To disentangle the effects of patch size and isolation, Fahrig recommended that researchers statistically control for the amount of habitat when measuring landscape fragmentation. Because isolation may depend on such critical factors as the species or landscape under study, there has been much discussion in the literature over how best to measure isolation (and its inverse, connectivity), specifically with regard to how species perceive and move through different landscapes (e.g., Tischendorf and Fahrig 2000, 2001; Moilanen and Nieminen 2002; Bender, Tischendorf, and Fahrig 2003; Tischendorf, Bender, and Fahrig 2003; Calabrese and Fagan 2004).

Essentially there are two categories of isolation measurements, focused either on structural or functional aspects of connectivity (Calabrese and Fagan 2004). *Structural connectivity* measures aspects of the spatial characteristics of the landscape, independent of the movement abilities of species that inhabit the landscape. *Functional connectivity* includes both physical landscape features and the movement abilities of the species found in that landscape. According to Calabrese and Fagan's (2004) useful classification scheme, functional connectivity comes in two flavors: *potential*, which incorporates the predicted dispersal abilities of species, and *actual*, which takes into account the measured movement pathways of individuals. To illustrate the difference between structural and functional connectivity, consider the isolation of two small towns near where I live in the mountains of Colorado. The structural connectivity (expressed as the straight-line distance) between the towns of Jamestown and Nederland is only about 20 km. But the functional connectivity for a human in an automobile, following paved roads, is closer to 40 km. And if that human were adventurous enough to ride a mountain bike between the two towns, the functional connectivity would be about 25 km.

Most of the research cited in this chapter measured structural connectivity, or isolation in relation to the physical features of the landscape. The method typically used in these studies is to locate a point in patch A (either at the patch center or at an edge) and a point in patch B on a map and then measure the shortest distance between the two points. With multiple patches in a landscape, the usual method is to measure the nearest-neighbor, straight-line distance, that is, the shortest distance that one particular habitat patch is from another patch of suitable habitat. A recent review of papers that reported connectivity measurements found that 44% of the studies used some type of nearest-neighbor distance measurement to estimate isolation (Moilanen and Nieminen 2002). For example, Soulé et al. (1988) studied bird species richness in chaparral fragments in Southern California and measured isolation in two ways: first, as the distance from one chaparral fragment (canyon) to the nearest canyon that contained suitable habitat and species composition, and second, as the distance to the nearest canyon that was of an equivalent size to or larger than the focal fragment. Similarly, Enoksson, Angelstam, and Larsson (1995) used distance-based criteria to define deciduous forest patches in Sweden as isolated or aggregated. *Isolated* patches were surrounded by very few other deciduous patches within a 4 km radius (the average distance between patches was 1.4 km), whereas *aggregated* deciduous forest patches were those that were situated in close proximity to other deciduous forest patches (here the average distance between patches was 363 m). Turchi et al. (1995) incorporated both inter-patch distance and patch area into four measures of isolation for aspen stands in Colorado's Rocky Mountains. The reasoning behind the area-weighted distance measures was that a forest fragment would effectively be less isolated if it is 1 km from a large forest patch than if it is 1 km from a small one (recall Fahrig 2003). Finally, Petit et al. (2004) developed three measures of isolation for ancient British woodland patches: the total area of woodland within 500 m of a forest patch, the total number of woodland patches within 500 m of the focal patch, and the length of hedgerows and lines of trees within a 1 km square surrounding the focal woodland fragment. All of these isolation measures describe structural connectivity (*sensu* Calabrese and Fagan 2004) in that they account for isolation based on habitat features present in the landscape but do not explicitly consider the ease with which plants or animals move among the fragments.

Fragment Isolation and Species Richness

Having considered the theoretical underpinnings of patch isolation and the practicalities of measuring isolation in the field, the question now becomes, How well do observations of habitat fragments that are isolated to varying degrees fit predictions of ecological theory? In the case of island biogeography theory and its modifications, observational studies often, but not always, show inverse relationships between species richness and patch isolation. Many of the ones that examine both patch size and isolation in relation to species richness or population density show negative effects of isolation that are equal to or greater than the positive effects of patch size on species richness (Grashof-Bokdam 1997; Estrada, Coates-Estrada, and Meritt 1998; Kehler and Bondrup-Nielsen 1999; Kolozsvary and Swihart 1999; Deng and Zheng 2004; Ficetola and De Bernardi 2004; Piessens et al. 2004; Parris 2006). Other research finds a primary effect of patch size and a secondary or negligible effect of patch isolation on species richness (Bellamy, Hinsley, and Newton 1996; Bruun 2000; Fernández-Juricic 2000; Brotons and Herrando 2001; Krauss et al. 2004; Watson, Whittaker, and Dawson 2004; Harcourt and Doherty 2005). The lack of consistency in effects of isolation among these and other studies may be at least partly due to the way isolation is defined (Fahrig 2003). As an example of the relatively stronger effects of patch isolation than of size, Piessens et al. (2004) examined plant species richness in remnant heathland patches in Belgium and found that species richness increased with area, but was even more strongly influenced by five different measures of patch isolation. In particular, the results provided evidence for both spatial and temporal rescue effects. Species appeared to be able to disperse easily between closely spaced patches, preventing extinctions and maintaining relatively high species richness in less isolated patches. Additionally, species with short-lived seeds that do not remain for extended periods in the soil seed bank were more sensitive to isolation (i.e., they were absent or less abundant in more isolated patches) than species with longer-lived seeds (table 3.2). This is evidence for a temporal rescue effect or *storage effect* (*sensu* Warner and Chesson 1985), with the persistence of a species

TABLE 3.2

Effects of heathland fragment isolation, measured in five different ways, on the occurrence of plant species with low or high seed longevity, where high seed longevity refers to species with a more persistent seed bank. Connectivity was an area-weighted distance measure that takes into account all other heathland patches. *Low seed longevity* refers to categories 1 and 2 of Piessens et al. (2004), and *high longevity* refers to their categories 3 and 4. Correlations between the number of species of each category in heathland patches and patch isolation measures are as follows: *** represents $P < 0.001$, ** represents a *P*-value between 0.001 and 0.01, * represents a *P*-value between 0.05 and 0.01, NS equals not significant, + represents a positive correlation, − represents a negative correlation. Compiled from Piessens et al. (2004).

Seed longevity	Low	High
Distance to nearest patch	*** (−)	NS
Mean distance to nearest five patches	*** (−)	NS
Area of heathland within 100 m	*** (+)	* (+)
Area of heathland within 500 m	*** (+)	** (+)
Connectivity	*** (+)	*** (+)

over time occurring via survival in the soil seed bank, even in more isolated patches.

The effects of isolation may be most severe for relatively sedentary species, such as vascular plants, and for species that require two or more habitat types to complete their life cycle, such as amphibians. Herbaceous plants of ancient woodland patches in Britain (Petit et al. 2004) showed distribution patterns similar to plants in the heathland patches described above (Piessens et al. 2004). The study focused on British ancient woodland indicator species, which are species known to occur primarily in old-growth forest habitats that are relatively weak dispersers. For 218 woodland patches located in the British lowlands, Petit and his colleagues observed the strong effects of two measures of fragment isolation on herbaceous plant species richness in woodland fragments. The total length of linear wooded features (hedgerows and lines of trees) within 500 m of a forest patch had a strong positive influence on species richness within the patch, and the total area of woodland within 500 m of the fragment secondarily influenced species richness. Thus these wooded linear features in the landscape may serve as refugia or as dispersal corridors for forest-specialist plant species.

Amphibians typically require aquatic habitats for breeding and for their larval life stages, and terrestrial habitats for their adult life stages. This life-history feature may render them particularly sensitive to habitat isolation, since they must move between these habitat types several times during their lives (Cushman 2006). Parris (2006) studied amphibian assemblages in ponds situated in public parks within the urban context of Melbourne, Australia. As a measure of isolation, Parris quantified the area covered by paved roads within 500 m of the center point of each pond. This measure assumes that roads form at least partial barriers to the movement of amphibians across this landscape and that higher densities of roads will cause greater isolation. Interestingly, this isolation of ponds by roads had the greatest effect on amphibian species richness within the ponds, compared to the effects of pond area and local habitat variables. Ponds surrounded by high road cover contained only a small fraction of the species that occurred in ponds surrounded by low road cover. Parris surmised that with increasing urbanization and road-building, these ponds may become even more isolated, so that the rescue of species from extinction or the recolonization of ponds following local extinction may become increasingly unlikely.

In a rather extreme instance of habitat isolation, the tropical forest reserve that is now Barro Colorado Island (BCI) is a land-bridge island that was continuous with the mainland of Panama prior to construction of the Panama Canal and Lake Gatun in 1914. It has been estimated that approximately 65 of an original

209 bird species that were present prior to isolation have now gone extinct on this 1600 ha island (Willis 1974; Karr 1990; W. Robinson 1999). Studies of a suite of forested islands in Lake Gatun created at the same time revealed similar patterns of species loss in relation to habitat isolation. Tropical tree species richness was lower on six islands than in comparable areas on the mainland (Leigh et al. 1993).

In other cases, isolation may play a secondary or even negligible role in determining species richness within fragments. For example, Turchi et al. (1995) studied bird communities associated with aspen stands in Rocky Mountain National Park, Colorado, and found no significant effects of isolation on bird species richness, despite using four different measures of isolation. For birds in the Mediterranean basin, Brotons and Herrando (2001) observed that fragment area explained 70% of the variance in species richness among pine fragments in an agricultural matrix, and that fragment isolation contributed relatively less to determining species composition. Similarly, species richness and the abundance of dung beetles in tropical forest fragments in French Guiana were positively related to fragment size but not significantly related to measures of fragment isolation (Feer and Hingrat 2005), and Veddeler et al. (2005) observed no effects of isolation on tropical butterfly species richness in Indonesia, at least for forest fragments separated by up to 1700 m.

Fragment Isolation and Population Occurrence

As with relationships between species richness and patch area, relationships between species richness and patch isolation ultimately represent the collective responses of particular species to the spatial structure of landscapes. Shifts in population occurrence may be driven by changes in species interactions in isolated habitats, but most studies of these phenomena either are of naturally patchy systems or use experimental approaches, so they will be considered elsewhere in this book. Several studies have specifically examined the responses of individual species (primarily species occurrence rather than abundance) to fragment isolation across a range of habitat types and life forms. For example, three of six forest-dwelling bird species in south-central Sweden appeared with lower frequency in deciduous forest patches isolated by coniferous forest than in deciduous forest patches that were more closely aggregated (Enoksson, Angelstam, and Larsson 1995). In New South Wales, Australia, the decline in the presence of the brown treecreeper in fragmented forests is thought to be due primarily to disrupted dispersal caused by habitat isolation (Walters, Ford, and Cooper 1999). The incidence of Cabot's tragopan in forest fragments of southeastern China was

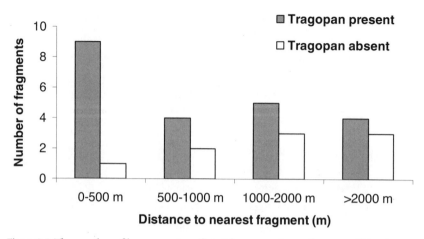

Figure 3.2. The number of lower montane forest fragments in southeastern China in which Cabot's tragopan (*Tragopan caboti*) was present (gray bars) and absent (white bars) in relation to that fragment's distance from the next-nearest fragment, calculated for four distance classes. Thirty-one fragments were sampled, which ranged in size from 2.5 to 48.5 ha. Redrawn from Deng and Zheng (2004).

positively affected by fragment size (mentioned above), and negatively affected by isolation (Deng and Zheng 2004; see fig. 3.2). These birds were found with greater frequency in forest fragments that were within 500 m of the nearest suitable habitat than in forest fragments that were more than 1000 m apart. For three of four ranid (frog) species studied in fragmented wetland habitats in the midwestern United States, habitat isolation had a strong negative effect on occurrence (Kolozsvary and Swihart 1999). Kehler and Bondrup-Nielsen (1999) found significant effects of isolation on the presence of a fungivorous forest beetle, *Bolitotherus cornutus*, in fragmented and continuous forests. They compared beetle occurrence patterns using isolation measures that included the distance to forest patches occupied by the beetles and forest patches unoccupied by them, but found no difference in beetle occurrence using these two types of measures.

Shifts in occurrence patterns of a particular species may be explained by the behavioral attributes of the species. For example, the isolation of remnant shrublands in western Australia alters dispersal behavior and social interactions of the white-browed babbler (*Pomatostomus superciliosus*) (Cale 2003). These birds are relatively sedentary, cooperative breeders, and both males and females may disperse from their natal habitat to find new breeding opportunities. Male birds were hesitant to disperse to patches more than 1 km from their natal patch, which resulted in smaller social groups in isolated patches and perhaps lower levels of productivity in these patches.

CONSERVATION CONTROVERSIES

The ecological relationships between species richness and fragment size and isolation are key components of theories in population and community ecology. They were featured prominently in the early principles of reserve design (Willis 1974; Diamond 1975; E. Wilson and Willis 1975) and have become cornerstones of modern conservation science. Ecologists and conservation biologists have debated—with considerable fervor—three particular issues that arise from these relationships. Each of these ecological concepts is essentially a hypothesis to be tested in the real world of conservation. First, given the consistent positive relationship between island or habitat area and species richness, one reserve design principle is that a large reserve is better than a small reserve. Similar reasoning led to the principle that a single large reserve would better protect biodiversity than several small reserves whose total area equaled that of the large reserve (Diamond 1975; Terborgh 1975; E. Wilson and Willis 1975). Second, species-area relationships have been used to predict rates of species extinctions given the expected rates of habitat loss (E. Wilson and Willis 1975; E. Wilson 1992). Third, given that isolated habitat fragments often have fewer species compared to more continuous habitat, another reserve design principle is that extinction rates will be lower when reserves are connected by corridors rather than when they are isolated (Diamond 1975; E. Wilson and Willis 1975). These will be discussed one at a time, starting with SLOSS.

Single Large or Several Small (SLOSS)?

Species-area curves show that the number of species typically increases with habitat area. Thus it is reasonable to assume that large reserves would contain more species than small reserves, and that fragmentation of a large reserve into smaller chunks of equal total area would diminish species richness. But the degree to which a set of small reserves harbors fewer species than a large reserve depends on how similar the small reserves are to one another in species composition (Quinn and Harrison 1988). E. Wilson and Willis (1975) stated that their design principles applied to preserves in "a homogeneous environment" (p. 529). This is a key qualifying statement, for if small patches are relatively homogeneous and all tend to support the same suite of species, then a collection of small reserves will cumulatively contain fewer species than a single large reserve. But if the small patches vary substantially from one another in species composition, then the set of small reserves is likely to contain cumulatively more species than the single large reserve. Diamond (1975) similarly foresaw the importance of habitat heterogeneity in his original discussion of this reserve design principle: "Separate reserves in an inhomogeneous region may each favour the survival of a different group of species; and . . . even in a homogeneous region, separate reserves may save more species of a set of vicariant similar species, one of which would ultimately exclude the others from a single reserve" (p. 144).

In fact, the typical observation for most habitat islands that have been surveyed is opposite to the "a single large preserve is better" principle. Collections of small reserves tend to contain more species than a single large reserve, which suggests that habitat heterogeneity must play a key role in determining the spatial structure of species richness patterns, as E. Wilson and Willis (1975) and Diamond (1975) expected. For example, Simberloff and Gotelli (1984) examined plant species richness in remnant prairie and forest fragments from five different data sets gathered in the midwestern United States (148 total fragments). They found that where the total area was equal, groups of small sites tended to have more species than single sites. A similar pattern was found for animals in U.S. national parks (Quinn and Harrison 1988); vascular plants in mires (Virolainen et al. 1998); grassland butterflies (Tscharntke et al. 2002); pond-dwelling plants and animals (Oertli et al. 2002); plants in urban woodlots (Godefroid and Koedam 2003); and vascular plants, brophytes, day-active butterflies, and grasshoppers in calcareous fens (Peintinger, Bergamini, and Schmid 2003).

The community property of nestedness has also been discussed widely in the context of the sloss debate (e.g., Cutler 1991; R. Cook 1995; Boecklen 1997). A species assemblage that is *perfectly nested* is one in which the collection of species in species-poor sites is a subset of species that occur in species-rich sites. If there is a positive relationship between species richness and area, and the biota is perfectly nested, then small patches would be comprised of a subset of species that occur in the large patches. With such a pattern, rare species, which are usually the species of conservation concern, would be present in large patches, but they would be less likely to occur as patch size declined. Using this logic, if an assemblage is perfectly nested, then a single large patch should always be a preferable reserve design to several small patches.

Some research has examined the degree of nestedness within assemblages, with varying results. Boecklen (1997) reviewed a series of published studies on species distributional patterns and observed that nestedness varied substantially among groups and study systems, making it difficult to generalize nestedness patterns to the sloss discussion. In fact, he concluded that "nestedness says little about optimal reserve design and management, and appears to be a weak conservation tool" (p. 123). Others have observed highly nested species assemblages in at least some of the communities that they have studied, ranging from plants to butterflies to vertebrates (Honnay, Hermy, and Coppin 1999; Fleishman and MacNally 2002; Berglund and Jonsson 2003; Godefroid and Koedam 2003; Donlan et al. 2005). Fischer and Lindenmayer (2005) perhaps best summarize the relevance of nestedness to reserve design by highlighting the distinction between perfectly nested assemblages and significantly nested ones. Their point is that unless an assemblage is perfectly nested, then a single large reserve will not necessarily be preferable for the maintenance of biodiversity than several small reserves.

But whether a single large reserve is better than several small ones also depends on what metric is being used to measure "better." Some species may simply not be able to persist in small habitat fragments, due to their body size, home range requirements, or dispersal mode. A collection of small reserves may often maximize species richness compared to a single large reserve, but large reserves may minimize the probability of species extinctions, especially for mammal species (Picton 1979; Newmark 1986, 1987, 1995, 1996; McCarthy and Lindenmayer 1999). Newmark (1987) analyzed local extinctions of large mammals in North American national parks since the parks were established, using a comparison of historical records to present-day distributions. The analysis very clearly showed that larger parks maintained a higher number of species over time than did smaller parks, supporting the notion that large areas are essential to minimize extinctions of particular species. These analyses were repeated for a larger suite of mammals in North American parks and for six national parks in Tanzania, and similar patterns were found: mammalian extinctions were inversely related to park size (Newmark 1995, 1996).

So the resolution to the sloss debate seems to depend on one's conservation goals. If the goal is to maximize species richness in reserves, all other things being equal, then a collection of small reserves will likely harbor more species, particularly if the habitat is relatively heterogeneous, than a single large reserve. If the goal, however, is to minimize species' extinctions or the extinction of a particular species, then a large reserve will almost always be preferable to a small reserve. In all likelihood, both goals will be important in reserve design, and some combination of protected areas of varying sizes in varying habitat types will be the most suitable arrangement for biodiversity protection (e.g., see Peintinger,

Bergamini, and Schmid 2003). Recent modeling efforts show that—depending on the objectives and the distribution of diversity among habitats—a combination of reserve sizes may be appropriate (McCarthy, Thompson, and Possingham 2005; Wiersma and Urban 2005). Rarely, however, do we have the opportunity to design a reserve system from scratch, so decisions must typically be made in the context of existing protected areas.

Species-Area Curves and Extinction Rates

Species-area curves display distinct relationships between habitat area and species richness. They are typically thought of as describing the *increase* in species richness with increasing area, but one can also slide down this line to calculate the extent to which species richness *decreases* with a decrease in area (fig. 3.3). Thus species-area curves should be useful in estimating the rates of species loss from expected rates of habitat loss. The magnitude of the effect of species loss will depend on the z-value, the slope of the line that describes the change in species richness with area (see chapter 2). These z-values generally range between 0.15 and 0.40, and, by using a z-value of 0.30, a general rule emerges that reducing habitat area by 90% reduces species richness by 50% (E. Wilson and

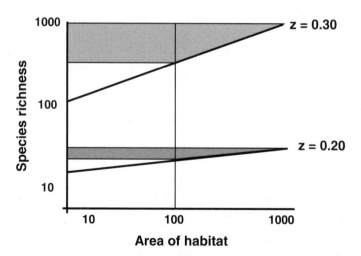

Figure 3.3. A conceptual representation of estimated species loss with reduction in habitat area, using species-area curves. When $z = 0.20$, the number of species lost as habitat area is reduced (represented by the lower gray polygon) is relatively low compared to when $z = 0.30$, where the number of species lost with the same reduction in area is much greater (represented by the upper gray polygon).

Willis 1975; E. Wilson 1992). Based on historic and current rates of habitat destruction, and using a conservative z-value of 0.15, in 1992 Wilson estimated that human activities are committing 27,000 species per year to extinction, which is 1000 to 10,000 times the background extinction rate estimated from the fossil record.

Critics have argued that these dramatic estimates of species extinctions have not been realized. We are simply not seeing the number of species extinctions that these estimates suggest we should be seeing. May, Lawton, and Stork (1995) and Pimm (2002) noted that there are likely to be lengthy time lags in species' responses to habitat loss. So even for species that have not yet gone extinct at the expected rates due to habitat loss, the number of populations and individuals are still in decline, and the species is "committed to extinction" (May, Lawton, and Stork, p. 16). Thus we are likely to see the end results within the next century. This time-lag phenomenon in species extinctions has also been referred to as the *extinction debt* (e.g., by Vellend et al. 2006).

Briefly, then, z-values derived from species-area curves strongly influence extinction rate estimates, and z-values vary substantially among species groups and habitats. Drakare, Lennon, and Hillebrand (2006) performed a quantitative meta-analysis of species-area curve studies and found extensive and systematic variation in species-area curves related to species' traits, geographic location, and habitat type. Their review suggested that z-values are "strong indicators" of species' sensitivities to habitat loss. Additionally, the shape of species' geographic ranges (Ney-Nifle and Mangel 2000) and non-random patterns of habitat loss (Seabloom, Dobson, and Stoms 2002) may cause extinction rates to even exceed the estimates derived from classic species-area curves.

Habitat Corridors

In addition to reserve size, the isolation of nature reserves was considered in shaping the principles of reserve design (E. Wilson and Willis 1975, Diamond 1975). Willis's view of habitat isolation derived from his studies of birds on Barro Colorado Island: "Limitation of human use of space will be most effective in preserving natural biotas if natural areas are not isolated islands in lakes or seas of humanity but instead are linked by corridor zones" (Willis 1974, p. 167).

Habitat corridors are linear strips of protected habitat; in biological conservation, they are proposed as a way to moderate the negative effects of habitat isolation on animal movement and species persistence. Habitat corridors that structurally link otherwise isolated habitat remnants are supposed to increase landscape connectivity by facilitating the movement of organisms between habitat fragments, thereby reducing rates of species loss, enhancing probabilities of colonization, and increasing overall species richness in connected fragments (Simberloff and Cox 1987; A. Bennett 1990; Saunders and Hobbs 1991; Lidicker and Koenig 1996).

The ultimate test of this design-principle hypothesis would be to directly compare the species richness or population abundance of a particular species in habitat fragments that *are* connected with corridors to fragments that are unconnected. I first reviewed this topic over ten years ago (Collinge 1996) and concluded at that time that there was insufficient evidence to support or refute the hypothesis that corridors enhance species richness. Rather surprisingly, there is still relatively scant observational evidence that habitat fragments with corridors have higher species richness or a higher abundance of particular species than isolated fragments. Direct comparisons have not been done very frequently, and perhaps this is due to the difficulty in coming up with reasonable sampling designs in heterogeneous landscapes. As reviewed above, dozens of studies have been conducted on how size and isolation (measured as distance or area-weighted distance) influence species richness and population abundance. But finding replicated examples of connected and unconnected fragments for controlled comparisons is definitely more difficult.

That said, there are a few studies in which fragments with corridors have been compared to isolated fragments. Soon after the flurry of activity on design principles in the mid-1970s, a group of researchers studying songbirds in deciduous forests in Maryland noticed that one of their forest fragments supported a higher number of bird species than would be expected from its relatively small size (MacClintock, Whitcomb, and Whitcomb 1977). Closer examination and further surveys revealed that this 35 acre forest remnant was connected by a "disturbed corridor" to an adjacent 400 acre forest patch. They concluded that the presence of this corridor explained the higher species richness in this woodlot compared to other fragments of similar size that were more isolated. Despite the fact that the authors observed only one example of the corridor phenomenon, this study was widely cited for several years as evidence of the beneficial effects of corridors. A handful of additional observational studies provide support for the notion that species richness is enhanced by corridors, and all come from tropical forest ecosystems. Bat species richness in the forests of French Guiana was positively affected by the presence of forested corridors (Brosset et al. 1996). In Costa Rica, agricultural windbreaks connected to forests had significantly higher forest tree species richness than windbreaks that were not connected (Harvey 2000). And the diversity of small mammals within fragments of Atlantic coastal forest in Brazil was lower in isolated fragments as compared to those connected by corridors (Pardini et al. 2005).

Several observational studies of the effects of corridors on species composition in fragmented habitats have been discussed here. The bulk of evidence on the utility of corridors in conservation, however, comes from experimental studies, which will be covered in the next chapter. These are ultimately more powerful than observational studies, since they normally involve surveys of species before and after habitat isolation occurs, and usually attempt to experimentally control other factors that influence species abundance and distribution, including fragment size, shape, context, and habitat heterogeneity. Other evidence comes from research that had slightly different goals than measuring species richness in connected and unconnected fragments. For example, if corridors are to be used in linking protecting areas, then we need to know whether animals and plants find corridors to be suitable, so many papers have focused on the use of corridors as habitat and movement pathways by different species (Lindenmayer, Cunningham, and Donnelly 1993; Lindenmayer 1994). Modeling studies may also extend research on habitat use or movement by simulating the likely effects of corridors for various species (e.g., the white-footed mouse in Fahrig and Merriam 1985).

SYNTHESIS

If you remember only one thing from this book, it will probably be that larger chunks of habitat support more species than smaller chunks of habitat. Ecological theory predicts this positive species-area relationship, and a vast collection of studies confirms this prediction. Because species richness is so tightly linked to habitat area, it should be no surprise by now that the major reason for species decline in the past 50 years is because rates of habitat loss are unprecedented. Some species are particularly vulnerable to habitat loss, due to certain life-history traits such as low mobility, rarity, and large body size. Generalist species tend to be much more able to withstand the consequences of shrinking habitats.

This chapter also emphasized the effects of fragment isolation, since both size and isolation are derived from ecological theory and have been very frequently cited in fragmentation studies. Isolation can be measured in a variety of ways and ultimately is in the eye of the beholder. Whether a forest fragment surrounded by an agricultural field is considered to be isolated depends on the perspective of the species that live in the forest. If the agricultural field is a place where predators lurk and desiccation is highly likely, then the forest patch probably represents the entire universe for that species. Conversely, for species that are able to move freely through agricultural fields, the forest fragment may just be one of many habitat patches used by them. So isolation may critically affect some species but not others, depending on their behavioral and physiological traits.

Despite the importance of fragment size and isolation, many authors have also observed that these two features of habitat fragments are not necessarily the whole story. Size and isolation may influence species composition and population abundance by interacting synergistically with habitat quality (e.g., J. Thomas et al. 2001), fragment age (Soulé et al. 1988; Veddeler et al. 2005), edge effects (Parker et al. 2005), landscape context (Forman 1995; Ewers and Didham 2006), and, importantly, with each other (Fahrig 2003). Moreover, size and isolation measured at the local scale of fragments must be considered in view of the composition and configuration of the broader landscape (Andrén 1994; Fahrig 1997, 2003). For these reasons, the next generation of empirical and theoretical studies incorporates multiple factors that describe habitat spatial structure and their ecological consequences.

CHAPTER FOUR

Experimenting with Fragmentation

In principle, fragmentation experiments could provide a
rich testing ground for theories and methodologies dealing
with spatiotemporal dynamics.

—*Diane M. Debinski and Robert D. Holt (2000)*

Previous chapters extensively covered the theories that relate to habitat loss and
fragmentation and considered ecological observations related to changes in patch
size and isolation. As many of these early observations were appearing in the
published literature, however, skepticism surfaced in the scientific community.
Critics asked whether the observed ecological patterns represented the real responses of species to habitat loss and fragmentation, or if they were reactions to
some other, confounded environmental variables. Particular questions arose.
Can cause and effect relationships between landscape change and species' responses be established? And do we know anything about the mechanisms responsible for the observed responses? In many cases, the answer to these two
questions was a resounding "maybe," but not an unequivocal "yes." Field observations are essential to understanding species' responses to habitat spatial structure, but the most obvious way to resolve cause and effect, and to determine
mechanisms, is to conduct experiments where habitat size, isolation, and fragmentation are properly controlled and where ecological responses are measured.
Researchers responded to these knowledge gaps by designing creative field and
laboratory experiments to specifically test ecological theory and determine the

mechanisms responsible for species' responses to habitat loss and fragmentation.

There are several benefits to be gained by conducting experiments. First, researchers can directly test predictions of ecological theory using real organisms that occur in real habitats. If results are consistently incongruous with theoretical predictions, then theories are revised or expanded, and our scientific knowledge creeps forward. Second, a significant advantage of experiments is that researchers can take measurements prior to as well as after experimental manipulation, in order to directly assess the effects of a particular intervention. If the experiment is well designed and controlled, then the ecological responses can be directly attributed to the experimental manipulation(s). Third, because researchers are present both before the experiment starts and as the experiment proceeds, they can witness the dynamics of ecological responses over time. This may be an especially meaningful way to understand the mechanisms responsible for particular ecological responses.

An additional advantage of manipulative experiments of habitat loss and fragmentation is that, if properly designed, they can disentangle the effects of loss versus fragmentation per se. As discussed in chapter 1, the term habitat fragmentation has come to mean many things; some have argued that its meaning has been lost in its overuse and misuse (e.g., Haila 2002; Fahrig 2003; Lindenmayer and Fischer 2006). Some of the experiments discussed here focus on the ecological effects of variation in patch size and isolation, while others concentrate on the influence of habitat subdivision, or fragmentation in the strict sense. The former experiments are technically not tests of the effects of fragmentation per se, but they do provide insights into how habitat loss and isolation influence ecological processes. In practice, both types of experiments are useful for and relevant to the conservation of biological diversity, but it is critically important to understand the differences between the impacts of loss versus fragmentation and then to interpret experimental results accordingly.

Debinski and Holt (2000) surveyed literature in ecology and conservation biology published from 1984 to 1998, explicitly seeking terrestrial, field-based, experimental fragmentation studies. They found 20 such studies that met their criteria and noted the rapid increase in fragmentation experiments in recent years (they counted three ongoing experiments in 1988 versus 14 ongoing studies at the time of their literature review). This chapter expands the scope of that review to include fragmentation studies in laboratory as well as field settings, and in marine as well as terrestrial ecosystems. For ease in discussing this topic, the experiments are divided into three categories, which, for simplicity, will just be called small, medium, and large. But this categorization is certainly not intended to imply that small experiments are less meaningful than large ones. Experiments in each category have their advantages and disadvantages, but all contribute to the body of knowledge regarding the ecological effects of habitat loss and fragmentation.

Small experiments are those conducted in laboratory microcosms, typically in an aquatic medium with planktonic organisms, but they also include species such as fruit flies (*Drosophila* spp.). These microcosm experiments are essentially extensions of theoretical models (see chapter 9), yet they are more realistic than models in that they allow tests of real processes (e.g., predator-prey interactions) using real organisms in real systems (e.g., bacteria and protozoa in aquatic habitats). The main advantage of microcosm experiments is that it is relatively easy to design an experiment with a high number of replicates, facilitating rigorous statistical analyses and interpretation. A further gain is that dynamics can be followed over many generations of the study organisms (since bacteria and protozoa reproduce very quickly), allowing an examination of both short-term and long-term responses to experimental manipulation. *Medium experiments* are those that move outside the laboratory and typically involve manipulations of relatively short-lived species, such as small mammals, insects or other microarthropods in terrestrial ecosystems, or other invertebrate groups, such as crabs and shrimp in marine ecosystems. Such experiments are usually conducted at spatial scales that also allow easy manipulation of key factors. Generally, however, it is not possible to conduct an experiment with as many replicates as in small experiments, and the organisms are observed over fewer generations, so there is less capacity to view and understand long-term dynamics. Finally, *large experiments* usually have the distinct advantage of allowing the examination of larger organisms that move over broader spatial scales (such as birds or primates), which may be particularly useful when it comes to incorporating experimental results into conservation planning actions. In many cases, large experiments are so time- and labor-intensive to establish that they have been continued for at least one or two decades in order to observe both short- and longer-term ecological dynamics. But generation times for larger organisms are much longer than for organisms in microcosms, so there are still limitations to understanding the ecological effects over many generations.

This chapter reviews the experimental designs and major findings of selected studies that fall along the gradient from small microcosm experiments to medium mesocosm (or *microlandscape*) experiments to large forest and grassland fragmentation experiments. Because an explicit discussion of species interactions is featured in chapter 7, this chapter concentrates primarily on experiments that examine the dynamics of single species populations or aggregate community responses, such as those of richness or diversity to experimental loss and fragmentation. Along the way, the chapter will highlight the tradeoffs among these different experimental approaches; point out whether these experiments focus specifically on habitat loss, isolation, or fragmentation; discuss how results from different experimental systems relate to one another; and consider the types of insights gained from observational versus experimental studies in understanding the ecology of fragmented landscapes.

MICROCOSM EXPERIMENTS: SIZE SMALL

Most graduate students in ecology are encouraged to read about the experiments of Huffaker (1958); indeed, this is one of the classic papers in ecology (presented in Real and Brown 1991). Huffaker's clever laboratory experiments with oranges, rubber balls, and mites tested theory regarding the role of patchiness in promoting the persistence of species interactions, in this case, between predator and prey species of mites. Most ecology students go on to learn that his experiments were pseudo-replicated, so there are limits to the inferences that can be made. But they are nevertheless provocative experiments, ones that were novel and provided insights into the mechanisms of species coexistence, a major problem in ecology. The key result from Huffaker's experiments was that habitat patchiness promoted the coexistence of predators and prey. Despite its limitations, Huffaker's work is relevant to understanding species' responses to fragmentation, and it paved the way for future studies of the role of spatial heterogeneity in ecological dynamics.

Forney and Gilpin (1989) chose fruit flies in the genus *Drosophila*, the quintessential experimental organism, to test whether habitat loss and fragmentation would cause species extinctions. Although their study used juice bottles and fruit flies, their focus was explicitly oriented toward conservation (the paper was published in one of the early volumes of the journal *Conservation Biology*). They considered their study "a model of population processes relevant to extinction" (p. 46). The researchers used plastic juice bottles containing laboratory medium to configure fly "habitats" that varied in total size and connectedness. They then examined the amount of time until population extinction for flies in three different configurations: *separated* (two patches that were isolated from one another), *large* (a patch twice the size of the separated ones, with frequent movement of flies within the patch), and *connected* (the same size as large, but with a very limited movement of flies within the patch). The laboratory setting allowed for 15 replicates of each configuration for each of two *Drosophila* species, and the experiment lasted 18 weeks (six to eight *Drosophila* generations). For their study species, *Drosophila pseudoobscura*, extinction rates were lowest for the large systems, next lowest for the connected systems, and highest for the separated systems. Hence the inference was that both larger and connected patches are significantly less vulnerable to extinction than small, isolated patches. The authors concluded by suggesting that this model system was quite useful for determining patterns and mechanisms of species' responses to habitat loss and isolation: "Our model can be used to deepen our understanding of the extinction process . . . the laboratory can be an important venue for exploring and testing theoretical ideas on species extinction and other issues of interest to conservation biology" (p. 50). This experiment helped researchers understand the population dynamics of a single species in the absence of other interacting species, such as competitors or predators.

The theory of metapopulation dynamics was discussed in previous chapters, and one of its predictions is that dispersal among spatially distinct habitat patches (local populations) should enhance the overall persistence of the regional metapopulation. Holyoak and Lawler (1996) studied the population dynamics of two aquatic ciliates (protists); one a bacterivore (*Colpidium*) and the second a predator of *Colpidium* (*Didinium nasutum*). They devised experiments to specifically investigate the effects of habitat subdivision, or fragmentation in the strict sense, using either undivided habitats (plastic bottles) of four different total sizes, or divided habitats, which were arrays of bottles linked by tubes to allow dispersal (fig. 4.1). The researchers were able to observe population dynamics in the experimental system for an impressive 602 prey generations and 437 predator generations (in only 130 days)! The key findings from this study that are relevant to ecological theory regarding habitat loss and fragmentation were that (1) the predator-prey interaction persisted for a much longer period of time in larger versus smaller undivided habitat patches, (2) the prey species exhibited frequent extinctions in the small bottles of the arrays, but individuals were able to recolonize bottles by dispersing via connecting tubes, and (3) predator populations displayed rescue effects in the arrays via dispersal through connecting tubes, as predicted by J.H. Brown and Kodric-Brown's (1977) modification of island biogeography theory. Thus these experiments showed that predators and prey in isolated habitat patches that are connected by dispersal pathways may persist for several hundred generations, much longer than in unconnected habitats.

A similar laboratory culture system was used by Burkey (1997) to study the

Figure 4.1. A photograph of a microcosm array used in studies of predator-prey interactions in spatially subdivided habitats (Holyoak and Lawler 1996). The entire 9-bottle array shown here is about 30 cm × 30 cm. Photo courtesy of Marcel Holyoak, University of California–Davis.

effects of habitat loss and fragmentation on the extinction of a community of bacteria and protozoa. Burkey also explicitly considered habitat loss and fragmentation per se by manipulating the size of laboratory jars (to mimic habitat loss), the subdivision of laboratory jars (to mimic habitat fragmentation), and the presence of connections between laboratory jars (to mimic movement corridors); he then observed the time to extinction of the top predator in the system (either *Didinium nasutum*, as above, or *Euplotes aediculatus*). For each experiment, the top predator always went extinct sooner in a small versus large habitat, and in a fragmented versus unfragmented habitat of the same total size. Here the effect of dispersal corridors was counter to theoretical predictions, as well as to the earlier experimental results of Holyoak and Lawler (1996); the linked bottles reached extinction sooner than the unlinked bottles. Burkey surmised that the dispersal links (corridors) served to synchronize the population dynamics of the bottles so that the decline to extinction of a single bottle reverberated throughout the linked system, causing overall population extinction. One lesson learned from these two sets of bottle experiments is that for persistence to occur, the amount of dispersal among habitat patches must be like the temperature of the porridge in the story of *Goldilocks and the Three Bears*: "just right." With too much dispersal, population dynamics are synchronized among patches, rendering the entire metapopulation vulnerable to extinction from a single stochastic event. With too little dispersal, the rescue of populations from extinction and recolonization of extinct populations occurs too infrequently to positively affect metapopulation dynamics. Species' responses to habitat patchiness in laboratory experiments have indeed provided valuable insights into the effects of habitat loss and fragmentation in the outside world—the stuff of medium and large experiments to be considered next.

MICROLANDSCAPE EXPERIMENTS: SIZE MEDIUM

The majority of manipulative experiments on fragmentation have been conducted at places outside of the laboratory and at scales that are intermediate in time and space to the small and large studies discussed in this chapter. The experimental designs of these studies typically emphasize either habitat loss and isolation, or habitat subdivision or fragmentation per se, depending on the researchers' questions (fig. 4.2). The first explicit field study to test ecological theory regarding habitat loss and fragmentation was Dan Simberloff's dissertation research on small mangrove islands in the Florida Keys (Simberloff 1969; Simberloff and Wilson 1969, 1970; the last is another classic paper in ecology). Simberloff concentrated his research on testing one of the main tenets of MacArthur and Wilson's (1967) island biogeography theory—that islands close to a mainland source of colonists will experience higher colonization, lower extinction, and thus higher species richness at equilibrium than islands that are more remote. So, in the strict sense, this was an experiment to test the effects of habitat isolation, but not a fundamental test of fragmentation. To create empty islands ripe for colonization, the researchers fumigated six small mangrove islands with an insecticide to remove all terrestrial arthropods, then observed the recolonization of the islands over a two-year period. Consistent with expectations from island biogeography theory, after two years of colonization the most remote, isolated island hosted the fewest number of insect species, the intermediate islands had intermediate numbers of insect species, and the island closest to the mainland had the most species (Simberloff and Wilson 1970).

Many field experiments have followed Simberloff's research and have focused on asking questions regarding the effects of habitat patchiness on population and community dynamics. Quinn and Robinson (1987) and G. Robinson and Quinn (1988) reported results from "the first direct experimental examination of the effects of habitat subdivision *per se* on extinction . . . and overall species diversity" (G. Robinson and Quinn 1988, p. 79). At their study site near Davis, California, they created 42 isolated grassland patches of three sizes (32 plots of 2 m², 8 plots of 8 m² and 2 plots of 32 m², similar to the left-hand example in fig. 4.2a) by mowing the vegetation around the patches; they then monitored the percent cover of plants in each plot over three full growing seasons. As predicted from island biogeography theory, the number of plant species was positively related to plot area; the largest plots contained about 50% more species than the smallest plots. Also as predicted, over the course of the study the small plots exhibited higher per-species extinction rates than the large plots. But when the researchers compared plant species richness in areas of equal size that varied in the amount of habitat fragmentation (e.g., plant species richness in 32 small plots combined versus 8 medium plots versus 2 large plots), they found the highest overall species richness in the most subdivided (combined 2 m²) habitat patches. Moreover, there were no detectable effects of habitat subdivision on extinction. The authors inferred from their results that spatially distributed reserves may be effective in protecting species, especially those that are quite patchily distributed. This research highlights a key distinction between experimental studies focused on habitat loss and isolation and those focused on subdivision or fragmentation (some of which are illustrated in fig. 4.2). The former studies concentrate on the characteristics of local patches, such as size or isolation, and the latter tend to highlight regional collections of habitat patches and the dynamics of regional

Figure 4.2. (opposite) Representative experimental designs of fragmentation experiments in the medium size category. In each scheme the dark gray areas denote native habitat, the white areas denote cleared areas, and the black lines are outlines of sample plots. (a) An experimental design that allows examination of the effects of fragment area and subdivision. The collection of small plots ($N = 16$) is equal in total area to the collection of medium plots ($N = 8$) and the collection of large plots ($N = 2$). Based on Quinn, Wolin, and Judge (1989) and Dooley and Bowers (1998). (b) An experimental design that allows examination of the effects of habitat area. Equal numbers of replicates ($N = 5$) of large, medium, and small fragments are arranged in blocks. Based on Collinge (2000). (c) An experimental design to assess the efficacy of habitat corridors in species persistence. Replicates of equal-sized plots are isolated (left), connected by corridors (center), or located in continuous habitat (right). Based on La Polla and Barrett (1993) and Aars, Andreassen, and Ims (1995).

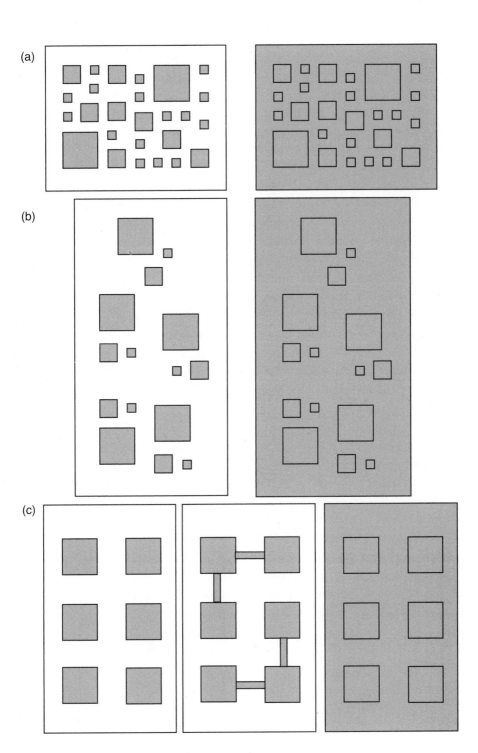

metapopulations. Thus there are different inferences that can be made from these varying experimental designs, which will be discussed in greater depth at the end of this chapter.

Small mammals in grassland or old-field habitats have frequently been used in experimental systems designed to understand the effects of habitat loss and fragmentation (e.g., La Polla and Barrett 1993; Aars, Andreassen, and Ims 1995; Andreassen, Halle, and Ims 1996; Dooley and Bowers 1996, 1998; Aars, Johanne-sen, and Ims 1999; Andreassen and Ims 2001; Orrock and Danielson 2005). For example, Dooley and Bowers (1998) experimentally manipulated patch size and fragmentation to examine the population responses of the herbivorous meadow vole, *Microtus pennsylvanicus*. The researchers compared population density, growth rate, survivorship, and recruitment of vole populations at two spatial scales. The local-scale comparison was among abandoned agricultural field frag-ments of three different sizes, including five small patches (0.06 ha), four me-dium patches (0.25 ha), and four large patches (1.0 ha). The second, landscape comparison was between the 13-patch, 20 ha fragmented landscape just de-scribed and an equally-sized adjacent, unfragmented landscape (fig. 4.2a). They found no effect of fragment size on population parameters, but there were significant effects of landscape fragmentation on population density, growth rate, and recruitment. The results, however, were opposite to theoretical predictions: population density, growth rate, and recruitment (but not survival) were higher in the fragmented landscape than the unfragmented one. The authors suggested that the fact that no patch size effects were discerned may have been due to a lack of statistical power, small differences in the amount of edge habitat among the three sizes of patches, or opposing processes that essentially cancelled each other out and precluded their detecting any net effect of patch size. The authors attrib-uted the significant landscape response to enhanced habitat quality resulting from fragmentation—voles appear to have higher individual growth rates in edge habitats, perhaps due to the higher nutritional quality of new plant growth along the edges, so the greater amount of edge habitat may translate into higher reproductive rates. There is a cautionary note here, however; because of the over-all reduction in habitat area, the fragmented landscape maintained a lower ab-solute population size of voles than did the unfragmented landscape.

Habitat isolation may be reduced by the presence of corridors (linkages be-tween habitat patches), and there are now several published studies that experi-mentally test the role of corridors in species persistence (fig. 4.2c; see chapter 6 for a detailed discussion of movement corridors). F. Gilbert, Gonzalez, and Evans-Freke (1998) designed an innovative experiment with microarthropods (mites, ticks, and springtails) that occur in moss patches on rocks to ask whether corridors influence species loss from otherwise isolated patches. The researchers chose six large, moss-covered rocks in a forest in the United Kingdom and imposed four different experimental treatments on each rock, with each treatment comprised of four, 10 cm-in-diameter, circular patches. The *mainland treatment* involved sampling the four circular patches on the rock for microarthropods, but no moss was removed, so this was effectively one large continuous patch. The *corridor treatment* consisted of four circular patches connected by narrow linear corridors; the *broken corridor treatment* was identical to the corridor treatment but with a 5 cm gap in the middle of each corridor (this treatment controlled for the effect of the increased area provided by the corridors). The *insular treatment* was comprised of four circular isolated patches. Microarthropod samples were collected prior to experimental manipulations and again three and six months after the treatments were imposed. The time length of the experiment allowed for several generations of the microarthropod species that occur in the moss habitat. The results clearly showed that corridors had a strong effect on species persistence—both the overall number of species and the percentage of predator species were significantly higher in patches connected by corridors than in isolated patches. As was the case in the laboratory microcosm experiments, the authors here also argued that the higher species richness in connected patches was due to the enhanced movement of microarthropods among the patches via the corridors.

Corridors may moderate the negative consequences of habitat isolation, but the relative effect of corridors may depend on habitat area. In one experiment from my own dissertation research (Collinge 1998, 2000), I manipulated the size and connectivity of grassland patches to ask whether insect populations and communities were affected by habitat loss and isolation. I took advantage of a popular treatment used in these types of experiments (mowing) to manipulate habitat structure. The patch sizes used in my experiment were 1 m², 10 m², and 100 m² (fig. 4.2b). Each was crossed with three connectivity treatments—continuous (control), corridor (the presence of a narrow linear corridor connecting the patch to continuous grassland), and isolated (no connections between the patches)— and replicated six times across the 6 ha study area near Boulder, Colorado. I sampled aboveground insects in each plot prior to mowing, and then re-sampled the plots monthly during each of three summer field seasons. The effects of corridors that I observed in my experiment were not as dramatic as those observed by F. Gilbert, Gonzalez, and Evans-Freke (1998), and they appeared to be contingent upon several factors. In particular, an influence of corridors was apparent only for the intermediate-sized (10 m²) plots, but not for the small and large plots. Corridor effects were stronger in the driest of the three years, suggesting that perhaps the effects of isolation were exacerbated by low resources due to environmental variability. Moreover, corridors appeared to positively affect some species and not others—specifically, insects grouped as "low mobility" that were relatively common showed positive responses to corridors. This experiment demonstrated, however, that by far the strongest influence on insect species richness was habitat fragment size, with connectivity having much weaker effects.

These two field experiments with arthropods (in the United Kingdom and in Colorado) primarily examined the effects of habitat connectivity by asking whether corridors influenced species richness in habitat patches. In a second experiment for my dissertation (Collinge 1998; Collinge and Forman 1998), I examined whether different patterns of land transformation, mimicked once again by mowing patches of grassland, affected insect species assemblages. I imposed mowing treatments that simulated four different spatial configurations of shifts from 100% to 25% habitat area. These configurations were four (see fig. 1.1) of the several proposed by Forman (1995); a comparison of two of these treatments allowed us to examine the direct ecological effects of habitat subdivision. One treatment, *shrinkage*, involved a single patch of habitat that was reduced in area from 100% to 25% during successive steps of the experiment. For a second treatment, *fragmentation*, the first step involved dividing the initial grassland patch into nine equal-sized patches that were made progressively smaller. Each mowing treatment was imposed every four days, and the insects were subsequently sampled. We sampled insects again five weeks after the final treatment to assess longer-term responses to habitat spatial configuration. In this grassland fragmentation experiment, spatial configuration did influence insects significantly. The first of its two main results were observations of increased insect densities in fragmented versus unfragmented habitats. This crowding effect has been noticed in the early phases of other experimental studies, and the inference is that animals move from the modified habitat to the remaining undisturbed habitat, which results in higher densities in small fragments. Second, large rare insect species were equally abundant in the fragmented and unfragmented sequences. This may be due to the fact that these species perceived the entire 9-patch fragmented system as one large habitat patch, given that inter-patch distances were relatively small (from 1 to 2.5 m).

Most fragmentation experiments have been conducted in terrestrial ecosystems, but habitat loss and fragmentation may clearly affect the behavioral, population, and community ecology of marine organisms as well. Several studies have independently manipulated habitat patch size and the degree of habitat subdivision or fragmentation in marine settings to examine the relative impacts of each on population and community dynamics. Quinn, Wolin, and Judge (1989) studied intertidal snails to test more general ideas about the probability of extinction of local and regional populations in relation to local patch size and the degree of habitat subdivision (fragmentation per se.). They used laminate plates attached to plywood as habitat for the predatory snail, *Nucella emarginata*, in the sand flats of Bodega Bay, California. Their experiment involved the subdivision of 1 m² square plates into progressively smaller square patches (to 1/2, 1/4, 1/8, and so on down to 1/64th m²). Replicate plates of each size (64 × 1/64 m²; 32 × 1/32 m²; and so on) were affixed to the sand substrate and allowed to be colonized naturally by barnacles (prey for *Nucella*); then an equal number of *Nucella* (128 individuals) were introduced to each experimental plate. The researchers followed these populations over two years, so that population extinction could be examined in relation to patch size and fragmentation. As predicted, there was a steady increase in the probability of local population extinction as patch size decreased—none of the populations on large patches went extinct, but about 80% of the populations on small (1/64 m²) patches did so during the study. When the combined plates of each size were compared, however, there was no discernible effect of habitat subdivision on extinction (*Nucella* had not gone extinct from any of the combined treatments by the conclusion of the experiment). As in the study of California annual grasslands cited above (G. Robinson and Quinn 1988), local extinction was negatively related to patch area, but regional extinction was unaffected by habitat subdivision or fragmentation.

In a different marine setting, Caley, Buckley, and Jones (2001) examined the relative effects of coral habitat degradation and fragmentation on coral commensal invertebrate communities (particularly those of the crab *Trapezia cymodoce*, and the shrimp *Palaemonella* spp.) on the Great Barrier Reef, Australia. Their experimental protocol involved degrading (killing) and fragmenting (dividing) small coral colonies in a 50 m × 50 m experimental plot, and then observing the composition of coral commensal species 1–2 days, 1 month, and 2 months after the treatments were applied. Species richness and overall abundance were not affected significantly by fragmentation; however one species, *T. cymodoce*, increased in response to the fragmentation treatment. The major effect detected in this experiment was that habitat degradation had a much greater impact on species richness and overall abundance than did the fragmentation of coral colonies.

Similarly, Goodsell and Connell (2002) independently manipulated habitat loss (the number of habitat patches) and habitat isolation (the distance between patches) in a marine environment, in this case, kelp forest habitat in South Australia. They recorded invertebrate (polychaetes, bryozoans, crustaceans, and echinoderms) species richness, species composition, and abundance over a two-month period. The researchers cleared a 1 m² area of kelp forest to create either nine or five kelp habitat patches, and habitat isolation was manipulated by removing kelp plants either within 10 cm of an experimental plant or at a distance greater than 20 cm. The resultant kelp spacing was 10 cm for the proximate treatment and 20 cm for the distant treatment. This 2 × 2 factorial experimental design was replicated six times across the study area and allowed for an examination of the independent and interactive effects of habitat loss and isolation on species assemblages. The separate effects of the amount of habitat and isolation were not significant, but these two factors interacted to influence invertebrate species composition. In particular, the effects of habitat loss on species composition and relative abundance were lessened when habitat patches were in close proximity. Although they did not directly examine the movement of individuals among patches, the authors suggest that the effects were due to movement among patches rather than to changes in population growth rates in different experimental treatments. This implies that perhaps the rescue effect (*sensu* J.H. Brown and Kodric Brown 1977) alleviated the negative effects of habitat isolation on species persistence.

One of the strengths of medium studies is that the spatial and temporal scales are generally appropriate for the organisms being studied, which are often insects but also include other terrestrial or marine invertebrates or small mammals. Further, these experiments have an advantage over microcosm experiments in that they are conducted in field settings, so they are subject to realistic levels of environmental variation. As with microcosm studies, one limitation of medium studies is that they are often difficult to scale up. It is hard to directly compare the responses of an insect predator to grassland fragmentation to a mountain lion's response to changes in landscape structure. The next section will look at experiments that are conducted at scales appropriate to these larger animals.

BROAD-SCALE FRAGMENTATION EXPERIMENTS: SIZE LARGE
Brazilian Amazon Rainforest: BDFF

Probably the most well known of the large fragmentation experiments is the Biological Dynamics of Forest Fragments (BDFF) project in the Amazon rainforest

north of Manaus, Brazil (table 4.1). The experiment was initiated in 1979 as a collaborative effort between the Smithsonian Institution and the Brazilian Institute
for Research in the Amazon (INPA) and is now the "world's largest and longest-
running experimental study of habitat fragmentation" (W. Laurance et al. 2002,
p. 606). Originally called the Minimum Critical Size of Ecosystems project, the
study was designed to assess a minimum size for rainforest reserves that would
effectively conserve biological diversity. The results and conclusions from this experiment have contributed profoundly to our understanding of the ecological
consequences of forest habitat loss, isolation, and landscape context. In their
2002 review, for example, William Laurance and colleagues referred to an ex-
traordinary 340 publications and theses that have been produced from this broad-
scale experiment. Several syntheses of this experiment have been published
since the study's inception and provide useful detailed summaries of the ecological consequences of Amazonian forest loss and isolation (Lovejoy et al. 1984,
1986; Bierregaard et al. 1992; W. Laurance et al. 1997; Gascon and Lovejoy 1998;
W. Laurance et al. 2002).

The study design included the establishment of 11 forest fragments (five of 1
ha, four of 10 ha, and two of 100 ha) that were isolated in the early 1980s by distances of 80–650 m from surrounding intact forest as a result of forest clearing
for cattle pasture in this region of the Amazon basin. At the same time, 12 forest
reserves (three of 1 ha, four of 10 ha, two of 100 ha, and three of 1000 ha) were
established in intact forest and left undisturbed as experimental controls. Technically, this experiment formally tests the effects of habitat area (see the example
in fig. 4.2b), and not different levels of isolation or fragmentation on ecological
dynamics. Since the experiment was initiated, cattle-ranching activities have
been largely abandoned in this area, allowing a growth of secondary forest to surround the fragments. Thus the researchers have periodically cleared 100 m wide
swaths of secondary forest to maintain the experimental fragments (W. Laurance
et al. 2002). Prior to isolation, a large number of taxa—including plants, butterflies, beetles, amphibians, birds, and primates—were surveyed in the fragments-
to-be as well as in the forest reserves, and these areas have been repeatedly surveyed for the almost three decades following isolation.

The manipulation of fragment size that was imposed in this experiment has
resulted in major impacts on forest biota and dynamics. As predicted from island
biogeography theory, small forest fragments contain many fewer species than
large continuous tracts of rainforest. This area effect has been noted for several
species of primates (K. Gilbert and Setz 2001), understory birds (Stouffer and
Bierregaard 1995; Ferraz et al. 2007), bees (Powell and Powell 1987), ants (Gas-

Table 4.1
Details of broad-scale (large) fragmentation studies

Experiment	Location	Habitat	Fragment sizes	Controls	Connectivity	Replication	References
Biological Dynamics of Forest Fragments (BDFF)	Brazil: 3° S, 60° W	Tropical rainforest	1, 10, 100 ha	1, 10, 100, 1000 ha continuous forest	Not explicitly included in design	5 of 1 ha, 4 of 10 ha, and 2 of 100 ha fragments	W. Laurance et al. 2002; Ferraz et al. 2007
Wog Wog	Australia: 37° S, 149° E	Native *Eucalyptus* forest	0.25, 0.875, and 3.062 ha	Same plot sizes in continuous forest	Not explicitly included in design	4 replicates of each fragment size, with 2 replicates in continuous forests	Margules 1992; Davies, Melbourne, and Margules 2001; Davies, Margules, and Lawrence 2003
Kansas succession study	United States: 39° N, 94° W	Abandoned wheat field undergoing succession to a mosaic of open prairie and deciduous forest	0.5, 0.028, and 0.0032 ha	Large (0.5 ha) patches are control for subdivided patches	Not explicitly included in design	6 large, 18 medium, and 82 small patches	G. Robinson et al. 1992; Holt, Robinson, and Gaines 1995; W. Cook et al. 2005
Calling Lake	Canada: 55° N, 113° W	Boreal mixedwood forest	1, 10, 40, and 100 ha	Same plot sizes in continuous forest	100 m wide corridors (riparian buffer strips)	3 of each patch size and treatment	Schmiegelow, Machtans, and Hannon 1997; Hannon and Schmiegelow 2002
Savannah River	United States: 33° N, 81° W	Pine plantation with clear-cut patches	1, 1.375, and 1.64 ha	Unconnected patches of equal size	25 m wide × 150 m long corridors	8 replicates	Haddad 1999; Tewksbury et al. 2002

con et al. 1999), beetles (Klein 1989; Didham et al. 1998), understory plants (Benítez-Malvido and Martínez-Ramos 2003), and epiphyllous bryophtes (Zartman 2003). One likely mechanism for lower species richness in smaller fragments is that observed rates of extinction have been higher in small versus large fragments for many of the above-cited taxa (Stratford and Stouffer 1999; W. Laurance et al. 2002; Ferraz et al. 2007). Although they did not control isolation experimentally, researchers have discovered that even relatively small forest clearings or gaps in the forest canopy are disruptive enough to sever connections between forest areas and limit animal movement. For example, several insectivorous birds have experienced local extinction in small fragments and have been unable to recolonize these areas, even though they are isolated from continuous forest areas by only 70–80 m (Stratford and Stouffer 1999). In contrast, other species or species groups have increased in small fragments or have been unaffected by changes in forest area, including generalist butterflies (K. Brown and Hutchings 1997), some frog species (Gascon et al. 1999), and small mammals (Malcolm 1997).

Researchers have also observed changes in key ecosystem processes in relation to fragment size, including sharp reductions in tree biomass (*biomass collapse*) in small fragments, due to higher rates of tree-fall at the fragment edges (W. Laurance et al. 1997, 1998), and lowered leaf litter and dung decomposition rates in small fragments compared to large ones (Klein 1989, Didham 1998). Interestingly, the increased mortality of large, old-growth, forest interior tree species along forest edges has resulted in a major shift toward dominance by early successional tree species (such as *Cecropia*, *Vismia*, and *Miconia*) along the edges of forest fragments (W. Laurance et al. 2006), which is likely to have major effects on ecosystem dynamics.

In addition to the discovery of significant fragment area effects—the investigation for which the experiment was originally designed—two major findings have surfaced from this experiment. Neither of these was predicted directly by island biogeography theory, however. The first is that the most pervasive and damaging effects on biota in relation to habitat modification come from edge effects (summarized in W. Laurance et al. 2002; see also chapter 5). The increased temperature, decreased soil and air moisture (Kapos 1989), and increased wind turbulence (W. Laurance et al. 1997) at forest edges modify the environment so much that certain organisms are no longer able to persist in forest fragments. These edge effects are most severe for the smallest fragments, given their higher edge-to-interior ratios. For example, W. Laurance et al. (1998) used a mathematical model to predict that edge effects would be especially severe as fragment size decreased to 100 to 400 ha (depending on fragment shape). Second, the matrix surrounding the forest fragments substantially affected species composition and ecosystem dynamics in the fragments (see also chapter 5). This became apparent as the forests began to regrow around the isolated fragments. If the surrounding forest had been cleared but not burned, regrowth was dominated by *Cecropia*, whereas if the surrounding forest had been burned and used for cattle pasture, the regrowth was dominated by *Vismia*. The composition of the surrounding matrix profoundly influenced the species composition of understory birds in forest fragments (Stouffer and Bierregaard 1995), as well as that of small mammals and frogs (Gascon et al. 1999).

To understand the mechanisms underlying species' responses to habitat loss and fragmentation, it would be ideal to study every aspect of the life cycle and identify key stages that are particularly vulnerable to habitat modification. Within the context of the BDFF study, Emilio Bruna performed a series of innovative experiments and conducted multi-year population censuses to assess the effects of forest fragment area on the demography of a perennial understory herb, *Heliconia acuminata* (Bruna 1999; Bruna and Kress 2002; Bruna et al. 2002; Bruna 2003; Bruna and Oli 2005). This body of research is especially valuable because it involved a systematic study of each aspect of the plant's life cycle in both fragments and continuous forest, and it employed demographic modeling to assess critical life-cycle stages and estimate population growth rates. Bruna began at the beginning of the life cycle with seeds and seedlings. He collected seeds of *H. acuminata* from continuous forest and planted them in 1 ha and 10 ha fragments, as well as back into the continuous forest. Seeds planted in the forest fragments were much less likely to germinate than those planted in continuous forest (Bruna 1999). To assess relative plant growth rates in fragments versus continuous forest, Bruna et al. (2002) reciprocally transplanted *H. acuminata* plants from continuous forest sites to fragment sites, and from fragment sites to forest sites, with appropriate controls. Plant growth strongly differed in these two landscape settings: plants in the fragments actually had fewer shoots and leaves 32 months after transplanting than when they were first transplanted. In other words, these plants shrank, while plants in continuous forest sites increased slightly in size. The reduction in plant size in fragments was likely due to heat and water stress associated with the modified microclimate of small fragments with pervasive edge effects, and to the concomitant loss of leaves and shoots. Consistent with these data on growth rates from the reciprocal transplants, multi-year surveys revealed that populations of *H. acuminata* in the fragments were dominated by individuals in smaller size classes than populations in continuous forest, and there was a non-significant trend toward a higher proportion of individuals flowering in continuous forest versus forest fragments (Bruna and Kress 2002).

These multi-year census data of over 5000 individual plants were used to construct a matrix population model to compare expected population growth rates with observed growth rates over two transition years (Bruna 2003). The modeling effort revealed that projected population growth rates were substantially lower than observed growth rates, based on yearly censuses. This discrepancy between projected and observed growth rates led Bruna and colleagues to re-examine model assumptions and ultimately invoke an additional mechanism to explain why the observed numbers of seedlings in the study plots were greater than expected. They reasoned that because their model assumed that most seed dispersal was localized around parent plants, it did not account for longer-distance seed dispersal into the forest fragments from the surrounding forest. Thus this unmeasured component of demography (immigration) likely explained the higher observed number of seedlings than the model projected. Using an expanded data set (5 years of census data), Bruna and Oli (2005) concluded that population growth rates in continuous forest were significantly higher than in forest fragments. Taken as a collection, these detailed field and modeling efforts provide unparalleled information on both the patterns and the demographic mechanisms for a species' response to habitat loss and isolation.

Australian Eucalyptus Forest: Wog Wog

Started in 1985 in southeastern New South Wales, Australia (table 4.1), the Wog Wog fragmentation experiment was designed to test the hypotheses from island biogeography theory that (1) a reduction in the area of native *Eucalyptus* forest would result in reduced species richness at equilibrium in forest fragments, and (2) the reduction in species richness would be greatest for small fragments (Margules 1992). The experimental design consisted of six replicates of each of three fragment sizes (0.25 ha, 0.875 ha, and 3.062 ha). Four *Eucalyptus* fragments of each size (for a total of 12 fragments) were isolated when the 80- to 100-year-old *Eucalyptus* forest surrounding these plots was cleared for a pine plantation (fig. 4.3). Two other replicates were established in uncleared, continuous *Eucalyptus* forest as control plots. As in the BDFF experiment just described, floral and faunal surveys were conducted for two years prior to the experimental isolation of the forest patches, and at least annually following the experimental treatments. The research effort included sampling many ground-dwelling invertebrate groups (e.g., beetles, spiders, ants, amphipods, and scorpions), as well as bats, birds,

Figure 4.3. Aerial photographs of the Wog Wog fragmentation experiment in south-eastern Australia. *Top: Eucalyptus* fragment (0.875 ha), immediately after clearing to create experimental fragments. *Bottom:* fragments of three sizes, seven years after clearing to create experimental fragments. Photos courtesy of Chris Margules, CSIRO, Australia.

mammals, and skinks. Permanent sampling sites were established both at eight different locations within each fragment or each control plot and in the exotic pine plantation matrix surrounding the native *Eucalyptus* forest fragments. Sample sites were located in either slope or drainage sites within fragments, as well as either near fragment edges or in fragment interiors, allowing for detailed, spatially explicit patterns of species abundance and distribution across the study area.

The beetle fauna sampled in these forest fragments is highly diverse (over 655 beetle species captured in the first five years), and it is the group that has been analyzed most extensively in the context of forest loss and isolation (Davies and Margules 1998; Davies, Margules, and Lawrence 2000; Davies et al. 2001; Davies, Margules, and Lawrence 2004). Both population and community responses have been examined, and within-patch as well as between-patch processes have been explored to explain species occurrence patterns. Several key conclusions can be drawn from these studies. First, particular traits appear to influence significantly whether a species will be positively or negatively affected by habitat loss and isolation (Davies, Margules, and Lawrence 2000). Specifically, for the 69 beetle species that were sufficiently abundant to be subjected to statistical analyses, those that were relatively rare, those that were confined to the *Eucalyptus* forest and did not regularly occur in the pine matrix, and those at higher trophic levels (predators) all declined in abundance in small, isolated patches. Further analyses indicated that these traits may interact synergistically (Davies, Margules, and Lawrence 2004); species that were both rare and specialized were especially vulnerable to extinction compared to other species. Second, community measures of species richness, species composition, and relative abundance all appeared to be more strongly affected by within-patch processes than between-patch processes. In particular, changes in patch area and isolation appeared to influence community structure primarily via changes in the microclimate at the edges (*edge effects*), not because of differences in extinction or colonization, as was predicted by island biogeography theory. This study provides an important link between population and community responses to fragmentation, as well as between studies that focus primarily on local processes, such as per-patch extinction rates, and broad-scale processes that encompass multiple patches, such as dispersal and colonization among habitats.

Grasslands: Kansas Old-Field Succession

Given that ecological theory predicts that rates of extinction and colonization should be affected by patch spatial characteristics, patterns of ecological succes-

sion that involve changes in species distributions over time are likely to be influenced by habitat area and fragmentation. In 1984, researchers at the University
of Kansas designed a field experiment to test the effects of habitat subdivision
(fragmentation) on plant secondary succession in an abandoned wheat field in
eastern Kansas (G. Robinson et al. 1992; Holt, Robinson, and Gaines 1995). The
study design involved the establishment of three patch sizes, with each patch isolated by a distance of 15 m from surrounding patches through continued mowing of the matrix areas. Clusters of small patches were equal in area to clusters of
medium patches and to large patches (similar to the example in fig. 4.2a; each
large patch was 0.5 ha).

Succession was allowed to proceed on each of the patches, and researchers followed the fates of above-ground arthropods, small mammals, vascular plants,
and snakes; they also measured soil moisture and nutrient concentrations. After
six years of succession (and fragmentation), there were no effects of habitat subdivision on soil properties, nor on aggregate measures of species richness for any
of the taxa that were studied (G. Robinson et al. 1992). In other words, species
richness values were similar for large patches and for clusters of small and
medium patches. However, there were effects of fragmentation on particular
populations or species groups. For example, clonal plant species, which reproduce primarily by vegetative growth and therefore move gradually across the
landscape, were more likely to persist in large fragments than in smaller fragments, where colonization was probably limited by habitat disruption created by
mowing between the patches (G. Robinson et al. 1992; Holt, Robinson, and
Gaines 1995). And populations of two of the three most common species of small
mammals studied (*Microtus ochrogaster* and *Sigmodon hispidus*) persisted much
longer in large patches than in small ones (G. Robinson et al. 1992; Diffendorfer,
Gaines, and Holt 1995).

The relatively modest, short-term effects of fragmentation on species abundance and distribution became more pronounced as succession proceeded
(W. Cook et al. 2005). After 18 years of succession, woody plants were more dense
in large versus small patches, overall plant species richness was higher in large
patches than small ones, and species turnover was lower in large patches. Furthermore, the spatial position of the experimental patches relative to native forest cover differed, and position affected successional dynamics significantly. In
particular, patches close to native forest had higher woody plant density, higher
species richness, and lower species turnover than patches that were more distant
from native forest. This spatial position effect highlights the important role of
plant propagule immigration from outside the study area, a key factor noted in

Bruna's (2003) study of plant demography in the fragmented Amazon forest. Overall, both larger patches and patches close to native forest had accelerated secondary succession and greater temporal stability (lower turnover) relative to smaller fragments (W. Cook et al. 2005). Clearly, there is great value to the long-term, repeated sampling of this experiment, since it revealed patterns nearly two decades after the experiment began that were not evident early in the successional trajectory.

The three large experiments just discussed manipulated fragment size and subdivision and asked how species and communities responded to these changes in habitat spatial structure. The next two experiments, conducted in two very different study systems, ask whether habitat corridors that connect otherwise isolated patches may influence population and community processes.

Canadian Boreal Forest: Calling Lake, Alberta

We shift now to the boreal forest and the Calling Lake fragmentation project, which was established in 80- to 130-year-old aspen-dominated forests in north-central Alberta, Canada, in 1993 (Schmiegelow, Machtans, and Hannon 1997; Hannon and Schmiegelow 2002). This experiment was specifically designed to assess whether habitat corridors would mediate the effects of habitat isolation, and it included three replicates each of 1, 10, 40, and 100 ha areas, which were either isolated by clear-cutting around the forest patch or connected via a 100 m wide riparian corridor to one other patch (table 4.1). Control plots identical in size to each isolated fragment were established in the nearby continuous forest (approximately 4000 ha in size). Pre-isolation data were gathered in 1993, and post-isolation data were collected up to five years after the forest harvest, as the surrounding forest regenerated. Researchers concentrated on the responses of breeding songbirds (both residents and neotropical migrants) to changes in habitat area and isolation. Measurements included species richness, the relative abundance of all species, responses of forest species versus generalist species, and the movement of both adults and juveniles along the riparian corridors connecting otherwise isolated patches.

Changes in forest area and isolation that were imposed in this boreal forest experiment affected bird species composition significantly following the isolation treatments (Schmiegelow, Machtans, and Hannon 1997). Overall, there was no change in bird species richness across treatments in the first two years following isolation, but, relative to the continuous forest, neotropical migrant birds were less abundant in both isolated and connected patches, and resident species declined in isolated versus connected patches. These results appeared to be relatively consistent as the experiment progressed. In a subsequent analysis, Hannon and Schmiegelow (2002) categorized species as either forest species or generalists, in addition to their migratory status. Consistent with their expectations, they found that forest species were more abundant in the forest reserves (control sites) than in the isolated or connected forest patches, but habitat generalist species were equally abundant across treatments. Forest species on the whole, however, were not affected by the presence of corridors—the abundance of species in this group was similar between connected and isolated forest fragments. As a group, resident species (versus neotropical migrants) were more abundant in control sites and patches with corridors than in isolated patches, but no individual species showed significant and consistent positive effects from corridors.

These data on species occurrence in isolated versus connected patches suggests that some bird species, but not all, may be using corridors to move between patches. Movement studies were conducted in the context of this experiment to help explain patterns in species responses (Machtans, Villard, and Hannon 1996; Robichaud, Villard, and Machtans 2002). Taken together, these studies showed that in the first two years following isolation, some songbird species appeared to use the riparian corridors as preferred movement pathways between patches, since they were more frequently captured in mist nets set up in riparian corridors than in the adjacent clear-cuts. This pattern was especially strong for juvenile birds (Machtans, Villard, and Hannon 1996). As the forest regenerated, however, the differences in physical structure between riparian corridors and the surrounding forest began to lessen, and bird movement behavior changed. Several years after isolation and regeneration, bird movement was equally frequent along riparian corridors and in adjacent, regenerating forest surrounding the patches (Robichaud, Villard, and Hannon 2002).

The Calling Lake experiment emphasized the efficacy of habitat corridors in ameliorating species loss from habitat fragments, and so most of the published results focus on isolation, rather than on patch size effects. Based on their multi-year observations of songbird responses, the authors concluded that corridors may not be all that effective for these boreal forest bird species, perhaps because the birds have evolved in the context of both fine- and broad-scale forest disturbances, such as wind-throw and periodic fires. Moreover, they suggested that this experiment occurred within "a landscape that was still ◻67% forested" (Hannon and Schmiegelow 2002, p. 1464), so perhaps most songbirds that were studied were not severely habitat-limited. The authors suggest that, given their results for forest-specialist birds, perhaps forest management efforts in this region should focus on enhancing the size of large blocks of intact forest rather than spending limited resources on protecting movement corridors that may not ultimately mitigate for habitat loss.

Southern Pine Forests: Savannah River, South Carolina

The Savannah River fragmentation experiments are unique compared to others discussed in this chapter in that they are the reverse image of the other studies. Two consecutive experiments conducted at this site (Tewksbury et al. 2002; Haddad et al. 2003) focus on cleared patches and corridors within southeastern United States pine forest plantations composed of loblolly (*Pinus taeda*) and slash pine (*P. elliotii*), rather than on forest remnants within a cleared matrix. Because the target habitat in this study is cleared forest, Nick Haddad and colleagues have concentrated their research on species that occur in open or early-successional habitats in this region. The initial experiment began in 1994 and addressed the effects of habitat isolation on animal movement by manipulating both the presence of habitat corridors and the distance between isolated patches (Haddad 1999). Patch size was held constant (1.64 ha) but the linear distance between patches varied (four distances, ranging from 64 to 384 m), and some patches were connected by linear corridors. Haddad initially focused on the movement of two butterfly species among patches, to assess whether corridors had positive effects. Indeed, for both species, the rates of movement between patches with corridors were significantly higher than for isolated patches, and the enhancement of movement increased as the distance between the patches increased (Haddad 1999). The higher rates of movement probably produced higher densities of butterflies in connected versus isolated patches (Haddad and Baum 1999). Not only did corridors positively affect butterflies in this experiment, but they also appeared to direct the movements of small mammals and bird-dispersed plants (Danielson and Hubbard 2000; Haddad et al. 2003).

A second experiment, initiated in 1999, was also designed to examine the effects of corridors on a wide variety of taxa, but these experimental treatments controlled for the effect of an increased area of patches that are connected by corridors (Tewksbury et al. 2002). In each replicate of the experiment, a 1 ha target patch was connected to another 1 ha patch with a 25 m wide × 150 m long corridor. The increase in patch area provided by the corridor means that the total area of the target patch was actually 1.375 ha (patch plus corridor). To control for this increase in patch area provided by the corridor, the experimental treatments include the 1 ha target patch surrounded on all sides by three additional 1 ha patches. Besides the 1 ha patch connected to the target patch via the corridor just described, a second, rectangular patch of 1.375 ha was located 150 m from the target patch, a third 1 ha patch had two wings that were 25 m wide × 75 m long (together they equal the area of the patch plus corridor), and a fourth patch was either another rectangular patch or another winged patch. Butterfly movement was again followed in this experiment, and the results showed that both butterfly species were more likely to move between connected patches than isolated ones, even after controlling for the effect of the increased area of patches with corridors. Moreover, pollination and seed dispersal were enhanced in connected versus isolated patches (see chapter 7), and in research conducted by Damschen et al. (2006), plant species richness was significantly higher in patches connected by corridors than in isolated patches. However, exotic plant species richness did not differ among the various connected and unconnected patches, so the corridors did not appear to facilitate the invasion of non-native species in this study system.

FRAGMENTATION EXPERIMENT UPDATE

By now there are dozens of fragmentation experiments that have been conducted in the laboratory and in the field, with a wide range of organisms and at multiple spatial and temporal scales. The common advantage of these diverse studies is that researchers have the ability to measure ecological characteristics before the treatments are imposed, and to follow the dynamics over time to assess both short- and longer-term responses to habitat loss and fragmentation. From experiments, we can observe the direct responses of species to landscape change and, often, the mechanisms that produce those responses. At the end of Debinski and Holt's (2000) review of fragmentation experiments, they identified five suggestions for future studies: (1) an increased focus on understanding the mechanisms behind observed community and population-level patterns, including dispersal and movement, (2) a better understanding of species interactions, such as plant-pollinator interactions or competition in fragmented landscapes, (3) an emphasis on species responses to the matrix (the habitat surrounding fragments), (4) an examination of how connectivity influences community dynamics, and (5) an analysis of the effects of fragmentation on genetic variation within and among populations.

In the ten years since their review (they included papers published through 1998), significant progress has been made in filling some of these information gaps. For example, we now have a clearer understanding of how dispersal and movement may influence species' responses to habitat loss and fragmentation (e.g., Haddad 1999; Robichaud, Villard, and Machtans 2002; Tewksbury et al. 2002). Movement is clearly crucial in species' responses to fragmentation. Several studies have shown that differences in movement may retard or accelerate extinction in small, isolated habitat patches; these differences may also stabilize interspecific interactions. We have also learned more about how species interactions—such as pollination, parasitism, seed predation, seed dispersal, and competition—may be affected by fragmentation (e.g., Tewksbury et al. 2002; see also chapter 7). The BDFF project, in particular, has shown the critical impact of surrounding matrix habitat on species composition in habitat fragments. And we know more about how connectivity, in the form of habitat corridors, affects populations and communities (fig. 4.4). At least some of the species' responses to corridors appear to generalize across scales; for example, experiments with habitat corridors typically show significant effects on the persistence or abundance of some species, but not others. We also have a greater understanding of the distribution of genetic variation in fragmented populations (see also chapter 6).

SYNTHESIS

Several key issues emerge from this review of fragmentation experiments, ones that were perhaps not as apparent from observational studies of habitat loss and isolation. First, manipulative experiments have allowed us to partition the effects of habitat loss, isolation, and fragmentation in the strict sense. In an observational study, it is virtually impossible to control the amount of habitat fragmentation while holding habitat area constant. Yet this is a relatively straightforward process in an experimental setting, and several well-designed studies at each of the spatial scales considered here have contributed greatly to our understanding of these separate effects (e.g., Quinn, Wolin, and Judge 1989; G. Robinson et al. 1992; Holyoak and Lawler 1996; Dooley and Bowers 1998). In observational studies (see chapter 3), researchers focused on the characteristics of the patches themselves, such as patch area or isolation. But in studies that manipulate habitat subdivision, we can examine both local and regional population dynamics in relation to the composition of the landscape. These studies show us that in some (but not all) cases there are no differences in regional (metapopulation) extinction in fragmented habitats, even though local extinction may be strongly influenced by patch size and isolation.

Second, experiments have helped to show the critical effects of the surrounding matrix on ecological processes within habitat remnants (see also chapter 5).

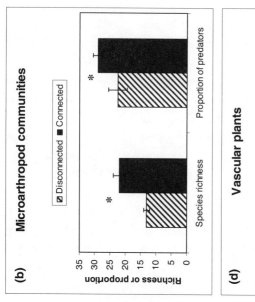

(a) *Drosophila* population persistence

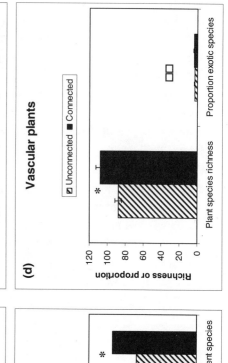

(b) Microarthropod communities

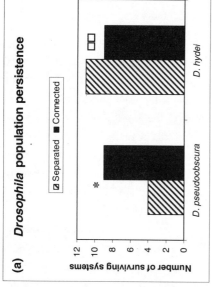

(c) Boreal forest birds

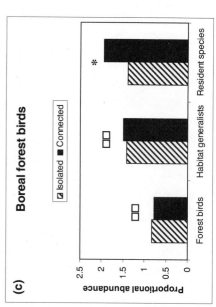

(d) Vascular plants

Some experiments have revealed only weak or neutral effects of habitat loss and isolation on population and community dynamics (e.g., Collinge 2000; Rantalainen et al. 2005), perhaps because the patch and matrix habitats were relatively similar to one another in physical structure or in the availability of resources. In other studies, the effects of the matrix have been revealed over time as regrowth occurs around the habitat fragments (Hannon and Schmiegelow 2002; W. Laurance et al. 2002).

Third, experiments at multiple temporal and spatial scales have shown consistent effects that represent generalized responses to habitat loss and fragmentation. For example, there are substantial positive effects of habitat area on local population persistence, as predicted by ecological theory; these have been shown for at least some species in nearly every experimental study that has been conducted. These positive effects on populations often translate into higher aggregate community measures of richness or diversity in large versus small habitat remnants. Further, microclimatic shifts at habitat edges clearly have negative effects on some species and positive effects on others. And experiments have shown that corridors promote connectivity (fig. 4.4), at least in landscapes where preferred habitat is limited, but not necessarily where it is dominant (e.g., compare the results of the Calling Lake experiment to those of the Savannah River site). This phenomenon has also been explored in simulation models (e.g., Fahrig 1997, 1998; With and King 1999), which have revealed thresholds in species' response to habitat connectivity, depending on the amount of available habitat in the landscape (see chapter 9 for a detailed discussion).

A related final point is that experiments have focused on a wide variety of taxa, which allows us to begin to make generalizations about species traits (e.g., trophic level, body size, rarity; see Davies, Margules, and Lawrence 2000; Davies, Melbourne, and Margules 2001; Davies, Margules, and Lawrence 2004) that may influence responses to habitat loss and fragmentation. This may help us to overcome the major challenge of relating results among small, medium and large experiments. We can do laboratory experiments on plankton, but what does that tell

Figure 4.4. (opposite) Species' responses to experimental isolation in four different experiments. NS denotes non-significant, * represents a significant effect of isolation. Similar patterns emerge from study systems at different spatial and temporal scales. (a) *Drosophila* population persistence. Redrawn from Forney and Gilpin (1989, fig. 2). (b) Microarthropod communities. Redrawn from F. Gilbert, Gonzalez, and Evans-Freke (1998, fig. 1). (c) Boreal forest birds. Redrawn from Hannon and Schmiegelow (2002, figs. 2 and 3). (d) Vascular plants. Redrawn from Damschen et al. (2006, fig. 2A).

us about species such as wolves and grizzly bears, which are often the focus of conservation efforts? As with models, controlled laboratory experiments often tell us what *can* occur, but not necessarily what *does* occur in the outside world of grass and trees or rocks and ocean. They can lead us to an expanded view of the possible mechanisms driving species responses and help us interpret field observations—they may give us ideas that we might not have considered. Experiments may also guide formal exercises in ecological scaling (e.g., Levin 1992; Petersen and Hastings 2001) and generate the discovery of ways to relate scaling characteristics of species (e.g., home range size, movement rates, or generation times) to their responses, allowing us to formally compare outcomes at multiple scales.

CHAPTER FIVE

Fragment Context and
Edge Effects

In landscapes and regions context is usually more important
than content. That is, the surrounding mosaic has a greater
effect on patch functioning and change than do the present
characteristics within the patch.

—*Richard T. T. Forman (1995)*

Many ecological studies of habitat loss and fragmentation have focused on how
ecological patterns or processes are affected by the characteristics of habitat fragments, such as size and isolation. For example, observational studies (see chapter 3) clearly show that larger patches support a higher diversity of species and
that isolation may have detrimental effects on species persistence. And most experiments (see chapter 4) have manipulated the size and isolation of habitat fragments, or fragmentation per se, to directly examine their effects on ecological
processes. This emphasis on characteristics such as size and isolation makes
sense. After all, these features are easily measured, and there is a good deal of
ecological theory to suggest that these two factors may largely explain variation in
species richness among habitat islands or fragments. But as the title of Daniel
Janzen's 1983 paper graphically suggests, "no park [or fragment] is an island."
Ecologists have learned that there is often a lot of action at the interface where
fragments end and the rest of the landscape begins. This chapter explores the dynamic changes that occur at the boundaries of fragments and examines how
these changes are affected by the characteristics of the surrounding landscape. It

reviews an abundant quantity of literature on species responses to edges, and a growing body of research on varied landscape spatial arrangements, to suggest ways in which understanding the effects of fragment spatial position in the landscape relates to prior studies of fragment size and isolation. This discussion emphasizes a major contribution of landscape ecology, which is to take an expanded view—to look beyond characteristics of local habitats or fragments and to consider the ecological dynamics of the broader landscape mosaic (e.g., Forman 1995).

This expanded view stresses how landscape context influences ecological processes. Think of landscape context as a form of ecological peer pressure. In the painfully familiar form of peer pressure that most teenagers experience, opinions and actions are often shaped by the behaviors or habits of people in their immediate surroundings. The same thing can happen in habitat fragments that are embedded in complex landscapes. Like those teenage peers, the land uses and natural habitats that surround a fragment may exert strong influences on that patch. For example, a forest fragment surrounded by fertilized agricultural fields is likely to experience different pressures and inputs than one surrounded by urban development. Or a prairie dog colony surrounded by continuous grassland is likely to be quite different from one surrounded by heavily trafficked roads and industrial development (fig. 5.1; see the section on prairie dogs at the end of this chapter).

Before delving into detail on the topic of fragment context, however, it is necessary to first mention edge effects. The study of edges is intimately linked with landscape context, for it is at these boundaries between fragments and their surroundings where the most immediate imprint of landscape context is detectable. Put simply, *edges* are the borders between two habitat types, such as the edge between a river and its bank, or one between a meadow and a forest. Most conservation ecologists are concerned with edges that are created as a result of human activities, such as the edge between a forest and an agricultural field, or one between old-growth forest and a clear-cut. Accordingly, this discussion focuses primarily on the latter type, that is, on human-induced edges.

Discussions of ecological edges often refer back to Aldo Leopold's (1933) treatise on game management, where he noted that wildlife in the midwestern United States, such as deer, quail, and grouse, tended to be most abundant in landscapes rich with edges between deciduous forests and agricultural fields. Stimulated by this observation, wildlife managers spent the next few decades championing this "edges are good" paradigm, since edges were considered favorable for wildlife.

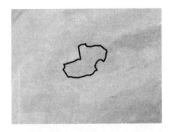

Figure 5.1. The landscape context of two black-tailed prairie dog (*Cynomys ludovicianus*) colonies within grassland parcels near Boulder, Colorado. The colonies are represented as polygons outlined with a solid black line in each photo. *Top:* colony surrounded by continuous grassland. *Bottom:* colony surrounded by roads, housing, and industrial activity. Digital aerial photos from 2000, obtained from the City of Boulder Open Space Department.

Leopold's concept began to be questioned when ecologists realized that not all species were favored by edge habitats. In fact, many of the rare species valued by conservationists, such as migratory songbirds, appeared to be positively associated with intact, continuous forests; they showed marked declines in abundance at habitat edges (Brittingham and Temple 1983; Wilcove 1985; Yahner 1988). Ecological research over the past several decades has shown that many species do decline at habitat edges, while other species increase or are unaffected by these human-induced edges (reviewed in Paton 1994; Murcia 1995; Ries et al. 2004). The response of particular species to edges may critically determine patterns of overall species composition in habitat fragments. So it is imperative to understand the magnitude and direction of edge effects in order to gain a more comprehensive view of species' responses to habitat loss and fragmentation.

SUMMARY OF EDGE EFFECTS

Imagine taking a walk from the dark dampness of an old-growth forest into an adjacent field recently cleared to make way for an agricultural crop. You would notice several distinct changes en route, especially that the cleared field is much brighter, warmer, and drier than the old-growth forest. Many edge studies emphasize variation in these microclimatic variables—such as light intensity, temperature, and relative humidity—across forest-field boundaries (Kapos 1989; Chen, Franklin, and Spies1992; Newmark 2001). The amount of light reaching plants is obviously higher at the edge of a forest fragment than in the forest interior, which generally means a higher temperature and lower relative humidity at the forest edge. Moreover, wind velocities are typically greater at edges than in forest interiors (Harris 1984; W. Laurance et al. 1998). The changes in light, moisture, temperature, and wind, which are most pronounced at the fragment edge, may alter significantly the plant and animal communities that occur there. This microclimatic edge effect may not only influence the environment at the edge of the fragment, but may also permeate the habitat remnant for tens of meters (e.g., Ranney, Bruner, and Levenson 1981; Harris 1984; Kapos et al. 1997). Edge effects have been primarily studied in forests, probably because of their dominant vertical organization and the dramatic changes in the structural characteristics of the habitat that occur as a result of forest clearing (Chen, Franklin, and Spies 1992; Murcia 1995; W. Laurance et al. 1997; Newmark 2001; Driscoll and Donovan 2004).

Because studies of edge effects often produce idiosyncratic results—at least partially due to variation in study design, methodology, and the temporal dynamics of edges—it has been difficult for ecologists to generalize about the consequences of increased edge habitat in fragmented landscapes and to invoke the mechanisms underlying those effects (Paton 1994; Murcia 1995; Newmark 2001; Chalfoun, Thompson, and Ratnaswamy 2002). Recent conceptual developments focused on boundaries and edges have been particularly useful in overcoming these difficulties. For example, Cadenasso et al. (2003) devised a framework for understanding the structure and function of ecological boundaries and for generalizing across different ecosystems and spatial scales. They emphasized how boundary architecture may influence multi-directional flows across boundaries. And Ries et al. (2004) developed a resource-based, predictive model to account for variability in edge effect studies, one that is based on estimates of habitat quality and whether the resources used by organisms across edges are supplementary or complementary. In a review of empirical literature on edge effects, they showed that edge responses are largely predictable if examined in terms of their mechanistic model of habitat quality and resource distribution. Both of these contributions are giant steps toward a more general understanding of edge dynamics.

With regard to the wide range of results gathered from empirical studies of edge effects, and a considerable amount of discussion regarding the mechanisms responsible for observed effects, there are at least four straightforward generalizations that can be drawn from this multifarious literature. First, due simply to geometry, small, irregularly shaped patches have more edge for a given area, and therefore they experience greater edge effects than do larger, compact patches. Second, edge aspect may influence the magnitude of edge effects, which has been shown in north and south temperate-zone studies, but typically not in tropical studies. This is based on the relative input of solar radiation. Third, the distances over which edge effects permeate range from a few meters to over several hundred meters, and abiotic effects tend to occur over shorter distances than biotic effects. Fourth, the magnitude and character of the edge effect may depend significantly on landscape context. More time will be spent discussing these last two issues, because they are more complex and warrant a closer look.

Patch Shape

The first issue occurs simply because small, irregular shapes have a higher perimeter-to-area ratio than do large, compact shapes. Small habitat fragments have more edge per unit area than do larger patches (Collinge 1996). If all things associated with edges are mostly bad, then small patches will experience proportionately greater negative effects than will large ones. Benítez-Malvido and Lemus-Albor (2005) found that pathogen damage on woody seedlings in tropical forests in Chiapas, Mexico, was three times greater on plants at forest edges than in forest interiors. They did not report plant mortality, but they did suggest that if pathogen infestation resulted in increased plant mortality, then disease spread at forest edges may threaten tropical forest vegetation. Similarly, Donoso, Grez, and Simonetti (2003) observed higher seed predation by rodents at the edges than in the interiors of *Nothofagus* forest fragments in central Chile. In both instances, these negative effects at edges are likely to have disproportionately greater effects in small versus large fragments.

This geometric fact is probably a vital piece in explaining the declines of native species in small versus large habitat patches noted in chapter 4, but it is sometimes difficult to disentangle area effects from edge effects. For example, in a recent review of studies of bird nest predation at forest edges, Parker et al.

(2005) asked whether there are fewer birds in small patches because of patch area, or because edge effects are more pronounced in small patches. They concluded that

> because almost no one has examined the effect of forest patch area independent of the effect of distance from forest edge, *the question of whether effects of area exist that are distinct from effects of edge remains unanswered.* Designing an area effect study that completely controls for edge effect may be impossible because even if the distance to the closest edge is controlled, edge distances in other directions are bound to vary with varying patch area . . . Regardless of which mechanisms are at work, the result remains that for many species, occurrence rates are lower in small forest fragments. Thus, although it is important to test alternative hypotheses to explain this phenomenon, this should not preclude management to limit further forest fragmentation. (pp. 1164–1165; italics mine)

The corresponding issue of edge effects in relation to patch shape has also been raised in discussions about reserve design. Diamond (1975) advocated the selection of large circular reserves because they would be more beneficial to native species than small linear reserves, based on the reasoning that negative edge effects would be less pronounced in protected areas with more compact shapes. Game (1980) argued that in some instances, however, a circular reserve may not be optimal, particularly if it is desirable to enhance the rates of immigration into the reserve. This assumes that there are individuals of desired species moving from the surrounding landscape into the protected area, and the linear shape of the reserve will likely increase the movement rate of organisms into the reserve. A few studies have shown the effects of patch shape on movement patterns (Collinge and Palmer 2002; Tanner 2003), but the key issue for designing reserves to protect biodiversity is ultimately whether the benefits of enhanced movement into reserves supersede the negative effects of abiotic and biotic changes at the edges (e.g., Bogaert et al. 2001; J. Williams, ReVelle, and Levin 2005).

Edge Orientation

Second, and perhaps not surprisingly, the extent to which edge effects penetrate habitat fragments may be influenced significantly by the aspect or orientation of the edge (Wales 1972; Ranney, Bruner, and Levenson 1981; Brothers and Spingarn 1992; Brothers 1993; Matlack 1993a; Young and Mitchell 1994; Turton and Freiburger 1997). In the Northern Hemisphere, higher inputs of solar radiation to south-facing edges generally mean that they are warmer, drier. and wider than north-facing edges; the opposite is true in the Southern Hemisphere (Young and Mitchell 1994). There are only a few studies that explicitly examine north-facing versus south-facing edge effects in tropical systems, but the expectation is that because there are not strong seasonal differences in the inputs of solar radiation to north- or south-facing edges in equatorial ecosystems, there would not be distinct effects of edge orientation on microclimatic variables.

Changes in the structure and composition of the existing plant communities are associated with changes in light, temperature, moisture, and wind conditions at forest edges. For example, in the deciduous forest patches of southeastern Wisconsin studied by Ranney, Bruner, and Levenson (1981), forest edges typically contained more pioneer and xeric plant species than the interior—higher densities of shrubs and herbaceous ground-layer vegetation—an effect that penetrated for several meters into the forest; the edges also had higher species richness than the interior. In tropical forest remnants in northeast Australia, the number of rainforest seedlings was higher in the forest interior relative to the edges, presumably because germination declined at the edges due to greater light intensity and soil surface temperatures (Turton and Freiburger 1997). Higher species richness in forest edges may result from the invasion of exotic plant species (Brothers and Spingarn 1992). Edge orientation may also influence species composition; south- and west-facing edges contained more xeric plant species than did north- and east-facing edges (Wales 1972; Ranney, Bruner, and Levenson 1981), due to variation in light and moisture conditions. Ranney, Bruner, and Levenson (1981) predicted that the composition of these forest patches would eventually change as a result of increased seed dispersal from the edge to the interior, with shade-intolerant edge species—e.g., hickory (*Carya* spp.) and hawthorn (*Crataegus* spp.)—replacing shade-tolerant plants of the interior—e.g. sugar maple (*Acer saccharum*) and American beech (*Fagus grandifolia*).

Similarly, they found that the windward edges of forest patches tend to be warmer, drier, and wider than leeward edges. Moreover, trees growing along the edge of a forest fragment may be exposed to greater wind velocities than those occurring in the forest interior. For example, in the United States, in the Douglas fir forests of the Pacific Northwest, higher wind velocities were recorded from the forest edge to a distance 60 m into the forest interior (Chen, Franklin, and Spies 1992). Increased wind velocities at the forest edge may result in an increased incidence of mortality due to tree-fall, especially for shallow-rooted tropical trees. In an isolated 10 ha forest patch in Brazil, the overwhelming majority of tree mortality along the margins was on the windward margin of the patch; annual

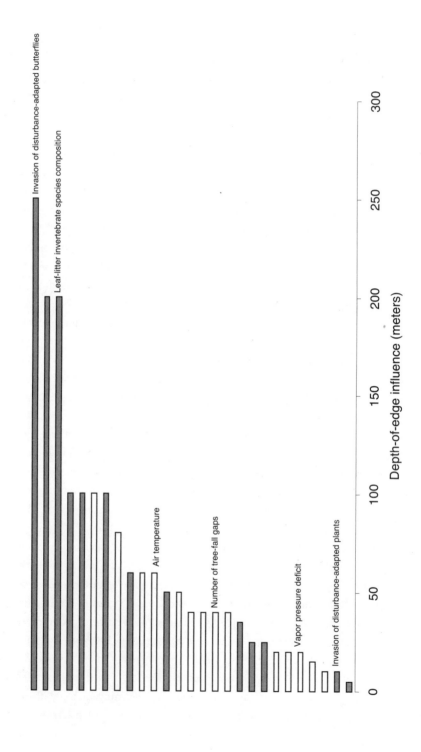

Invasion of disturbance-adapted butterflies

Leaf-litter invertebrate species composition

Air temperature

Number of tree-fall gaps

Vapor pressure deficit

Invasion of disturbance-adapted plants

Depth-of-edge influence (meters)

tree mortality rates were estimated to be 2.6% in isolated patches versus 1.5% for continuous forest (Lovejoy et al. 1984; Bierregaard et al. 1992).

Depth-of-Edge Influence

A third generalization that arises from edge effect studies is that the distance over which edge effects extend (often called *depth-of-edge influence*) is determined not only by the aspect or orientation of the edge, but also depends significantly on what particular variable is being measured across the habitat boundary. The attenuation of light at forest-field edges, for instance, may occur for only a few meters inside the forest fragment, whereas shifts in soil moisture may extend much farther. For example, old-growth forest remnants studied in central Indiana exhibited significant increases in light levels and temperature and a marked decrease in humidity at the forest-field edge (Brothers and Spingarn 1992). These microclimatic differences ceased to exist beyond a distance of 8 m into the forest for these relatively small, 8–23 ha forest fragments. Similarly, temperature and light decreased within relatively short distances in eastern deciduous forest fragments in Pennsylvania and Delaware (Matlack 1993a), but humidity and leaf-litter moisture continued to change 50 m into the forest interior.

In their intensive, long-term study of Amazonian forest fragments (highlighted in chapter 4), W. Laurance et al. (1997) measured the edge effect distances of many different ecological phenomena (fig. 5.2). The beauty of this study is that multiple variables were measured across the same forest-pasture boundaries, which makes it easy to compare edge effect distances for a variety of variables. These measurements reveal a quite interesting pattern, which is that abiotic variables (such as relative humidity, the number of tree-fall gaps, and air temperature) tend to penetrate from 10 to 70 m into the forest, but shifts in biotic factors (such as the species composition of leaf-litter invertebrates and the invasion of disturbance-adapted butterflies) extend much farther, from 100 to 250 m into the forest.

Other edge effects may occur at even broader spatial scales (W. Laurance 1991, 2000). W. Laurance (2000) reviewed two different studies of edge effects in South-

Figure 5.2. (opposite) Edge penetration distance, or depth-of-edge influence, for 27 response variables measured at the Biological Dynamics of Forest Fragments (BDFF) project in the Brazilian Amazon. Distance is measured in meters. The white bars denote abiotic variables, the dark bars denote biotic variables. Three abiotic and three biotic variables are noted as examples. Redrawn from W. Laurance et al. (1997).

east Asian forests (Curran et al. 1999; H. Peters 2001), both of which suggested much greater depths-of-edge influence than previously recorded. Laurance surmised that at Curran et al.'s (1999) study sites in western Borneo, the influx of seed predators from degraded forests to intact dipterocarp forests, and their subsequent negative impact on seedling establishment, must have occurred over distances of 10 km. In peninsular Malaysia, a greater abundance of wild pigs, due to an increase in their food supply in agricultural areas, caused shifts in the invasion of exotic species in a protected forest, even though the oil-palm plantations that supplement the pigs' diets occur more than 2 km outside the forest reserve.

So the answer to the question, How wide is the edge? really depends on what feature of the edge effect is being considered. The distance is relatively small for microclimatic effects, but shifts in animal abundance and distribution may occur over several kilometers. Moreover, scientists involved in the Amazonian study found edge effects to be so pervasive and to have such profound impacts that they concluded that "the Theory of Island Biogeography has not lived up to its initial promise of serving practical conservation biology. Recent knowledge from BDFFP [Biological Dynamics of Forest Fragments Project] . . . suggests that much of the ecological degradation can be accounted for by the influences of *edge effects and the surrounding matrix*, neither of which is addressed by Island Biogeography Theory" (Gascon and Lovejoy 1998, p. 273; italics mine).

Edge Effects in Varied Landscapes

A fourth generalization that can be noted from studies of edge effects is that the extent and character of edge effects depends on the features of the matrix surrounding a habitat patch. This issue of landscape context will be considered in broader terms in the following section, but the present discussion emphasizes the explicit differences in edge effects in varied landscapes. A key aspect of landscape boundaries is *contrast*, which refers to the degree to which habitat types on either side of the boundary differ from one another. Thus boundary contrast is directly related to the characteristics of the surrounding matrix. Boundaries may be described as abrupt, sharp, and hard (Wiens, Crawford, and Gosz 1985; Forman 1995); as soft and gradual; or anywhere in between these two extremes. The adjectives "sharp" and "gradual" clearly reflect human perception of the structural differences across boundaries. We still know relatively little about how these structural attributes (boundary *contrast*) translate into functional aspects (boundary *permeability*), such as the movement of animals or materials across boundaries (Holmquist 1998; Collinge and Palmer 2002; Cadenasso et al. 2003; see also chapter 6). A field study of montane forest edges in the highlands of Chiapas, Mexico, compared edge effects for boundaries with high contrast between forest and field (cleared pastures) versus low contrast (old fields) (López-Barrera et al. 2007). Oak acorn dispersal by small mammals extended farther from remnant forests into fields in places where the boundary between the two habitat types was soft (low contrast) than for high-contrast boundaries. Thus the extent of edge effects (seed dispersal across forest-field edges) was greater where the contrast between the two patch types was less.

In addition to variation in the distance over which edge effects occur across different patch types, the character of edge effects may change depending on surrounding land use. For example, in addition to microenvironmental changes, forest fragments in suburban areas experience altered conditions due to human activity. Matlack (1993b) observed that suburban residents dumped grass clippings, Christmas trees, and building rubble in nearby forest remnants. They also gathered firewood, pruned limbs, and built tree houses in these habitats. In this northern Delaware suburban landscape, 95% of these human activities occurred within 82 m of the forest edge. Thus changes in forest conditions due to human activities may permeate the forest interior as far as or farther than microclimatic changes. The influence of a suburban development on adjacent native habitat may be visually subtle, but it may carry far beyond property-line boundaries. In other words, suburban developments may have an ecological "aura" that extends into adjacent native habitats. Such significant influences must be an important consideration in the design and planning of suburban developments in continuous native habitat.

As mentioned above, tree mortality may be higher at forest edges, but this mortality at the edges may also vary significantly, depending on what type of habitat surrounds the fragment. For example, Amazonian old-growth forest fragments that were part of the BDFF project (chapter 4) and studied by Mesquita, Delamonica, and Laurance (1999) were surrounded either by cattle pastures or by one of two types of second-growth forests. The second-growth forests were dominated either by the pioneer tree species, *Cecropia*, or a second pioneer tree species, *Vismia*. Tree mortality at the forest edges was higher in fragments surrounded by cattle pastures than those surrounded by second-growth forests. Edge effects also appeared to extend slightly farther (by about 20 m) into the forest fragments bordered by pastures than those surrounded by second-growth forest.

Many studies of edge effects have focused on bird predation at forest edges. Early investigators (e.g., Gates and Gysel 1978) noted higher predation on bird nests at forest edges than those in intact interior forests, and this emerged as a major mechanism to explain the declines of birds in highly fragmented forests (Paton 1994). Further field studies and meta-analyses of nest predation have concluded that nest predation at edges may be mediated by landscape context (Andrén 1992; Donovan et al. 1997; Hartley and Hunter 1998; Lahti 2001; Chalfoun, Thompson, and Ratnaswamy 2002; Driscoll and Donovan 2004; Tewksbury et al. 2006). Two separate meta-analyses noted that in general, nest predation increased with proximity to the edge, but the patterns were stronger in agricultural or non-forested landscapes than in forest-dominated ones (Hartley and Hunter 1998; Chalfoun, Thompson, and Ratnaswamy 2002). In introducing their own field studies of nest predation in wood thrushes, Driscoll and Donovan (2004) asserted: "There is an emerging pattern in investigations of habitat fragmentation that the occurrence of edge effects, especially across forest-field edges, depends on the composition of the landscape in which the edge is embedded. Yet the generality of this pattern—that edge effects are context-dependent—is still uncertain" (p. 1331).

LANDSCAPE CONTEXT

These observations—that edge effects may differ depending on the composition of the surrounding landscape—provide some background for a broader and more detailed discussion of landscape context. As defined above, *landscape context* refers to the characteristics of the landscape surrounding a fragment (the *matrix*), and it may have profound effects on ecological patterns and processes within the fragments. Compared to studies of fragment size and isolation, relatively few researchers have examined how the landscape context of habitat remnants affects population and community dynamics. A number of published studies reveal striking ecological patterns among remnants that differ in their surroundings (e.g., Åberg et al. 1995; Stouffer and Bierregaard 1995; Sisk, Haddad, and Ehrlich 1997; Ricketts et al. 2001; Guirado, Pino, and Roda 2006; Ohlemüller, Walker, and Wilson 2006; Rand and Louda 2006). As an example of a strong effect of landscape context, Sisk, Haddad, and Ehrlich (1997) found that bird species richness and abundance in northern California remnant woodlands differed significantly in patches surrounded by chaparral compared to those surrounded by grassland.

It is worth taking a moment here to discuss how researchers describe and study the effects of landscape context. After all, landscapes are complex, heterogeneous amalgamations of various habitat types and land uses, so it is a difficult task to develop appropriate measures of context. As a useful and relatively easy first pass, many researchers describe the landscape surrounding a patch by first demarcating a defined area (usually circular or square in shape) that borders a habitat patch on all sides, usually by using aerial photographs or satellite images. They then calculate the percent cover of that defined area (excluding the patch itself) that is occupied by a particular habitat type or land use, such as grassland, forest, urbanization, or agriculture. This is repeated for several to many patches in the landscape, which are usually specifically chosen to represent the desired range of conditions present in the surrounding landscape. The dependent variable of interest (e.g., bird species richness, herbivory on a particular plant species, the prevalence of a particular pathogen, the abundance of an endangered mammal, or nutrient input to a lake) is measured in all of the focal patches. Then it is related to the independent variable, the percent cover of the different habitat types or land uses in the landscape. This description of context can also be done for multiple spatial scales by drawing concentric circles or squares of ever-larger dimensions around the patch (e.g., Roland and Taylor 1997; Langlois et al. 2001; Steffan-Dewenter, Munzenberg, and Tscharntke 2001) and then calculating the percent cover of every single habitat type or land use in each of these nested sampling grids or areas. This is a particularly useful approach for identifying the spatial scale at which species or ecological processes respond to landscape context (e.g., Roland and Taylor 1997; Bock et al. 2002).

Species' Responses and Interactions

Many studies that have measured landscape context in this way have revealed striking patterns of variation in ecological parameters (such as species richness, species occurrence, population abundance, animal movement patterns, and species interactions) associated with variation in landscape context. Others have noted no such effects of landscape context, citing local patch variables as being more critical influences on ecological parameters than landscape context.

The abundance of individual species, as well as the number of species present in fragments, may vary significantly with variation in landscape context. For example, in a study of spruce forests in Norway, both species richness and the abundance of individual species of mycetophilids (small flies that develop in fungi) were positively associated with the amount of old-growth spruce forest in the surrounding landscape (Økland 1996). Ricketts et al. (2001) surveyed moths within agricultural sites near a large forest fragment in southern Costa Rica and found that moth species richness and abundance were higher in sites near the forest fragment than in those farther away from the fragment. Species composition also differed among sites, depending on how close sites were to remnant forests. These authors suggested that forest fragments form a halo by influencing species composition in adjacent agricultural habitats up to 1.4 km from the forest edge. The invasion of alien plant species into indigenous forest fragments in New Zealand was related significantly to the amount of indigenous forest cover in the surrounding landscape (Ohlemüller, Walker, and Wilson 2006); alien species richness was lower in fragments that were surrounded by a greater cover of native forest. In all three of these cases, landscape context significantly influenced species composition within fragments.

Whether a species even occurs in a given habitat fragment may depend on landscape context. Occurrence and abundance of the rare western ringtail possum, or ngwayir (*Pseudocheirus occidentalis*), was associated significantly with several features of the landscape surrounding its native *Eucalyptus* forest habitat in southwestern Australia (Wayne et al. 2006). The authors observed a positive association with ngwayir occurrence and the control of fox predators in the surrounding landscape, and found negative associations with the amount of forest fragmentation and the distance to non-native vegetation types (agricultural fields and tree plantations). In a study of birds in urban forest fragments near Tokyo, Japan, certain species, such as the Oriental turtle dove (*Streptopelia orientalis*) occurred more frequently in woodlands surrounded by agricultural areas than in woodlands surrounded by urbanized areas (Morimoto et al. 2006). The authors suggested that the occurrence patterns were probably related to movement abilities; the species that displayed this pattern could perhaps move easily across agricultural areas, but not urban areas, so they were found with greater frequency in woodlands surrounded by agricultural areas.

The differential abilities or affinities of species to travel across particular habitat types or land uses may ultimately explain variation in occurrence patterns in conjunction with landscape context. For example, hazel grouse (*Bonasa bonasia*) in Swedish forest fragments dispersed over much greater distances when the surrounding matrix was forested habitat than when it was open fields (Åberg et al. 1995). In agricultural landscapes, grouse appeared to disperse up to about 100 m between suitable forest habitats, but in predominately forested landscapes, grouse dispersed much farther, up to 2000 m. This link between matrix composition and animal movement will be explored more fully in the next chapter, but it is important to note here that variation in movement may help explain relationships between species occurrences in habitat fragments that are found in different landscape settings.

Because different landscape contexts may shift species composition in habitat fragments, interactions among species that occur in those fragments may also be altered. Several recent examples show how key interactions among species— including parasitism, herbivory, pollination, and seed dispersal—may shift in relation to landscape context. Three of four fly parasitoids that infest an economically important forest herbivore, the forest tent caterpillar, had higher rates of attack in contiguous rather than fragmented aspen forests near Edmonton, Alberta, Canada (Roland and Taylor 1997). Similarly, in eastern Canada, the spruce-budworm-caused mortality of balsam fir was greater in conifer stands surrounded by extensive conifer forest than in those surrounded by deciduous forest (Cappuccino et al. 1998). The number of visits by bee pollinators to *Centaurea jacea* (Asteraceae) flowers in central Germany was positively associated with the amount of semi-natural habitats (grasslands, orchard meadows, hedgerows, and garden land) within a 250 m radius of the focal plant (Steffan-Dewenter, Munzenberg, and Tscharntke 2001). The proportion of flower heads that were damaged by seed predators similarly increased in landscapes with a greater proportion of semi-natural habitats. Consequently, the net effect of landscape context on seed set was neutral; pollination was enhanced, but so was seed predation. The dispersal of seeds of the shrub *Solanum americanum* by frugivorous birds in Puerto Rico was enhanced significantly in plant neighborhoods that included a co-occurring fruiting species, *Cestrum diurnum*, than in monospecific experimental arrays that contained only *Solanum* (Carlo 2005). This study experimentally controlled the abundance and composition of fruiting plants, so the differences in frugivory and seed dispersal could be directly attributed to differences in the composition of fruiting landscapes. All of these studies show that species interactions are context-dependent and are not solely a function of the plant traits or inherent features of the species with which they interact.

A final example of species interactions that may be influenced by landscape context comes from a study of a biological control insect that has shifted plant hosts. *Rhinocyllus conicus* is a weevil that was introduced to the United States to control non-native noxious thistles, especially *Carduus nutans*. Rand and Louda (2004) investigated the distribution and abundance of *R. conicus* occurrence, specifically as a function of landscape context, on two species of native thistles (*Cirsium undulatum* and *C. flodmanii*) in central Nebraska. The intensity of herbivory by *R. conicus*, determined by the number of eggs deposited on flower heads of the native thistle (*C. undulatum*), was positively associated with the density of the noxious thistle (*Carduus nutans*) that was measured in fields surrounding the focal study sites (fig. 5.3). The risk of infestation by this herbivore

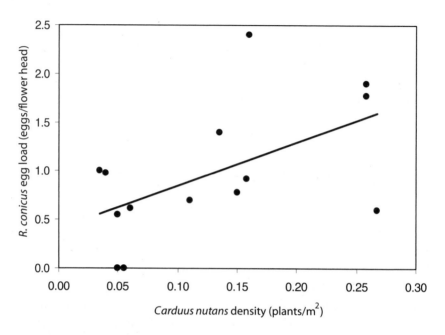

Figure 5.3. Top: The number of eggs of the herbivore *Rhinocyllus conicus* deposited on flower heads of the native wavy-leaved thistle, *Cirsium undulatum*, in relation to the abundance of the exotic thistle, *Carduus nutans*, in the surrounding landscape. Redrawn from Rand and Louda (2004). *Bottom:* Photograph of *Rhinocyllus conicus* pupae and larva inside the inflorescence of Platte thistle (*Cirsium canescens*) in Arapaho Prairie, Nebraska. Photo by Svata M. Louda.

Errata

p. 90, figure 4.4: the double squares above the bars in parts (a), (c), and (d) should read "NS."

p. 145, figure 7.1: the y-axis label should read "Probability of species interaction"; the x-axis label should read "Population abundance or density."

p. 148, figure 7.2: the part (b) x-axis label should read "*L. paralienus* nest density (residuals)."

p. 231, figure 10.3: the y-axis label should read "Relative species diversity, S_R/S_A"; the x-axis label should read "Proportion ancient forest, P_A."

on native thistles thus depended significantly on characteristics of the surrounding landscape.

Land Use and Aquatic Ecosystems

Thus far this chapter has surveyed species responses to landscape context and shifts in species interactions in different landscape types. However, the influences of varied adjacent habitat types and land uses on habitat patches is perhaps best understood for aquatic ecosystems, particularly lakes and streams (Likens et al. 1970; Peterjohn and Correll 1984; Pringle 1991; Sorrano et al. 1996; J. Allan 2004; Moore and Palmer 2005). The supply and flow of nutrients, materials, and energy within water bodies—as well as between lakes, streams, and the surrounding landscape—may differ substantially, depending on the adjacent land use or activity. For instance, in an agricultural landscape, the relative inputs of nitrogen and phosphorus into a stream varied according to the relative proportion of cultivated cropland versus riparian vegetation adjacent to the stream (Peterjohn and Correll 1984). The land-water interface can be seen as the ultimate hard boundary, since the contrast between aquatic and terrestrial habitats and organisms is usually quite sharp. Additionally, aquatic-terrestrial edges are usually only permeable in one direction, that is, nutrients, materials, and energy typically flow from land to water. A notable exception is during flooding events, where the aquatic environment essentially invades the terrestrial environment, reversing that flow of nutrients, materials, and energy.

Soranno et al. (1996) extended our understanding of how adjacent land uses contribute to water quality by developing a model for an agricultural-urban watershed in Wisconsin that incorporated landscape spatial configuration. The researchers calculated the percent cover of different habitat types and land uses in the watershed, as had been done previously, but they also calculated nutrient input to the lake by accounting for the spatial variation in topography and land use, both of which influence the rate of overland water flow. They used this model to identify major inputs of non-point-source phosphorus, as well as to predict the impacts of future land use changes on water quality in the lake. Subsequent analyses of dissolved organic carbon input to lakes in Adirondack Park in New York (Canham et al. 2004), and of nitrogen input to streams in the coastal plain of Maryland (R. King et al. 2005), revealed that the proportion of adjacent land use in different categories, as well as the spatial arrangement of land uses, can influence water chemistry and aquatic biodiversity.

As with studies of species' responses to landscape context, most studies of

landscape context effects in aquatic systems have primarily quantified the proportion of surrounding habitat types or land uses. Although it is a more difficult
proposition to characterize and compare the relative effects of distinct spatial
configurations of land uses on ecological processes, several published studies
now include such sophistication (e.g., Soranno et al. 1996; Canham et al. 2004;
R. King et al. 2005). In a recent review, M. Turner (2005) noted that "whether just
the composition of the uplands (i.e., the amount of different land uses) matters,
or if the spatial configuration is also important, remains unresolved because
studies have produced conflicting results" (p. 1970).

Local versus Landscape Effects

Although the foregoing discussion has featured a large number of ecological
studies that have demonstrated the importance of landscape context for ecological processes, some researchers have documented that local habitat characteristics determine ecological patterns and that landscape context plays a relatively
weak role (e.g., Edenius and Sjöberg 1997; Berry and Bock 1998; Lindenmayer,
Cunningham, and Pope 1999; Pharo, Lindenmayer, and Taws 2004). An extensive survey of mammals in 86 forest fragments in southeastern Australia demonstrated the variability in species' responses to landscape context. For several
mammal species, there were no significant differences in their presence or abundance in forest stands surrounded by native *Eucalyptus* forest versus those surrounded by non-native pine tree plantations (Lindenmayer, Cunningham, and
Pope 1999; Lindenmayer et al. 1999). The reasons for this lack of effect may stem
from the behavior and movement capabilities of the mammals in these forests.
Perhaps the animals move easily through the pine plantations, and so essentially
do not view their habitat as discontinuous. Alternatively, perhaps the forest fragments are large enough to support persistent populations of these mammal
species, so their dispersal from the fragments is unnecessary. Pharo, Lindenmayer, and Taws (2004) studied bryophyte distributions in these same remnant
forests, and similarly found that both liverworts and mosses were more strongly
affected by site and substrate factors than by the type of cover in the surrounding
landscape. In dry *Eucalyptus* forests in Queensland, Australia, the abundance of
several diurnal bird species was unaffected by landscape context, but the diversity and abundance of several glider species was altered significantly by landscape variables (McAlpine and Eyre 2002).

Given that ecological studies have produced variation in the responses of
species to landscape context, are there any generalizations that can be made re-

garding the importance of context for ecological processes? In an effort to discern general patterns, Mazerolle and Villard (1999) reviewed published studies that combined local and landscape effects. They asked whether local (patch) characteristics or landscape characteristics exerted strong effects on patterns of species abundance and distribution. At the time of their review, there were 61 studies appropriate for analysis, 49 of which were conducted in forests fragmented by agricultural activity. Their analysis revealed that the presence and abundance of birds, mammals, amphibians, and reptiles tended to be affected by both patch and landscape variables, whereas studies of invertebrates tended to show stronger effects of patch characteristics than landscape variables. The variation in the responses of species groups is probably attributable to body size and movement abilities, as well as to the scale at which species perceive landscape spatial structure (With 1994). The reasoning is that because larger animals have broader-ranging movement patterns than smaller, less mobile animals, the former should be more greatly affected by landscape context. Thus species' responses to landscape context may be dependent significantly on movement patterns. For example, individual Iberian lynx move at different rates and with different probabilities, depending on the habitat type and land use (Ferreras 2001).

Studies published since Mazerolle and Villard's (1999) review confirm and extend their findings. Recent research has shown the importance of landscape context to the presence and abundance of birds and mammals in a variety of habitat types (Villard, Trzcinski, and Merriam 1999; J. Miller et al. 2003; Dunford and Freemark 2005; Silva, Hartling, and Opps 2005; Martin et al. 2006). Moreover, these studies have shown that, similar to the aquatic studies mentioned above, the influence of landscape context is due not only to the percent cover of a particular habitat type or land use, but also to the spatial configuration (Villard, Trzcinski, and Merriam 1999) and intensity (Martin et al. 2006) of various land uses. In a study of forest fragments in southeastern Canada (Dunford and Freemark 2005), different land uses also appeared to exert effects on species at different scales; agricultural land uses affected species in forest fragments at a distance of up to 5 km, and urban land uses had impacts at both fine scales (up to 1.8 km) and at a broad scale (5 km).

Rapid Urbanization

An increasingly common landscape configuration is that of native habitat surrounded by urban and suburban development (Soulé et al. 1988; Theobald, Miller, and Hobbs 1997; Grimm et al. 2000). As noted above, fragments adjacent to urban developments and their associated features (such as roads and parks) may experience unique pressures compared to those surrounded by agricultural activities or forest plantations. In the western United States in particular, human population growth is as high as 13% per year in some regions (Riebsame 1997), so the juxtaposition of native habitat and urban development continues on a dramatic upward trajectory. In the past ten years, there has been increased attention to ecological studies in urban areas, with results that suggest unique and varied effects of urban landscape context on ecological patterns and processes. An ongoing, comprehensive study of this context near Boulder, Colorado, provides a useful case study for discussing this topic.

In 1994, Jane and Carl Bock, my colleagues at the University of Colorado at Boulder, began a study of biological diversity in relation to landscape context. The city of Boulder is located at the intersection of the western Great Plains and the eastern edge of the Rocky Mountain foothills (40°00′54″ N, 105°16′12″ W), and its human population has increased nearly five-fold since 1950, from 20,000 residents to a current population of about 100,000. The City of Boulder Open Space Department owns or manages over 10,000 ha of grassland habitat, which forms a nearly continuous green belt around the city (Bock et al. 2002; Collinge, Prudic, and Oliver 2003). These protected grasslands include tallgrass remnants and hayfields in lowland floodplains, and mixed grass and shortgrass prairies on upland slopes and mesas.

The Bocks established 66 circular, 200 m diameter study plots in grasslands near Boulder that were surrounded by varying levels of urbanization, from continuous grassland up to grasslands surrounded by about 30% urbanization. All of the study plots were placed within grassland habitats, with the plot centers at least 100 m from the nearest urban structures. Because of the relatively contiguous spatial configuration of these grasslands, characterizing them in terms of their surrounding context was more appropriate than including more typical measures such as patch size and isolation. All 66 study plots were located in one of four grassland types, including shortgrass prairie ($n = 13$), mixed-grass prairie ($n = 21$), tallgrass prairie ($n = 11$), and hayfields. Of the 66 plots, 30 were located near some form of human activity, such as residential or commercial development, and the other 36 plots were surrounded by native habitat (Bock et al. 2002).

For three consecutive field seasons, the Bocks and their field crews sampled the abundance and species composition of raptors, grassland birds, small mammals, grasshoppers, and flowering plants in each of these plots. I and my graduate-student colleagues followed with surveys of butterflies in each of the plots during

1999 and 2000, and with surveys of black-tailed prairie dogs in colonies surrounded by varying levels of urbanization in 2000 and 2001. To describe landscape context for each study plot, a satellite image was classified and converted to a land-cover map of the area (Haire et al. 2000). Then the percent of each habitat type and land use was calculated in each of five square windows, ranging in area from 6 to 400 ha. Our combined research efforts showed that some species are quite sensitive to relatively low levels of urbanization in the surrounding landscape, others are apparently unaffected, and the abundance of at least one species (prairie dogs) is positively associated with urbanization. Given the comprehensive scope of this combined research effort, we can directly compare responses among taxa that occur in the same grassland habitats.

Small Mammals

The abundance of the four dominant, native, small mammal species observed in this study was affected significantly by surrounding urbanization. Deer mice (*Peromyscus maniculatus*), prairie voles (*Microtus ochrogaster*), and hispid pocket mice (*Chaetodipus hispidus*; see fig. 5.4a) declined sharply in abundance in grassland sites surrounded by as little as 5%-10% cover of urban land uses (Bock et al. 2002), suggesting strong negative effects of adjacent urban development on native grassland species. The abundance of the only non-native species captured, the house mouse (*Mus musculus*), was not associated significantly with the amount of urbanization in the surrounding landscape. The authors suggested that one likely mechanism for reductions in small mammal abundance near urban edges may be the presence of rodent predators associated with human dwellings (e.g., house cats).

Songbirds

Increased urbanization near native grassland sites was also negatively correlated with the abundance of grassland nesting songbirds (Bock, Bock, and Bennett 1999; Haire et al. 2000). All seven songbird species studied declined in abundance as surrounding urbanization increased. For two species, the horned lark and the Savannah sparrow, no birds were observed in plots where 4%–7% of the surrounding landscape was urbanized, suggesting a high sensitivity to adjacent urban development. Grasshopper sparrows, which are mixed-grass and tallgrass specialists, declined abruptly in abundance at approximately 10% urbanization (fig. 5.4b).

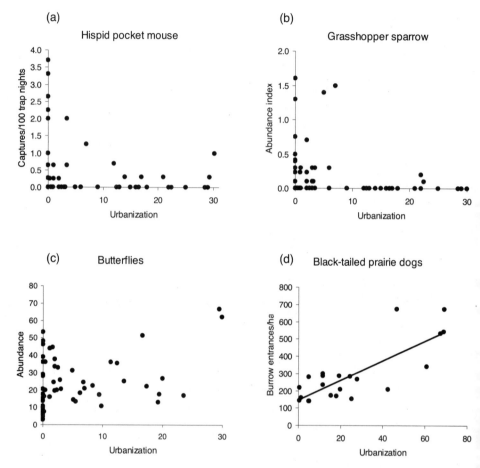

Figure 5.4. The effects of urban landscape context on native species abundance or richness in Boulder County, Colorado, grasslands. (a) Abundance of *Chaetodipus hispidis* (hispid pocket mouse). Redrawn from Bock et al. (2002). (b) Abundance of *Ammodramus savannarum* (grasshopper sparrow). Redrawn from Haire et al. (2000). (c) Sum of abundances of 57 species of grassland butterflies. Redrawn from Collinge, Prudic, and Oliver (2003). (d) Abundance of black-tailed prairie dogs. Redrawn from W. Johnson and Collinge (2004).

Raptors

Birds of prey forage over large areas and therefore might be expected to respond to landscape context at broader spatial scales than smaller organisms. Berry, Bock, and Haire (1998) used the same sampling plots near Boulder as described above, but merged closely spaced plots into single sampling units, so that the total number of plots sampled was 34, rather than the 66 plots used for the other taxa in this study. They counted 465 diurnal raptors of 11 species during three sampling seasons, including both winter and summer. Abundances of four of the wintering raptors (bald eagles, ferruginous hawks, rough-legged hawks, and prairie falcons) were negatively associated with the amount of urbanization in the surrounding landscape. As with small mammals and songbirds, abundances of these species declined sharply with as little as 5%–7% urbanization in the surrounding landscape. Three of these species (all except the prairie falcons) were also positively associated with another feature of landscape context—the presence of black-tailed prairie dog colonies. The abundance of these species increased in areas surrounded by prairie dog colonies, which are important prey resources. Red-tailed hawks were the most abundant raptors observed, and they were not associated significantly with urbanization. Six other raptor species were also not associated significantly with surrounding urbanization, but relatively few individuals of these species were found.

Plants

Barry Bennett (1997) surveyed vascular plants on each of the 66 plots for two field seasons and recorded the presence of each species along 200 m transects from each plot center to its boundary. These data were used to identify particular plant communities, and also to ask whether the surrounding urbanization influenced species composition or the presence of non-native plant species. Somewhat surprisingly, the sample plots did vary in the cover of non-native species, but not in relation to surrounding urbanization. Plots that contained a high cover of non-native plant species tended to be heavily grazed areas that were not necessarily in close proximity to urban edges. No distinct effects of surrounding urbanization on plant species composition were observed.

Grasshoppers

Grasshoppers are a highly diverse insect group in the western Great Plains, so they were chosen as a representative insect group for this biodiversity study (Craig et al. 1999). Grasshoppers were sampled in all 66 plots for two field seasons to examine their habitat associations, and their abundance and distribution in these grasslands were strongly affected by local habitat characteristics. The effects of surrounding urbanization on grasshopper abundance and distribution were not directly assessed in this study, but the authors did compare species' lists from their surveys with that of previous surveys conducted from 1949 to 1960 (Alexander and Hilliard 1969), when Boulder was much less heavily populated. This comparison with historical data suggested that grasshopper species composition has not changed considerably over the past 50 years in Boulder County (Craig et al. 1999).

Butterflies

Collinge, Prudic, and Oliver (2003) surveyed butterflies in the 66 grassland plots in 2000 and 2001. The response of butterflies to the urbanization of the Boulder Valley appeared to be similar to that of grasshoppers (Craig et al. 1999), but different from that of raptors, small mammals, and songbirds (Berry, Bock, and Haire 1998; Haire et al. 2000; Bock et al. 2002). We found that the grassland habitat type and quality, but not surrounding urbanization, were the key determinants of butterfly species richness and composition (fig. 5.4c). The lack of butterfly response to surrounding urbanization may be due to the fact that the centers of all circular, 200 m diameter, grassland plots surveyed in these studies were at least 100 m from the nearest urban edge. One explanation for the lack of a significant effect of landscape context for both grasshoppers and butterflies is that they may respond to urbanization on a much finer spatial scale than raptors, small mammals, and songbirds. Alternatively, the 5%-30% surrounding urbanization that we observed may not be sufficient to influence grasshopper and butterfly abundance and distribution. Although the direct effects of urbanization on butterflies were not apparent from our study, indirect effects—such as increased nutrient deposition, concomitant shifts in the distributions of exotic plant species (Lejeune and Seastedt 2001), and increased butterfly mortality from traffic (Ries, Debinski, and Wieland 2002)—may ultimately contribute to the long-term persistence of butterfly populations in urban landscapes.

Black-Tailed Prairie Dogs

Although mice and voles responded fairly negatively to surrounding urbanization in the Boulder grasslands (Bock et al. 2002), we observed quite the opposite effect in our study of black-tailed prairie dogs in this landscape (Johnson and Collinge 2004). Prairie dog density increased with increasing boundedness (a measure of urban landscape context) and roads (fig. 5.4d). The higher prairie dog density that we observed may be due to a refuge effect (or mesopredator release) in urbanized landscapes (e.g., Crooks and Soulé 1999). As noted above, Berry, Bock, and Haire (1998) showed that three species of raptors that are positively associated with prairie dog colonies (bald eagles, ferruginous hawks, and rough-legged hawks) are also negatively associated with the amount of urbanization in the surrounding landscape. If other potential prairie dog predators decline in urbanized landscapes, then prairie dog colonies may be able to achieve higher densities in these urban colonies. Similarly, Miller and Hobbs (2000) observed predation that was significantly higher on songbird nests at greater distances from recreational trails in Boulder County riparian areas, presumably due to human presence and declines in nest predators adjacent to recreational trails.

The combined results from the Boulder biodiversity study suggest that the response of grassland animals to urbanization in this region is scale-dependent, with larger animals (prairie dogs excepted) responding to landscape structure either at broader spatial scales or at lower levels of urbanization. For small mammals, songbirds, and raptors, there appears to be a critical landscape threshold (Andrén 1994; With and Crist 1995; see also chapter 10) that exists at 5%–7% urbanization of the landscape, a point where the abundance of these species sharply decreases (Bock et al. 2002). Because herbivorous insects such as grasshoppers and butterflies tend to be relatively host-plant specific, their occurrence at a site may depend largely on local habitat characteristics, such as the presence of their food resources. For example, we observed a rare species of butterfly (*Speyeria idalia*) at only two adjacent tallgrass prairie plots, one of which contained its native host plant, *Viola nuttallii* (Violaceae) (B. Bennett 1997; Collinge, Prudic, and Oliver 2003).

The results of our combined studies corroborate those in the Mazerolle and Villard (1999) review. They reported that for most invertebrate studies, patch characteristics explained a greater proportion of the variation in abundance and richness than did landscape characteristics; for vertebrate species, landscape characteristics explain a greater proportion of this variability, consistent with previous studies of vertebrate taxa in Boulder's grasslands (Berry, Bock, and Haire 1998; Bock et al. 2002; Haire et al. 2000). Black-tailed prairie dogs are a distinct exception, but their increased densities in urban areas may be due to a release from predation by larger animals, such as raptors, that tend to avoid grasslands surrounded by urbanization. Alternatively, prairie dogs may avoid dispersing from colonies that are surrounded by urbanization, allowing colonies to reach high densities. The next chapter will explore the effects of matrix composition and matrix quality on animal movement in more detail.

SYNTHESIS

This chapter has emphasized a broader view of fragmentation studies, one which goes beyond the fragments as islands paradigm reviewed in previous chapters. Clearly, the changes that occur at fragment edges have pervasive impacts for multiple species and processes in many landscapes. Smaller patches have greater edge effects due to their geometry, as edge effects tend to differ according to aspect and orientation (although this has not been well studied in tropical latitudes); biotic effects tend to permeate farther into interior habitats than do abiotic effects; and edge effects may differ depending on surrounding land uses. Further research on edges and edge effects should seek to identify the mechanisms that underlie these responses; they should also place a greater emphasis on more sophisticated analyses of the effects of multiple edges and edge types within landscapes (e.g., Ries et al. 2004; Fletcher 2005). Additionally, edge literature is still dominated by studies in forests. For a greater understanding of the generality of edge effects, we need more studies focused on edges between non-forested habitats and human-generated land uses. In non-forested habitats, edges are likely to be less abrupt and more diffuse, and therefore the character of the edge effects is likely to differ.

The composition of the landscape surrounding habitat fragments also clearly influences ecological patterns and processes within patches. It is increasingly recognized that it is impossible to fully understand the ecological aspects of fragmentation without considering the relevance of landscape context (e.g., Jules and Shahani 2003; Ewers and Didham 2006). Kupfer, Malanson, and Franklin (2006) recently referred to this as a "matrix-inclusive approach to fragmentation" (p. 15). Landscape context can exert an influence on everything from nest predation and nutrient flow to pollination and seed dispersal, so this concept is not only important to conservation biologists, but is clearly relevant to the entire science of ecology. The effects of context need to be considered, as well, especially in concert with other variables that are likely to affect ecological processes in fragmented landscapes, including patch size, shape, and quality.

Ecologists have made terrific progress in the past two decades in understanding how the composition of surrounding habitat types and land uses affects species richness, abundance, interspecific interactions, and the flows of nutrients and materials. Our knowledge now encompasses the spatial configuration of habitat types and land uses, which greatly enhances the precision with which we may be able to direct and manage landscape change for the benefit of biological diversity.

CHAPTER SIX

Animal and Plant Movement

The interplay between environmental heterogeneity and individual movements can have far-reaching consequences for the ecology of organisms.

—*Peter Turchin (1998)*

One of the most famous of Aesop's fables relates the story of the tortoise and the hare. After having been teased relentlessly by the hare about his rather plodding gait, one day the slow tortoise challenges the swift hare to a race. Most readers probably know how the story turns out. Along the route, the hare snacks on cabbages from a farmer's field and dozes off, only to wake up just in time to see the tortoise cross the finish line ahead of him. The story teaches children not to be boastful of their talents, but it also conveys the message that slow and steady will ultimately win over swift and hasty. The lesson is conveyed via the use of two animals moving through the landscape, at different paces and with different pathways to their destination.

Is there any biological relevance to this fable? Well, as an aside, hares and tortoises don't occur in the same habitat. But that doesn't really matter; it's a fable! In fact, where their movements have been carefully studied, tortoises and hares in the real world appear to both be similarly affected by anthropogenic landscape changes, particularly habitat loss and fragmentation, which alter their movement pathways (e.g., Clevenger, Chruszcz, and Gunson 2001; Wirsing, Steury, and Murray 2002; Edwards et al. 2004).

Why do individuals move, besides to race each other in children's stories? Animals typically move to forage, to find mates, to disperse to new colonies or populations, or to migrate (Armitage 1991; Dingle 1996; Bowler and Benton 2005). These different types of movements happen at different spatial and temporal scales. Short-term, fine-scale foraging movements occur many times within an animal's lifetime; they make fewer annual round-trip migrations or, usually, only a single once-in-a-lifetime dispersal trip to join a new population. Some species may use complex navigation mechanisms to reach their destination, while others may move more randomly. Most terrestrial animals move themselves, but the movement of aquatic animals may be determined largely by currents (Bilton, Freeland, and Okamura 2001). Plants are typically sedentary, but their products (pollen and seeds) or other plant parts are dispersed by wind, water, or animals (Chambers and MacMahon 1994; Ghazoul 2005). And parasites are typically dependent on their hosts for movement (Lion, van Baalen, and Wilson 2006). Species vary tremendously in their innate abilities and propensities for movement, and each of these types of movement may be strongly affected by landscape changes that alter the composition and configuration of habitats.

In fragmented landscapes, species often encounter obstacles in their attempts to travel safely from point A to point B, in other words, from one suitable patch to another. Because some species avoid certain landscape features, such as parking lots or corn fields, habitat loss and fragmentation may force species to navigate through complex patchworks of suitable habitat. This problem of species movement through fragmented landscapes motivated the potential solution of habitat corridors. Corridors have been widely proposed as a means to ameliorate the negative effects of habitat loss and fragmentation by linking otherwise isolated habitat patches (Saunders and Hobbs 1991; Harris and Silva-López 1992; Fahrig and Merriam 1994). *Corridors* have been variously defined as "narrow strips of land which differ from the matrix on either side" (Forman and Godron 1986, p. 123), "strips of a particular type that differ from the adjacent land on both sides" (Forman 1995, p. 38), "a linear landscape element that provides for movement between habitat patches" (Rosenberg, Noon, and Meslow 1997, p. 678), "a landscape element that plays a key role in connectivity" (A. Anderson and Jenkins 2006, p. 3), and "routes that facilitate movement of organisms between habitat fragments" (Hilty, Lidicker, and Merenlender 2006, p. 5).

These definitions emphasize both the structural (e.g., "linear elements") as well as the functional (e.g., "routes that facilitate movement") aspects of corridors. Despite the intuitive appeal of corridors as a resolution to the fragmentation conundrum, some researchers realized that these structural features may not always be the preferred movement pathways for plants and animals (Mann and Plummer 1995; Rosenberg, Noon, and Meslow 1997; Schultz 1998). This led to the broader concept of landscape connectivity. Corridors may be one type of landscape feature that promotes connectivity, but even landscapes in which habitat patches are not strictly linked by corridors offer some connectivity (Ricketts 2001; A. Anderson and Jenkins 2006; Crooks and Sanjayan 2006; Hilty, Lidicker, and Merenlender 2006).

The fact that we observe such profound changes in biological diversity in relation to habitat loss and fragmentation suggests that species vary widely in their tolerance for and their capability to move through modified habitats. In landscapes dominated by human activities, shifts in species composition and population abundance may be ultimately driven by altered plant and animal movement patterns (Kareiva 1987; Henein and Merriam 1990; Ims 1995; Andreassen, Hertzberg, and Ims 1998). Species' movement patterns may provide a critical link between individual decisions and population and community dynamics. For example, the behavioral avoidance of roads (e.g., Forman et al. 2003) or other modified habitats by animals may reduce movement rates between native habitat fragments relative to continuous habitats, resulting in higher probabilities of species extinction, lower rates of colonization, and diminished species richness (Wilcove, McLellan, and Dobson 1986; W. Laurance and Bierregaard 1997). Conversely, the presence of woody hedgerows or other linear landscape features may enhance movement rates between woodland fragments (Haas 1995; Tewksbury et al. 2002). The crucial question is, Which features of the landscape impede movement, and which features facilitate movement for a wide variety of organisms?

This chapter now turns to studies of plant and animal movement in fragmented habitats, in order to have a better understanding of these critical links between individual characteristics and population and community dynamics. It briefly reviews ecological theory related to movement and connectivity, discusses whether habitat corridors provide connectivity, introduces broader views of landscape connectivity, including a growing body of literature on landscape genetics, and suggests ways in which this basic understanding of individual movement patterns is essential to a synthesis of the effects of ecological fragmentation. Throughout the chapter, I emphasize the ecological aspects of short-term, relatively fine-scale movements, although there are a few examples of longer-term, broader-scale movements.

THEORY

The ecological theories of island biogeography (MacArthur and Wilson 1967) and metapopulation dynamics (Levins 1969) and, more recently, metacommunity dynamics (Holyoak, Leibold, and Holt 2005), discussed in chapter 2, provide conceptual frameworks for predictions regarding the role of habitat corridors in species persistence and community composition. The key prediction from island biogeography theory regarding habitat corridors is that immigration rates should vary inversely with the distance of a fragment from a source of colonists. For terrestrial habitat fragments, such a colonist source could be a continuous expanse of native habitat. J.H. Brown and Kodric-Brown (1977) added the rescue effect (fig. 2.2b), suggesting that island distance (in addition to size) may also influence extinction rates. Specifically, in closely spaced resource patches, immigration rates may be very high relative to extinction rates, shifting the balance of these two rates in favor of immigration and thereby rescuing small populations from extinction. If corridors reduce the effective distance between patches, then they should provide routes for an influx of colonists, which will increase the immigration rate and facilitate the rescue effect.

It must be abundantly clear by now that populations and communities often exist in patchy landscapes. If that were not so, then this book would not even be necessary! As in IBT, metapopulation theory suggests that corridors may modify species persistence (Hanski and Gilpin 1991; Doak and Mills 1994; McCullough 1996). Failed colonization of available habitat fragments may occur because of patch isolation (Hansson 1991; Fahrig and Merriam 1994), and the persistence of metapopulations may depend on rescuing the populations within fragments from extinction (*sensu* J.H. Brown and Kodric-Brown 1977) or recolonizing fragments following the extinction of the populations they contain (S. Harrison and Taylor 1997). Metapopulation theory, when applied to the dynamics of species in human-induced habitat fragments, suggests that corridors between fragments should increase the regional persistence of native species by reducing isolation and thus increasing the probability of colonization (Dunning et al. 1995; S. Harrison and Fahrig 1995; McCullough 1996). Further, interacting metapopulations form metacommunities. As was discussed in chapter 2, a metacommunity is broadly defined as a collection of communities connected by dispersal (Hanski and Gilpin 1991; Holyoak, Leibold, and Holt 2005). So, as with metapopulations, movement, or the lack thereof, among scattered habitat patches affects community dynamics and persistence significantly.

To summarize, ecological theory predicts that any mechanism that reduces isolation among habitat patches should also expedite the movement of organisms between patches, thereby reducing the rates of species loss and enhancing the probability of fragment recolonization. If corridors successfully reduce isolation, then habitat fragments connected by corridors should support more stable populations and, perhaps, a higher number of species than completely isolated fragments of equal size (E. Wilson and Willis 1975; J.H. Brown and Kodric-Brown 1977; Simberloff and Cox 1987; A. Bennett 1990; Saunders and Hobbs 1991). Dispersal and immigration are essential to understanding population and community dynamics in each of these theories. So we need to know what species characteristics and landscape features affect movement rates to be able to incorporate these into predictive models of population and community dynamics.

MOVEMENT AND CONNECTIVITY

Movement is intimately related to the concept of landscape connectivity. In fact, *connectivity* has been defined as "the degree to which the landscape facilitates or impedes movement" (P. Taylor et al. 1993, p. 571). To make sense of the multifarious literature on connectivity, Calabrese and Fagan (2004) reviewed and condensed measures of landscape connectivity into three types of metrics: (1) *structural* connectivity, which includes the size, shape, and location of habitat patches (and corridors) in the landscape, (2) *potential* connectivity, which includes the physical features of the habitat patches, as above, as well as the dispersal distances (usually averages) of various organisms, and (3) *actual* connectivity, which includes direct measurements of the physical features of the landscape and the movement pathways taken by plants or animals in those landscapes.

Most of the discussion of connectivity covered so far in this book has focused primarily on the first category in this classification scheme, structural connectivity. Previous chapters reviewed studies that feature descriptions of habitat patches and corridors in fragmented landscapes and examined species composition or population abundance in relation to those features. This chapter explicitly incorporates studies of plant and animal movement into this discussion, which moves us into the realms of potential and actual connectivity..

These broader views of connectivity arose from suggestions that corridors may not be just linear landscape features; they may often be diffuse, species-specific, and difficult to identify in spatially heterogeneous landscapes (Taylor et al. 1993; Gustafson and Gardner 1996). Researchers argued that landscape connectivity depends on how the movement patterns of particular species interact with the structural features of that landscape (S. Harrison and Fahrig 1995; Schumaker 1996). Movement patterns may thus serve as a record of the perception of and response to specific structural elements in the environment (Mader, Schell, and Kornaker 1990; Loreau and Nolf 1994; With 1994; Turchin 1998; Haddad 1999). Perception and response to spatial heterogeneity may also provide clues to the persistence capabilities of particular species (den Boer 1981). As humans continue to alter the spatial structure of native habitats, it is vital to understand and accurately predict how organisms will move through and be affected by these altered environments.

Ultimately, we would like to be able to trace the wanderings of plants and animals through landscapes, so that we may infer which landscape features facilitate connectivity. But studies of movement are notoriously difficult. If researchers have the time, patience, energy, and financial resources to do so, they can follow organisms in real time (or close to it)—using direct observations, radio-telemetry, or satellite-tracking—and record habitat use and behavior in relation to landscape features (Ims and Yoccoz 1997). However, researchers primarily employ indirect measures, such as (1) surveys of species in different habitat types using live-trapping, mist-netting, point-counts, spot-lighting, seed traps, and vegetation sampling, (2) mark-release-recapture studies of animals such as mammals, birds and amphibians, and (3) remotely-triggered cameras (*camera-traps*) that are heat- or movement-sensitive to capture images of larger animals, such as hyenas, leopards, raccoons, or alligators. More recently, genetic studies have been combined with analyses of underlying landscape structure. These genetic studies are proving to be quite illuminating, since an organism's genes essentially provide a historic record of movements over multiple generations. The disadvantage of genetic assessments, however, is that they can only measure movements that, depending on the organism, ended either in survival or reproduction, which does not give a full accounting of all movements.

DO CORRIDORS PROVIDE CONNECTIVITY?

In order to moderate the negative effects of habitat isolation on species persistence, one popular proposal is to safeguard linear landscape features that structurally link otherwise isolated habitat remnants (Diamond 1975; E. Wilson and Willis 1975; Forman and Godron 1981; Harris and Scheck 1991; Saunders and Hobbs 1991; Lindenmayer and Nix 1993). Biological conservation efforts aimed at the preservation of plant and animal diversity in fragmented landscapes have considered that including corridors is a critical strategy in reducing the detrimental effects of human disturbance (Arnold 1995; Beier and Noss 1998; A. Anderson and Jenkins 2006; Hilty, Lidicker, and Merenlender 2006).

Whether corridors do in fact provide crucial linkages in fragmented habitats has been debated vigorously (e.g., Hobbs 1992; Mann and Plummer 1995; Rosenberg, Noon, and Meslow 1997). Essentially, two key steps are required for corridors to do what they are intended to do. First, individuals must not avoid corridors—in other words, they must perceive them as suitable habitat. Second, individuals must move along corridors preferentially, rather than take random movement pathways across the landscape. Even if steps one and two are followed, however, it is uncertain whether population extinction would occur in the absence of corridors. Because of this complication, it has been difficult to amass scientific evidence to support or refute the role of corridors in enhancing movement and preventing population declines. Fragmentation experiments in controlled settings were discussed in detail in chapter 4—recall that studies directly examining the relationship between corridor presence and population persistence have shown positive effects of corridors on the persistence of some species (La Polla and Barrett 1993; F. Gilbert, Gonzalez, and Evans-Freke 1998; Haddad and Baum 1999; Tewksbury et al. 2002; Damschen et al. 2006), but not others (Collinge 2000; Hannon and Schmiegelow 2002).

Species Occurrence

As mentioned above, the first step to evaluating whether habitat corridors provide connectivity is to find out if plants and animals see them as suitable habitat. One of the earliest studies to document the presence of species in corridors was by MacClintock, Whitcomb, and Whitcomb (1977), who expressed surprise at finding forest-associated bird species in a "miserably disturbed" (p. 10) corridor in a wooded eastern deciduous forest landscape in the United States. They noted that this corridor appeared to provide breeding habitat for several species, and thereby reduced the negative effects of forest fragmentation. Their observations were compelling, but unfortunately there was no comparison of bird species richness in replicated fragments with and without corridors, so only limited conclusions could be drawn. However, this observation was followed by a flurry of studies over the next 25 years in which researchers surveyed habitat corridors for plants and animals. The collective results were that a wide variety of organisms could be found in many types of corridors, including mice and chipmunks in wooded fencerows (Wegner and Merriam 1979; A. Bennett 1990; A. Bennett, Henein, and Merriam 1994), carabid beetles in French hedgerows (Burel 1989), butterflies in British hedgerows (Dover and Sparks 2000) and South African grasslands (Pryke and Samways 2001), birds in wooded strips in grasslands in the United States (Haas 1995), herbaceous plant species in Great Britain (Petit et al. 2004), trees in Costa Rica (Harvey 2000), arboreal mammals in tropical Australian forests (S. Laurance and Laurance 1999), and arboreal marsupials in southeastern Australian forests (Lindenmayer, Cunningham, and Donnelly 1993; Lindenmayer 1994).

Corridors may serve as suitable habitat for some species but not others, even within the same landscape. For example, Bolger, Scott, and Rotenberry (2001) studied mammals and birds in "corridor-like landscape structures" in San Diego County, California, and observed that several rodent species spent as much time in remnant strips of coastal sage scrub habitat and revegetated highway rights-of-way as in remnant habitat patches. Based on information from prior studies, bird species were categorized as either fragmentation-sensitive or fragmentation-tolerant. As expected, the bird species that were more sensitive to fragmentation (and, presumably, habitat quality), occurred in remnant habitat strips, but not in the revegetated rights-of-way. However, these rights-of-way were acceptable to the fragmentation-tolerant bird species, and the authors noted that these landscape features have the potential to be effective habitat linkages.

Some animal species may actively avoid corridors because they are too narrow or too disturbed (e.g., Lindenmayer 1994), or because they are not floristically di-verse (S. Laurance and Laurance 1999). Animals' use of corridors may vary de-pending on their foraging patterns, body size, home-range size, degree of dietary specialization, mobility, and social behavior (R. Harrison 1992; Lindenmayer and Nix 1993), as well as on their cognitive abilities and navigation mechanisms (e.g., Gillis and Nams 1998; Caldwell and Nams 2006; Vuilleumier and Perrin 2006). Interestingly, Lindenmayer and Nix (1993) noted that linear remnants of mon-tane forest in Australia harbored several species of large arboreal marsupials, but smaller species were absent. These authors suggested that species occurrence in these corridors was largely determined by foraging and social behaviors, rather than by body size. The large animals foraged singly and fed on readily available leaves that were relatively evenly distributed, while the smaller species foraged in social groups and fed on more widely dispersed arthropods. Given the differing distributions of their food resources, the small marsupials appeared to be unable to find enough food in the relatively narrow corridors.

Because corridors are usually linear and have a relatively high perimeter-to-area ratio, they may be strongly influenced by edge effects (see chapter 5) and har-bor more generalist species than habitat patch interiors do. For example, in a study of the role of linear strips of Australian tropical rainforest in promoting animal dispersal (C. Hill 1995), individuals of two of the four forest interior species, a butterfly and a beetle, were observed in the rainforest corridor but not in the adjacent cultivated land. This suggests that dispersal for these two species may be enhanced by the presence of the corridor. However, two other forest interior species were not observed in the rainforest corridor, suggesting that the corridor's physical dimensions or habitat characteristics were insufficient to facilitate their spread. Hence the suitability of corridors as habitat and possible dispersal routes will vary among species. Similarly, shade-loving, forest interior plants may not thrive in corridors because of the altered microclimatic conditions found there (Fritz and Merriam 1993; de Blois, Domon, and Bouchard 2002).

Some species may not be present in corridors because they depend on other species that actively avoid corridors. For example, Norton, Hobbs, and Atkins (1995) noted that native hemiparasitic mistletoes (Loranthaceae: *Amyema miquelii*) that live on *Eucalyptus salmonophloia* (salmon gum) trees in Western Australia did not occur in wooded roadside corridors, even though these were considered suitable habitat. Their absence probably reflected the avoidance of these corridors by the birds that feed on mistletoe fruits and distribute their seeds.

More recently, Hilty and Merenlender (2004) used remotely triggered cameras to survey the use of riparian corridors by predators in Sonoma County, California, a region famous for its grape-growing and wine production. The authors categorized riparian corridors as degraded, narrow, or wide according to the characteristics of the riparian vegetation, and then surveyed predator presence in each type of corridor. They found that predators were far more likely to be detected in riparian areas than in adjacent vineyards, and that more species of predators were detected in wide corridors than in narrow or degraded corridors. These observations provide evidence that riparian corridors are perceived as habitat, and that they may be important in ensuring the persistence of native predator populations (Hilty and Merenlender 2004).

Therefore it appears that the first condition—corridors must be perceived as suitable habitat—is met for at least some species in some landscapes. Species may thrive in corridors if the requisite environmental conditions are met, or they may avoid using corridors as habitat based either on the physical characteristics of the corridor or on some aspect of their life history, behavior, or interactions with other species.

Movement Pathways

If plants and animals find habitat corridors suitable, then the next step is for them to use the corridors as preferred movement pathways between larger patches of suitable habitat. Some investigations have shown that corridors may function as *both* habitat and movement pathways (A. Bennett, Henein, and Merriam 1994; Pryke and Samways 2001). For example, resident chipmunks in an agricultural landscape used wooded fencerows as habitat, while transient chipmunks used fencerows to travel between forest patches (A. Bennett, Henein, and Merriam 1994). As noted above in the discussion of corridors as habitat, corridors have been found to function as movement pathways for animals of various taxa, including mammals (Merriam and Lanoue 1990; Lindenmayer and Nix 1993; Gehring and Swihart 2004), birds (Borgella 1995; Haas 1995), and insects (Haddad and Baum 1999; Collinge 2000; Pryke and Samways 2001).

To clearly demonstrate whether corridors are preferred movement pathways, it is essential to compare the movement patterns of organisms in the presence and absence of corridors. Experimental studies thus provide the strongest evidence for the relevance of corridors to plant and animal movement (e.g., Tewksbury et al. 2002). In their recent work on a rainforest bird species, the Chucao Tapaculo (*Scelorchilus rubecula*), researchers radio-tagged, translocated, and released 41 birds and followed their movement behavior in three different landscape settings: forest patches surrounded by pasture, patches surrounded by shrub cover, or patches connected to other patches with wooded corridors (Castellón and Sieving 2006; see fig. 6.1). Birds that were released in patches surrounded by pasture (open matrix) stayed in the patches longer, prior to their dispersal, than those released in patches either surrounded by shrub cover or linked by corridors to other forest patches. The researchers concluded that this behavior indicated that the birds were more resistant to crossing the open pasture habitat than the shrub or corridor habitats. This direct comparison of behavioral choices is particularly valuable for understanding the contribution of corridors to species' movement patterns.

Amphibian species may have special needs when it comes to corridors as movement pathways, due to the fact that many cannot stray far from water. Mazzerolle (2005) studied the movements of green frogs (*Rana clamitans melanota*) in a mined peat-bog landscape in eastern New Brunswick, Canada. These pond-breeding frogs used drainage ditches to move between ponds that were otherwise isolated by the peat-mining activities. The frogs did not use the ditches as breeding habitat, but their survival was high there, and they preferred ditches to the

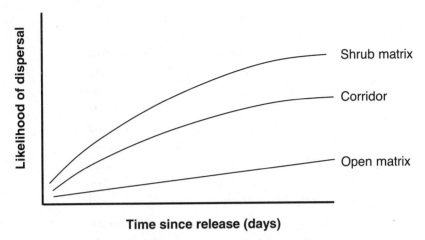

Figure 6.1. The likelihood of dispersal of a temperate rainforest bird, Chucao Tapaculo (*Scelorchilus rubecula*), in relation to the time since release, for forest patches surrounded by open pasture (*open matrix*), for forest patches connected to other patches via a wooded corridor (*corridor*), and for forest patches surrounded by dense shrubs (*shrub matrix*). Redrawn from data presented in Castellón and Sieving (2006).

more exposed peat-mined surfaces. Other amphibian species appear to have more diffuse movements across landscapes. Neither the abundance of newts in agricultural landscapes in France (Joly et al. 2001) nor the movements of red-legged frogs in California (Bulger, Scott, and Seymour 2003) were influenced by hedgerows or riparian corridors, respectively. Instead, researchers suggest that protection of these amphibians will require the maintenance of suitable habitat in the matrix surrounding breeding ponds (Joly et al. 2001; Bulger, Scott, and Seymour 2003; see also chapter 5).

Population Persistence in Connected Fragments

If corridors are to be useful in conservation, then ultimately they must not only facilitate the movement of individuals among isolated habitat patches, but that movement must translate into higher population densities or higher probabilities of persistence in fragments connected by corridors. Although theory predicts that corridors enhance the recolonization of habitat patches, and they are often assumed to do so (Sutcliffe and Thomas 1996), there is limited direct evidence to support it. Henderson, Merriam, and Wegner (1985) found that two deciduous forest patches connected by fencerows were recolonized by chipmunks following the experimental removal of resident individuals. There were no comparisons of recolonization in forest patches without corridors, however, so it was not possible to assess the relative effect of corridors on recolonization from their research. In my dissertation experiments with grassland insects, I found that corridors promoted the recolonization of grassland patches by less vagile species (Collinge 2000). A recent review of studies linking individual movements to population dynamics (Bowne and Bowers 2004) confirmed that our knowledge is still limited. Few authors reported the population effects of movement, yet when they did, the positive effects of movement on population persistence were mentioned in twice as many studies as were negative effects.

In the absence of direct evidence of population persistence in relation to movement via corridors, we can use observations of species richness in connected versus unconnected habitat patches to infer that corridors promote persistence. For example, Pardini et al. (2005) observed a higher diversity of small mammals in patches of Atlantic coastal rainforest in Brazil that were connected via corridors to larger forest remnants versus those found in patches that were isolated.

The Dark Side

In contrast to the evidence cited above, some ecologists have suggested that corridors may have neutral or even negative effects on individual movement and species persistence (Simberloff and Cox 1987; Simberloff et al. 1992; Lindenmayer, Cunningham, and Donnelly 1993; Schultz 1998). First, not all species may perceive and use linear strips of vegetation as movement pathways. For example, Schultz (1998) examined the dispersal movements of Fender's blue butterflies (*Icaricia icarioides fenderi*) in central Oregon and discovered that although the butterflies were mainly found in patches of their preferred host plant, lupine (*Lupinus* spp.), they did occasionally venture into areas that did not contain lupine. Based on detailed movement observations, Schultz surmised that these butterflies would likely benefit more from a series of stepping stones of habitat patches than from a linear corridor, which they were not likely to travel through. Second, depending on an organism's behavior, the narrow linear shape of corridors may make it difficult for animals to perceive them (Tilman, Lehman, and Kareiva 1997). In other words, the likelihood of finding corridors may be minimal for some species. A third, related issue is that because corridors are narrow, linear habitat strips, they have high perimeter-to-area ratios and may be susceptible to the negative effects of edges (Andreassen, Halle, and Ims 1996; Lidicker and Koenig 1996; see also chapter 5). A recent study demonstrated higher nest predation on indigo buntings in connected fragments with higher edge-to-interior ratios than on those in unconnected fragments (Weldon 2006). Fourth, corridors may have additional negative effects beyond those caused by edges, such as facilitating the propagation of disturbance or disease (Simberloff and Cox 1987; Simberloff et al. 1992; Hess 1994, 1996a, b). Hess's research used the results from simulation models to argue that under certain conditions, the connections between populations afforded by habitat corridors would enhance the spread of pathogens and increase the likelihood of population extinction (see also chapter 8). McCallum and Dobson (2002) expanded upon Hess's work by examining pathogen spread for diseases in which hosts develop resistance, as well as for ones such as Lyme disease or plague that have multiple reservoir hosts. They countered that in these scenarios, the benefits of corridors for population persistence exceed the potential costs of pathogen spread.

For many years, the lack of ecological field data on the contribution of corridors to population persistence generated much controversy concerning their implementation as a primary emphasis of conservation plans (Simberloff and Cox 1987; Hobbs 1992; Mann and Plummer 1995). We now have more data but, perhaps not surprisingly, no clear, one-size-fits-all solution has emerged. Given the uncertainties regarding corridor effectiveness, some have argued that either maintaining or enhancing large blocks of habitat or managing the matrix between habitats will contribute much more meaningfully to species persistence than preserving narrow corridors can (Fahrig 2001; Ricketts 2001). We still need to know more about the conditions under which corridors are likely to facilitate movement and diminish the negative effects of fragmentation (Collinge 2000), as well as about how behavioral decisions made by animals are integrated with landscape patterns (Chetkiewicz, St. Clair, and Boyce 2006).

BROADER VIEWS OF CONNECTIVITY

Corridors are conspicuous features of landscapes, and they undoubtedly affect habitat use and movement pathways for some species some of the time. But in other cases, such as the red-legged frogs mentioned above, animals appear to move across landscapes in mysterious and unexplained ways (i.e., what Pe'er, Saltz, and Frank 2005 call *virtual corridors*; see also fig. 6.2). Ultimately, our goal is to identify landscape features that impede or promote movement, so that we can understand how changes in landscape structure are likely to influence ecological dynamics. To reach this goal, researchers have used simulation models,

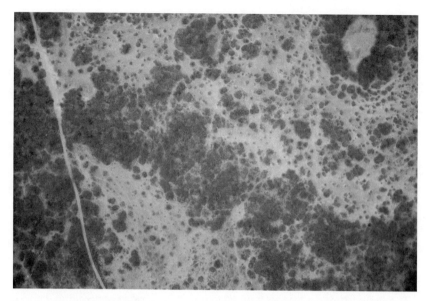

Figure 6.2. Movement pathways of African mammals as seen from the air. On the left side of the photo is an unpaved road for vehicles. Animal pathways show up as a white network around the dark gray patches of vegetation. Photo near Great Ruaha National Park, central Tanzania, by Peter Coppolillo, Wildlife Conservation Society.

direct studies of animal movement, and genetic analyses to go beyond corridors in order to describe or infer movement in complex landscapes.

Models of Movement

Models are especially useful in situations where we want to know what the expected outcome of a particular set of actions might be. Chapter 9 discusses more types of models in greater detail, but it is germane here to review a few modeling exercises that explicitly incorporate animal movement in diverse landscapes. For example, if a major landscape change is proposed, it would be nice to know beforehand whether the pattern and overall amount of the change is likely to have significant consequences for the biota. Boone and Hunter (1996) conducted such an analysis to determine whether proposed timber harvests were likely to affect grizzly bear (*Ursus arctos horribilis*) movement among large wilderness areas in the Rocky Mountains of Idaho, Montana, and Yellowstone National Park in Wyoming. The researchers used data gathered from experts on the habitat preferences of grizzly bears to develop an individually-based model of grizzly bear movement, combined with spatially-explicit (geographic information system or GIS) data layers on vegetation and land ownership. They assigned different permeabilities to each cell in the raster-based GIS, determined by land ownership and management as well as by vegetation type. For example, expert knowledge of grizzlies suggested that the bears would move relatively easily through designated wilderness areas, but that they were typically shot when they were encountered on private land. So Boone and Hunter set permeabilities of 99 (easy to move through) for wilderness areas and ≤ 50 (harder to move through) for private lands.

Based on these permeabilities, they simulated the movement of grizzlies through the landscape and then modified the model to examine how proposed harvests might influence movement patterns. Their simulations revealed some surprising results. First, they suggested that the proposed harvests would not affect movement from two wilderness areas into Yellowstone National Park, because grizzly movement along this route had already been severed by previous land conversion. However, they indicated that timber harvests would likely be detrimental to movement between two other wilderness areas in eastern Idaho. The authors concluded that road construction, both prior to 1990 and that proposed to facilitate the timber harvests, would negatively affect grizzly movement, since bears have been observed to avoid roads for a distance of up to 1 km.

Although mathematical models always require us to make assumptions that are our best guess at reality, they have been especially useful for exploring the possible effects of habitat loss and fragmentation before the damage is done. As Boone and Hunter (1996) noted, "there are many questions in landscape ecology that concern the movement of animals through space, and individually-based diffusion models allow identification of key variables. Once key variables are identified, other researchers, such as those using radio-telemetry, can focus on measuring these variables. With cooperation and feedback, modelers and field researchers can work toward a complete understanding of how land use affects wildlife" (p. 62).

Schumaker (1996) similarly focused on models of animal movement in realistic landscapes. His motivation was to test whether nine commonly used indices of landscape pattern were strongly correlated with the dispersal success of organisms. In other words, Schumaker wondered whether several *structural* metrics that are widely used in GIS analyses of landscapes—including the number of patches, patch area, core area, patch perimeter, nearest-neighbor distance, contagion, perimeter-area ratio, shape index, and fractal dimension—were useful in describing the *functional* aspects of landscapes, in this case, connectivity. Schumaker modeled movements of a territorial organism undergoing dispersal, based on the characteristics of the northern spotted owl (*Strix occidentalis caurina*), and asked which index or indices correlated most strongly with dispersal success. As in the grizzly study described above, Schumaker modeled the movements of individuals across a raster-based GIS landscape that reflected real patterns of old-growth forest vegetation for the Pacific Northwest. The noteworthy result from his analyses was that descriptions of landscape structure were only weakly correlated with functional characteristics of movement success. So he developed a new index called *patch cohesion*, which correlated very well with dispersal success. This new index combined an area-weighted perimeter-to-area ratio and a shape index, and it was relatively insensitive to small changes in patch perimeters. The main point from this exercise is that our human-biased measurements of landscape structure may not always accurately describe the view of the landscape from an animal's or plant's perspective. As Schumaker asserted, "to be as meaningful as possible, definitions of habitat connectivity should be predicated on animal movement because animal natural histories are fundamentally linked to landscape pattern through movement" (p. 1222).

These issues of functional connectivity continue to be explored and refined through innovative modeling efforts. Theobald (2006) combined estimates of landscape permeability with graph theory to construct landscape networks that describe likely movement pathways and critical habitat patches for Canada lynx (*Lynx canadensis*) recently introduced to southwestern Colorado. This involved constructing a map of habitat quality for lynx that incorporated vegetation or land cover type, proximity to a patch edge, and the distance to disturbances such as roads. Theobald then used GIS to cluster cells of similar quality to create habitat patches. He estimated permeability values for each habitat type (as in the grizzly bear example above), with the highest permeability for forested areas and the lowest for human-dominated land cover types. These permeability values were combined to compute a least cost surface that minimized the overall cost of moving across the landscape. A landscape network could then be analyzed to identify patches and movement pathways that were critical to maintaining landscape connectivity for lynx. This approach is especially useful for conservation planning efforts (see chapter 11), because it allows determining how various landscape-change scenarios are likely to impact the species that inhabit those landscapes. Although all of these models of landscape connectivity are abstractions of reality and may not incorporate all aspects of the landscapes and of species behavior, they are exceedingly useful in identifying likely movement pathways, and they provide insights into the probable consequences of landscape modification.

Observations of Movement

Studies of movement can tell us a lot about the choices that animals make as they move through landscapes. The three studies discussed below center on quite different organisms (butterflies, lynx, and turtles), but all involve observations of movement behavior in landscapes that are well characterized, so that movements and landscape structure can be effectively linked.

Butterflies are particularly suitable for movement studies, because they can be easily observed and marked, their recapture rates are relatively high, and the scale of their movements is usually easy for human researchers to follow. They have the added benefit of actively flying only on warm, sunny days, meaning that re-searchers can stay home on cold, rainy ones! Ricketts (2001) inferred movement patterns of 21 different butterfly species in a naturally fragmented alpine habitat in Colorado. Butterflies inhabited meadows (patches) that were surrounded by ei-ther willow or conifer habitat (the matrix), so Ricketts examined whether rates and patterns of movement between meadows differed, depending on the inter-vening habitat. He concluded that the matrix matters; depending on the species, butterflies were 3–12 times more likely to move between meadows embedded in willow habitat than in conifer habitat. The conifer matrix was more resistant to movement than the willow matrix, and it follows that meadows embedded in conifer habitat were more effectively isolated than those in willow habitat. Even though the flight pathways of butterflies were not directly observed, the number of individuals marked (6273 over two field seasons) and the recapture rates of the butterflies (25%) were relatively high, which suggests that these findings are ro-bust.

Additionally, when examining their resistance to movement through the ma-trix, there were differences among butterfly taxa that made sense when related to butterfly mobility. The most sedentary group stayed primarily within the mead-ows and moved relatively infrequently among patches; butterflies in this group were not affected significantly by the type of matrix habitat. Similarly, matrix type had no significant effect on the least sedentary group, which moved frequently among patches. The four groups with intermediate vagility, however, moved just enough among patches to be affected significantly by matrix type. This detailed study of local movement in alpine butterflies shows that matrix type can thwart or enhance movement, and that knowing this helps us to understand the impact of landscape changes on species abundance and distribution patterns.

Even though they are vastly different animals, lynx respond to landscape het-erogeneity in ways that are remarkably similar to butterflies. Long-term studies of the critically endangered Iberian lynx (*Lynx pardinus*), which occurs in southern Spain, provide an excellent example of detailed movement observations in relation to habitat types and human activities (Palomares et al. 1991, 2000; Ferreras et al. 1992; Ferreras 2001). For 15 years, a total of 65 lynx were fitted with radio collars and radio-tracked as they moved within and among populations (Ferreras 2001). The longer time scale and broader spatial scale of this study, compared to the research on alpine butterflies, means that the tracked movements constituted dispersal and migration in many cases, rather than just local movement patterns. In this study, connectivity was defined functionally as "the proportion of dispersers from a source that reached a given subpopulation" (p. 125) and was examined in relation to landscape features. Based on data obtained from the radio-collared animals, Ferreras estimated how much time lynx spent in each habitat type. He further reasoned that if a lynx spent little time in a particular habitat, say, croplands, then they perceived croplands as low quality habitat and incurred a high cost when traversing that habitat. He calculated this resistance to movement (or *friction*) of all six different habitat types found in the region, and found that marshes had the highest resistance to movement and scrubland had the lowest (i.e., scrublands were the preferred habitat and lynx moved most easily through them).

These measures of friction based on habitat selection were then incorporated into a GIS, where each cell in the GIS was assigned a measure of friction (recall the grizzly bears and the Canada lynx). Ferreras could then compare the movement of lynx among populations using both the simple straight-line distance and the effective distance, which accounts for the *viscosity* of the habitat to the moving lynx. The result was that *effective distance* (distance plus habitat quality in the intervening matrix) was more closely related to dispersal success than was straight-line distance. These studies have direct relevance to the conservation of this endangered species. These small lynx populations occur within and around Doñana National Park, which is surrounded by croplands, forestry plantations, marshes, and native scrubland. Lynx population persistence will likely depend on demographic exchange between scattered subpopulations, and the amount of exchange is directly related to the ability of the lynx to move through the landscape. Because we now know that certain habitat types are more resistant to movement than others, the landscape could be modified, through management and habitat restoration, to promote movement among populations and reduce the likelihood of extinction.

A third example of *actual* connectivity (*sensu* Calabrese and Fagan 2004) comes from a study of painted turtles (*Chrysemys picta*) in northern Virginia

(Bowne, Bowers, and Hines 2006). Bowne and colleagues used both mark-release-recapture and radio-telemetry to track the movement of turtles among nine ponds situated in a landscape of agricultural fields, woodlots, pastures, meadows, and development. Moreover, they categorized turtles into five age-sex classes to examine whether there were differences in movement at different life stages. Turtle movement appeared to be most strongly affected by habitat quality, in this case, the duration of the pond's inundation. Because some ponds dried out during the study, the researchers had the opportunity to observe vast differences in habitat quality and its effects on movement. Turtles moved to ponds that held water during the drought, regardless of their distance or the characteristics of the intervening matrix. For example, one pond (Rattlesnake Spring) was relatively unconnected during wet years, but had high actual connectivity during the dry period, when it was only one of three ponds available.

The temporal and spatial scales of movement by the turtles observed in this study were intermediate to those of the butterflies and lynx discussed above. Turtles moved beyond their home ranges but did not disperse over long distances. They often skipped low-quality ponds that were in close proximity in favor of high-quality ponds that were farther away. Yet their movements did not appear to be affected by landscape features, except for roads. A four-lane highway appeared to serve as a barrier to turtle movement, and at least one tracked turtle was killed by a car on this road. From their study, Bowne, Bowers, and Hines conclude that "purely structural connectivity values have little value if they are calculated without regard to the ecology of the organism" (p. 789). These authors argue strongly that the behaviors of organisms in relation to landscape features constitute the best estimates of connectivity, and that any conservation plans for imperiled species should incorporate such connectivity measurements.

Landscape Genetics

A particularly powerful method for understanding the consequences of fragmentation on movement is to examine population genetic structure in relation to landscape structure. This allows both an integrated look at individual movement patterns and the inference of gene flow over longer time periods than is possible with the direct measurements of animal movement discussed above. In animals, gene flow occurs via dispersal and subsequent breeding. In plants, gene flow can occur through pollen movement, seed dispersal, or the translocation of ramets or whole plants by physical vectors (wind or water) or by animals. Pollinators can move pollen from place to place, and their behavior may change in relation to habitat loss and fragmentation (Lennartsson 2002; Aguilar et al. 2006), affecting gene flow significantly. Thus this chapter concludes by highlighting recent advances in the study of landscape genetics. This research explores how we can effectively link genetic, demographic, and behavioral studies of movement in order to understand landscape connectivity over both short and long time scales.

For many years, population geneticists have sought to understand how the spatial location of individuals affects the probability of them interbreeding (*isolation by distance*). Typically, genetic analyses test for isolation by distance by correlating measures such as F_{st} (an estimate of genetic difference among populations) with the distance among populations. Often these analyses have revealed the expected result that populations in close proximity were more genetically similar than populations that were distant from one another. But as has now been shown, the straight-line distance between two points in the landscape may be an incomplete measure of the actual gap that is perceived by organisms that move through the landscape. For this reason, Manel et al. (2003) summarized the landscape genetics approach, which "promises to facilitate our understanding of how geographical and environmental features structure genetic variation at both the population and individual levels . . . [with] implications for ecology, evolution and conservation biology" (p. 189). Several relevant examples highlight how analyses of genetic structure can provide insights into animal and plant movement in complex landscapes.

To return to the question of whether corridors facilitate movement in fragmented landscapes, Mech and Hallett (2001) provided one of the first analyses of genetic structure in relation to the presence of corridors. They sampled two small mammal species, red-backed voles (*Clethrionomys gapperi*) and deer mice (*Peromyscus maniculatus*), in the Pacific Northwest in continuous forest, forest patches connected by corridors, and isolated forest patches, in order to see whether landscape structure contributed to population genetic similarity and inferred gene flow. For the forest-specialist red-backed vole, but not for the generalist deer mouse, genetic analyses revealed greater similarities in corridor landscapes than in isolated landscapes, suggesting that corridors do enhance movement and gene flow for this species (fig. 6.3).

Corridors also appear to enhance gene flow for certain plant species, as in the following case where it is mediated by seed dispersal rather than by pollination. Kirchner et al. (2003) studied the population genetic structure in an endangered species, *Ranunculus nodiflorus*, that lives in discrete, temporary ponds in the Fontainebleau forest in northern France. For these wetland plants, corridors were identified as "narrow paths of land, a few tens of centimeters wide without vege-

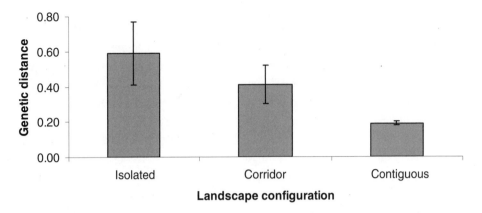

Figure 6.3. Genetic distances among red-backed voles (*Clethrionomys gapperi*) sampled at forest sites in three landscape configurations in northeastern Washington State: isolated forest patches, patches connected by corridors, and contiguous forest. Average (±1 SE) genetic distance was calculated as Nei's genetic distance for three pairs of sites in each landscape category. The ranking of site types by genetic distance was statistically significant ($P < 0.05$). Drawn from data presented in Mech and Hallett (2001).

tation . . . for seed dispersal between ponds" (p. 404). Plant populations in ponds that had these corridors were less differentiated from one another than those in ponds that were more isolated. Furthermore, ponds were more likely to be colonized if they were connected by corridors to other ponds with established *R. nodiflorus* populations. Studies of plants in fragmented landscapes are relatively rare compared to animal studies, so this is an important example of how corridors may be an important feature of connectivity and influence on population dynamics in this rare species.

Two additional examples show that corridors may influence gene flow among populations. Banks et al. (2005) measured the genetic structure of populations of the marsupial carnivore *Antechinus agilis* in fragmented *Eucalyptus* forest in southeastern Australia. Individuals in patches that were connected by riparian corridors of native *Eucalyptus* forest were more similar genetically than individuals that occupied forests separated by a matrix of exotic pine plantations. This suggests that movement of *A. agilis* was less likely to occur across pine plantations than along riparian corridors. In other words, pine plantations had lower permeability than did riparian corridors. Similarly, Dixon et al. (2006) used genetic analyses to evaluate whether a regional corridor that is under consideration for protection is currently an effective link between populations of Florida black bear (*Ursus americanus floridanus*) that occur in two national forests (the Osceola and the Ocala). This is a large corridor (30 km wide and 90 km long), comprised of mostly native vegetation. Results indicate that this regional corridor does provide connectivity between these two populations, which can then be used as the basis for formal, long-term conservation of these lands.

Movement and gene flow obviously affect population persistence, but they also have the potential to affect long-term evolutionary change among populations. In a unique study of landscape effects on song differentiation in a European bird, Dupont's lark (*Chersophilus duponti*), Laiolo and Tella (2006) recorded bird songs and linked their acoustic variation to landscape structure. Variation in bird calls recorded throughout Spain were examined in relation to landscape connectivity, as measured by both geographic distance and landscape unsuitability (i.e., akin to the permeability and friction measures mentioned in the studies discussed above). Landscape connectivity was a better predictor of call variation than the simple straight-line distance between individuals, which shows that the evolution of bird song may be influenced by the abundance and distribution of suitable habitat patches within the landscape.

This marriage of population genetics and landscape ecology continues to be a rapidly developing and innovative research theme (McRae et al. 2005; Orrock 2005; Holderegger and Wagner 2006; Jump and Peñuelas 2006; Riley et al. 2006; Storfer et al. 2007). New analytical techniques that consider molecular genetic data in the context of landscape networks are converting traditional *isolation by distance* measures of genetic differentiation to *isolation by resistance* measures (McRae 2006). These approaches facilitate an understanding of how landscape structure affects gene flow, and they also help to delimit populations in real landscapes based on genetic patterns rather than on a priori designations by researchers. In this way landscape genetics merges with studies of connectivity, in that it essentially allows the organisms to tell the researchers what the spatial extent of their movements and interactions is.

SYNTHESIS

The movement of organisms across complex landscapes reveals how species perceive landscape structure. Thus the study of movement is extremely useful in identifying which landscape features are barriers, which ones promote movement, and which habitat types and land uses effectively isolate populations from one another. Early studies of movement focused on corridors as structural elements, and research has expanded to express connectivity as a functional aspect of landscapes. Corridors may be part of this connectivity, but other features of the landscape also influence the perception and ability of species to move through the landscape. Understanding movement is essential, because it critically influences metapopulation and metacommunity dynamics, as well as microevolutionary processes.

CHAPTER SEVEN

Species Interactions

Although the first-order effects of the biodiversity
crisis—the loss of species—are dire, the second-order
consequences—the loss of species interactions—may be
more ominous.

—*Michael E. Soulé, James A. Estes, Joel Berger, and*
Carlos Martinez Del Rio (2003)

Most research in fragmented landscapes has centered on species' populations or
aggregate measures of community composition. Explanations for observed responses to habitat loss and fragmentation have frequently considered pervasive
microclimatic effects that occur at edges (chapter 5) or the random, demographic
and environmental variability associated with small populations in small habitat
remnants. But a rich history of ecological research tells us that the responses of
species to fragmentation are mediated not only by interactions with abiotic environmental factors, but also by interactions with other species within ecological
communities. Interactions among species are vital to survival, and they also provide critically important ecosystem services, such as pollination, seed dispersal,
and mycorrhizal associations (Kremen et al. 2007). Yet we still know less about
how species interactions are affected by habitat loss and fragmentation than we
know about overall changes in population size, species richness, or composition.

The influence of spatial heterogeneity on interactions among species (rather than
on individual populations) . . . represents an important future research direction.

Much of the research on how spatial pattern affects organisms focuses on how variables like patch size, edge-to-area ratio, and interpatch distances influence population presence or abundance, largely through effects on key demographic parameters. However, some changes in landscape patterns may have cascading influences among species . . . *and more work is needed on how spatial heterogeneity affects species interactions.* (M. Turner 2005, p. 1972; italics mine)

As has been evident throughout this book, habitat loss and isolation tend to reduce population sizes and diminish the functional connectivity of landscapes. To participate in interspecific interactions, individuals from one species must encounter individuals of another species. So any change in habitat spatial structure, and the concomitant shifts in critical dispersal pathways, may ultimately make it more difficult for some species interactions to occur, and less difficult for others. Pollinators must find pollen, predators must find prey, and birds must find berries; each of these searches may be more challenging, or sometimes perhaps easier, in fragmented landscapes. Conversely, organisms trying to escape their enemies or competitors may find that this is more difficult in fragmented landscapes, since both parties may be confined to an isolated fragment and unable to escape.

Currently, those who study habitat loss and fragmentation generally focus on theory that is based on single-species spatial models, such as metapopulation theory, or models of species richness, such as island biogeography theory. Both of these conceptual frameworks emphasize immigration and extinction as the dominant processes that influence ecological dynamics, and obviously both processes are affected by habitat loss and fragmentation. The emerging field of metacommunity dynamics shows us that the dynamics of multiple species in spatially divided habitats can be quite different, and that the dynamics of an individual species, when viewed in the context of the metacommunity, may differ from what single-species models predict. For example, single-species metapopulation models predict that habitat fragmentation (*subdivision*) can only be unfavorable, because large populations become increasingly small and isolated, making them increasingly vulnerable to local extinction via stochastic events. These small, isolated populations are much less likely to be recolonized following extinction, making their long-term persistence even more precarious.

But if we look at interacting species, the predictions for persistence in fragmented habitats may be different. Species that otherwise might not coexist may be able to do so in subdivided habitat, so that subdividing that habitat may actually reduce the risk of extinction for some species. Specifically, when species interact, especially in an antagonistic way via competition, predation, or parasitism, one species can drive the other to local extinction. In this case, diversity may actually be enhanced by habitat fragmentation. Metacommunity theory predicts that crucial factors in whether or not this actually happens are (1) the extent to which species interact, (2) how such interactions are modified by spatial dynamics (such as competition-colonization tradeoffs), and (3) how these pairwise interactions are situated within a more complex, multi-species community (Holyoak, Leibold, and Holt 2005).

Metacommunity theory is not yet developed to the degree where it makes specific predictions about species loss from fragmented landscapes (Holt, Holyoak, and Leibold 2005), but it provides a formal conceptual framework that describes how the spatial arrangement of patches may influence not only populations, but interacting species. Two issues are particularly relevant in this context. First, we know that the species that tend to be most vulnerable to extinction include large, rare species that have broad home ranges and tend to be specialized species of high trophic rank. Think of predators like tigers in Asia, jaguars in Central and South America, or the Florida panther. Gonzalez (2005) pointed out that because these are the first species likely to be lost from fragmented landscapes, this means that food chains and food webs are likely to disintegrate from the top down, causing cascading effects and, possibly, the extinctions of multiple species in the ecosystem (Terborgh et al. 2001; Soulé et al. 2003; van Nouhuys 2005).

A second issue highlights the consequences of altered species interactions more broadly. Because species may play critical functional roles via their interactions in ecological systems (e.g., Paine 1966; Redford 1984; Daily, Ehrlich, and Haddad 1993), the loss of these interactions will lead to declines in ecological function. If population abundance decreases below some threshold level, then not only is the focal species vulnerable to extinction—via random variation in demographic and environmental parameters, a process that Soulé (1987) calls the *extinction vortex*—but the probability of species interactions is diminished (fig. 7.1), reducing the functional role of that species in the ecosystem. A domino effect, similar to the extinction cascades noted above, may then result in additional species extinctions. For example, in Indonesia, a number of primate and bird species disperse the seeds of tropical trees (Kinnaird 1998; O'Brien et al. 2003). When these species decline due to overexploitation or habitat loss, their seed dispersal function is also lost, with major implications for tree recruitment in these forests.

Redford (1992) succinctly captured these phenomena in his description of the

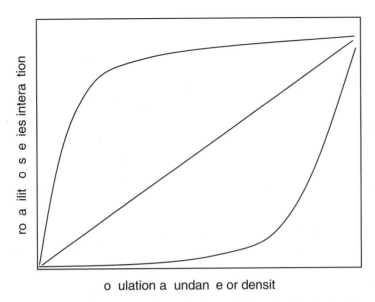

o ulation a undan e or densit

Figure 7.1. A conceptual diagram illustrating three possibilities for the relationship between population abundance or density and the probability of species interactions. *Abundance or density* refers to all species that participate in the interaction. In this diagram, the relationship may be linear, non-linear concave, or non-linear convex.

empty forest, which remains a key concept for conservation biologists. An *empty forest* (or grassland or ocean) is an ecosystem that appears to be intact, but is missing species and interactions that promote a fully functional ecosystem. So, what appears to be a forest, isn't—it is just a collection of trees. The corollary concept of conservation planning for *ecologically functional populations* goes beyond the low limits on population size set by population viability and considers the minimum population size for which a species remains a functional component of the ecosystem (Sanderson et al. 2002).

Because of the importance of species interactions for ecosystem function, Soulé et al. (2003) extended the idea of the empty forest, arguing that maintaining species interactions, not just species populations, should be a major goal of conservation biology. They defined *foundation species* as "highly interactive species that are often extremely abundant or ecologically dominant" (p. 1239), and included species such as bison, prairie dogs, termites, and the American chestnut. Their main message is that species interactions must be maintained in order to maintain functioning ecosystems, but that habitat loss and fragmentation may affect the abundance of foundation species and compromise their capacity within the ecosystem. Of particular relevance to this issue are two important questions. Which species are lost from the ecosystem? And in what order they are lost (Gonzalez 2005)? For instance, if an avian seed disperser declines to extinction in a habitat fragment, then the successful dispersal of the plant species whose seeds it scatters will cease. Had it happened the other way around, with the plant declining to extinction first, then the seed disperser might be able to continue to persist if additional suitable food resources were available.

In the context of habitat loss and fragmentation, metacommunity theory suggests that species interactions will be strongly influenced by the spatial arrangement of remnant habitats and the amount of dispersal among patches (Holyoak, Leibold, and Holt 2005). Studies of predation, pollination, seed dispersal, and parasitism in fragmented landscapes all show that these interactions are substantially influenced by habitat spatial arrangement, and this research has revealed mechanisms for individual species' responses to fragmentation. This chapter reviews the literature on species interactions in fragmented landscapes and discusses the importance of understanding these interactions for predicting future responses to fragmentation. The chapter is organized by the type of interaction, and looks at the evidence for effects of habitat loss and fragmentation on each of these interactions in turn. The interactions of hosts, vectors, parasites, and pathogens are treated separately in the following chapter, since there is a growing literature on disease emergence and pathogen prevalence in relation to anthropogenic landscape change (e.g., Collinge and Ray 2006).

NEGATIVE–NEGATIVE INTERACTIONS

Of all the interspecific interactions that ecologists have identified, there are probably more studies of competition than of any other interaction, yet in the context of habitat loss and fragmentation, it is perhaps the least well understood. On the one hand, habitat alteration may reduce population size and density, so one might expect competition to be less intense in fragmented landscapes, since fewer individuals are available to compete. But on the other, habitat loss and fragmentation may concentrate resources, individuals, and species into smaller and smaller areas, thereby increasing population density, so competition could be more intense. Many studies have examined changes in species richness or composition in fragmented landscapes (as noted in previous chapters), which may imply shifts in competitive interactions among species. Although there seems to be a dearth of studies that have directly examined the outcomes of competition, there are a few good examples that highlight shifts in competitive interactions in fragmented landscapes.

For instance, Suarez, Bolger, and Case (1998) studied the distribution of native ants in relation to the presence of the exotic Argentine ant (Hymenoptera: Formicidae: *Linepithema humile*) in the sage scrub fragments of San Diego County featured in previously discussed research (Soulé et al. 1988; Bolger, Scott, and Rotenberry 2001; Crooks et al. 2001; Crooks, Suarez, and Bolger 2004). Argentine ants were ubiquitous in small fragments and at the edges of larger fragments, but they did not pervade the interior of large fragments. Thus Argentine ants had a greater impact on native ants in fragments versus those in continuous sage scrub habitat. The displacement of native ants by the exotic species (fig. 7.2a) is apparently due to the loss of intraspecific aggression by the Argentine ant (Holway, Suarez, and Case 1998) and the related ability of these ants to surpass native ants at both interference and exploitative competition (Holway 1999). In other words, many native ant species spend a lot of time and energy defending their territories from invasion by conspecifics, but Argentine ants do not engage in this behavior, which effectively leaves them more energy to compete against other species. Another study of interactions among ant species produced similar results (Braschler and Baur 2005). Ants were studied in an experimentally fragmented grassland near Basel, Switzerland. The most abundant ant in these grasslands was *Lasius paralienus*, a generalist species that tends aphids and feeds on aphid honeydew and extrafloral nectar. In grassland fragments, *L. paralienus* density was negatively correlated with the species richness of other ant species, but there was no such relationship in control (unfragmented) plots (fig. 7.2b). So, similar to the Argentine ant study, this research suggests that fragmentation alters competitive interactions among species and magnifies the dominance of the most abundant generalist species.

Interspecific competition for resources among plants is widely viewed to be a dominant force structuring vegetative communities. But relatively little empirical information is available to examine whether competitive interactions among plants shift in response to habitat loss and isolation. In the interpretation of their results of field experiments on grassland fragmentation, Quinn and Robinson (1987) suggested that interspecific competition among plants may be responsible for their observation that subdivided habitats contained more species than did continuous grassland areas. Their reasoning went like this: in subdivided habitats, any one of several potentially dominant species has the opportunity to become established and preempt the colonization of other species in that partic-

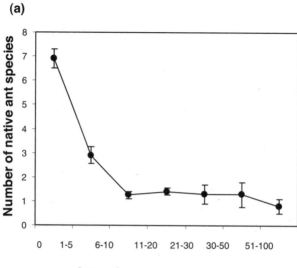

(a)

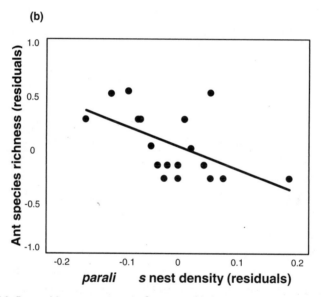

(b)

Figure 7.2. Competition among ants in fragmented habitats. (a) The number of native ant species present in sage scrub habitat fragments in San Diego County, California, in relation to Argentine ant abundance. All sites with Argentine ants had significantly lower native ant species richness than control sites with no Argentine ants. Native ant species richness increased with the distance from the fragment edge; Argentine ants were most abundant at fragment edges. Redrawn from Suarez et al. (1998). (b) Partial regressions of residuals from ANOVAs of *Lasius paralienus* nest density and the species richness of ants in fragmented grasslands near Basel, Switzerland. Redrawn from Braschler and Baur (2005).

ular fragment, which results in a random and diverse assortment of species among small patches. But in a large area, a single dominant species tends to suppress or even eliminate a number of species that would likely persist otherwise. Thus habitat subdivision is likely to increase diversity when interspecific interactions are strong, because less competitively dominant species are able to escape in space and time.

In essence, this hypothesized mechanism is the *competition-colonization trade-off*: good competitors are predicted to be lousy colonizers, and good colonizers do not compete very well (Levins and Culver 1971; Tilman 1994). Tilman et al. (1994) constructed a generalized model to examine the relationship between competition, colonization, and the probability of extinction for species within communities. They used the assumption of a competition-colonization tradeoff to argue that habitat destruction is likely to result in an *extinction debt*. Because good competitors are likely to be poor colonizers, then habitat loss will result in deterministic extinctions of dominant competitors, since they are unable to disperse to other, intact sites. As the competitive dominant is removed, competitive interactions shift toward dominance of the next-best competitor. As habitat loss proceeds over time, each species will go extinct one after the other, starting with the next-best competitor and continuing to the poorest competitor. When little habitat remains, the poorest competitor will exist by itself, but it will be unable to disperse to existing patches, since they are now so far apart, and will eventually go extinct, too. Thus this modeling exercise suggests that the competitive environment of a particular species may change as a result of habitat loss and fragmentation.

Changes in the strength of competitive interactions may also affect vertebrate abundance and distribution. Nour, Matthysen, and Dhondt (1997) tested the prediction from ecological theory that the niche dimensions for a species would expand or contract in relation to the presence of other, similar species. They reasoned that in small, isolated habitat fragments, some species would be absent, and thus other species would be expected to expand their niche. They examined this possibility by studying seven species of songbirds in the pariform guild (tits and chickadees) in mature oak forest in northern Belgium. They selected 18 forest sites, with 17 fragments ranging in size from 1 to 30 ha and a large forest site of approximately 200 ha. Careful observations of the foraging behavior of each species revealed that, in contrast to theoretical predictions, there were no differences in foraging height, niche width, and niche overlap in relation to fragment area. They concluded that changes in forest size and isolation did not seem to affect competition among similar species.

Habitat spatial characteristics may interact with other factors, however, to alter competitive interactions among species. For example, in Australia, the noisy miner (*Manorina melanocephala*) is a colonially nesting bird that aggressively defends its territories and appears to exclude other, less common birds from forest fragments (MacNally, Bennett, and Horrocks 2000; Ford et al. 2001). Small (20 to 30 ha) *Eucalyptus* woodlands in Tasmania were dominated by the noisy miner, and few other native songbirds were present (MacDonald and Kirkpatrick 2003). In larger woodlands, the bird community was dominated by small insectivorous birds, and noisy miners were relatively uncommon. It appears that the effects of fragmentation in this case follow a two-step process: a reduction in habitat area facilitates the occupation of small fragments by noisy miners, which then aggressively inhibit other species from gathering resources and maintaining viable populations.

POSITIVE–NEGATIVE INTERACTIONS
Predator-Prey

Previous chapters described Huffaker's (1958) classic experiments on predator-prey interactions. The noteworthy result from them was that spatial heterogeneity tended to stabilize predator-prey interactions. In other words, in patchy experimental arrays, predators and prey both persisted. Several field experiments have directly examined how habitat patchiness influences interactions between species where one species benefits and the other loses, such as those between predators and prey, or between parasitoids and hosts. For example, Kareiva (1987) directly field tested Huffaker's idea that habitat patchiness promotes the persistence of predators and prey by conducting an experiment with goldenrod (*Solidago canadensis*), herbivorous aphids (*Uroleucon nigrotuberculatum*), and predatory ladybird beetles (*Cocinella septempunctata*) in three different fields of goldenrod in Rhode Island. Kareiva created three replicates of patchy and continuous areas of goldenrod by establishing 1 m × 20 m plots of goldenrod, then mowing every other 1 m² patch of goldenrod in the patchy treatment and leaving the continuous treatment unmowed. During the summer months, aphids and beetles were censused in both patchy and continuous areas in all three fields for three consecutive years. Aphids were consistently more abundant in patchy versus continuous goldenrod, and they reached outbreak levels much more frequently in patchy goldenrod.

This result was counter to Huffaker's laboratory experiments, because aphids appeared to escape predation by beetles in patchy habitats, leading to a *lack* of coexistence of predators and prey in the fragmented treatment. In other words, predator and prey dynamics were less stable, not more stable, in patchy versus continuous habitats. Kareiva reasoned from both field and modeling evidence that the mechanism responsible for these contrary results was that patchiness disrupted the searching ability of ladybird beetle predators, making it more difficult for them to locate and aggregate near aphid prey. The end result was that fragmentation of goldenrod habitat promoted aphid population explosions, or *outbreaks*. This experiment was groundbreaking in that it allowed direct tests of the effects of patchiness on predator-prey dynamics, and also revealed a key behavioral mechanism responsible for the result.

One of the most commonly studied predator-prey interactions in fragmented landscapes is the predation of bird nests, especially along forest-field edges (e.g., Driscoll and Donovan 2004). Typically, nest predation is higher at edges than in forest interiors (see chapter 6), and this is usually attributed to increases in the population sizes of generalist predators in fragmented landscapes. Further, because large predators may be absent from fragmented landscapes, these medium-sized predators, or *mesopredators*, may escape their own predators.

When mesopredators evade their own predators, prey may suffer disproportionately. Mesopredator release poses an acute problem for native songbirds in the heavily modified urban landscapes of Southern California. Crooks and Soulé (1999) studied predator-prey interactions in 28 native sage scrub habitat remnants in San Diego County. In this system, coyotes (the top predators) were sensitive to habitat loss and isolation, so they occurred primarily in large remnants of sage scrub habitat, but not in small fragments. Coyotes prey on intermediate-sized carnivores such as striped skunks, raccoons, and grey foxes, as well as exotic predators such as opossums and even domestic cats. And these mesopredators feed primarily on native songbird eggs and nestlings, as well as on lizards and small mammals. As a result, scrub bird diversity was significantly lower in small fragments relative to large fragments, primarily because of mesopredator release due to the absence of coyotes in the small fragments. In this landscape, fragmentation affected multiple trophic levels, resulting in a *trophic cascade*. This study underscored the rather astonishing impact of domestic cats on biodiversity in urbanized landscapes. By surveying homeowners that lived next to sage scrub habitat remnants, Crooks and Soulé estimated that, on average, the "cats surrounding a moderately sized fragment (100 residences) return about 840 rodents, 525 birds and 595 lizards to residences per year" (p. 565). A similar case of mesopredator release was observed in a set of newly created islands in Venezuela (Terborgh et al. 2001). Small (0.25 to 0.9 ha) islands were too small to support predators, so the densities of herbivores (including rodents, howler monkeys, iguanas, and leaf-cutter ants) were 2–3 orders of magnitude greater than on the nearby mainland. In this case, the trophic cascade on small islands resulted in high herbivore pressure and subsequent sharp declines in the densities of seedlings and saplings of canopy trees.

Argentine ants in California appear to displace native ants through competitive interactions, as noted above (Suarez, Bolger, and Case 1998; Holway 1999), but they may also depress the abundance of other native species that they use as prey. For example, Huxel (2000) investigated relationships between the presence of Argentine ants and the threatened valley elderberry longhorn beetle (Coleoptera: Cerambycidae: *Desmocerus californicus dimorphus*), or VELB, along two different riparian zones in central California. Argentine ants are known to prey on eggs of another cerambycid species, but interactions with particular life stages of VELB are not known. The presence of Argentine ants was negatively correlated with VELB presence along one of the drainages (Putah Creek), but not along the other (the American River). Riparian vegetation along both of these drainages has been heavily modified and fragmented, which suggests that patchiness has not contributed significantly to the dynamics of predator-prey interactions in this instance.

As with other topics covered in this book, most research has focused on terrestrial systems, but there are some noteworthy studies of predator-prey interactions in patchy marine environments. Irlandi, Ambrose, and Orlando (1995) took advantage of naturally occurring variation in the patchiness of seagrass beds in coastal North Carolina to study predation on bay scallops in relation to varying habitat configurations. Plots of approximately equal areas of seagrass were established within the context of three different spatial patterns of seagrass bed vegetation (*very patchy*, *patchy*, and *continuous*). Equal numbers of juvenile bay scallops were introduced to each plot and tethered so that their fates could be followed throughout the experiment. After four weeks, a significantly higher proportion of scallops in the very patchy treatment were lost to predation, compared to the patchy and continuous treatments. This increase in predation associated with increased habitat patchiness may have been due to the larger amount of edge in patchy seagrass beds, as has been observed in terrestrial habitat fragments.

Within the same region, Micheli and Peterson (1999) conducted field-enclosure experiments and found that proximity to certain habitat types influenced the rates of predation by blue crabs. On reefs that were connected to other reefs by a substrate of seagrass beds, the rates of predation on clams by blue crabs were

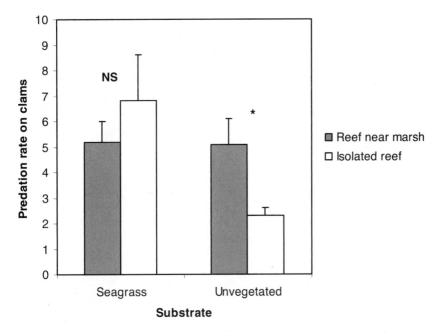

Figure 7.3. Blue crab predation (the number of clams crushed and missing) on clams occupying intertidal oyster reefs in estuarine habitats of coastal North Carolina. Oyster reefs occurred either adjacent to salt marshes or were isolated by 8–18 m from the marshes. The substrate was either seagrass or unvegetated. NS denotes non-significant, * represents $P < 0.05$. Redrawn from Micheli and Peterson (1999).

similar from one reef to another (left side of fig. 7.3). But where the marine substrate was unvegetated, predation rates were significantly lower on isolated reefs than on reefs adjacent to the salt marsh (right side of fig. 7.3). The seagrass substrate apparently increased the functional connectivity of this marine landscape by promoting the movement of blue crabs to otherwise isolated oyster reefs and by increasing the intensity of blue crab predation. Seagrass beds have been used extensively in studies of the effects of habitat fragmentation in marine systems, and a recent review highlights the relevance of these systems for understanding mechanisms of species' responses to habitat patchiness, including shifts in interspecific interactions (Böstrom, Jackson, and Simenstad 2006).

Ryall and Fahrig (2006) conducted a broad review of theory and empirical research related to habitat loss and fragmentation and predator-prey interactions. They suggested that despite much research in this area, there were still no clear generalizations that could be made regarding the effects of habitat loss and fragmentation on predator-prey interactions. They provided a very helpful summary of predator-prey theory that they hoped would guide future efforts of empirical researchers in illuminating these effects. Because empirical studies vary substantially both in their approaches and in the extent to which they directly test theoretical predictions, Ryall and Fahrig highlighted the importance of recording (1) the feeding habit of the predator in question (i.e., whether specialist or generalist), and (2) whether the matrix supports or does not support predator populations. With greater details on the feeding and habitat requirements of predators, we will hopefully be able to generate more robust conclusions regarding how habitat loss and fragmentation are likely to affect predator-prey interactions.

Herbivory and Seed Predation

Herbivory has been comprehensively studied, and it resembles parasitism in that most herbivores remove plant tissue but do not actually kill the plant. Seed predation is another form of herbivory, but it is more like predation than parasitism, in that the herbivore consumes and kills the seed. One other complexity is that there is often a fine line between seed predation and seed dispersal. For example, animals that cache seeds, including jays, nutcrackers, and kangaroo rats, retrieve and eat some cached seeds but do not retrieve others, effectively dispersing them (see the section on seed dispersal, below). For ease of discussion, these two processes, herbivory and seed predation, are lumped together here.

This continuum of interactions among animals and plants involving seeds was decisively demonstrated in a study based at Barro Colorado Island (BCI) in Panama (Asquith, Wright, and Clauss 1997). Researchers studied the seedling herbivory and seed predation of three large-seeded tropical tree species in four different site types: (1) mainland forest, which contained a relatively intact assemblage of mammalian predators and herbivores, (2) BCI, a large forested island in Lake Gatun, which supports the mainland assemblage minus two predatory cat species, (3) medium-sized forested islands in Lake Gatun that support small herbivorous mammals, and (4) small forested islands that support a highly reduced mammalian fauna (rats only). Seed predation was highest on the small islands, but on larger and larger islands, there was an increasing proportion of seeds that were dispersed instead of eaten. On small islands, seedling herbivory was also highest, compared to other forest types. As with the trophic cascades mediated by predator removal that were mentioned above, the depauperate mammalian fauna on small islands affected both seed viability and seedling success significantly. Shifts in animal species composition similarly affected interactions among seed dispersers, seed predators, and Spanish juniper (*Juniperus thurifera*) in central Spain (Santos and Tellería 1994). A common seed predator, the wood mouse (*Apodemus sylvaticus*), was nearly nine times more dense in small forest fragments than in large ones, and consequently consumed much higher quantities of seeds in the small fragments. Seed dispersers (thrushes, *Turdus* spp.) were more abundant in large forests. Junipers in small fragments thus experienced the "double-whammy" of having higher seed predation and lower seed dispersal than those in large fragments, which is likely to reduce juniper abundance in the small fragments.

Small fragments contain disproportionately higher amounts of edge, and edges may attract plant natural enemies. In their research on native New Zealand mistletoe (*Alepis flavida*), Bach and Kelley (2004) observed higher leaf herbivory and seed predation, but lower floral herbivory, on *Nothofagus* forest edges than in forest interior sites in New Zealand. The patterns depended on herbivore identity, as well as on the overall amounts of herbivory in a given year. For example, leaf herbivory by the introduced Australian brushtail possum (*Trichosurus vulpecula*) was significantly greater on forest edges than in the forest interior in a year of high possum herbivory, but not in a year with low damage levels. Leaf herbivory by insects did not differ between forest edges and the interior, but floral herbivory by a caterpillar was higher in the forest interior than on its edges, and seed predation was higher along forest edges. The net effect of all these types of herbivory was a higher vulnerability of mistletoe growing on forest edges, since the greatest amount of plant damage was by possums, who concentrated their feeding at forest edges.

A recent review of the effects of habitat loss and fragmentation on plant-herbivore interactions (Tscharntke and Brandl 2004) suggested that there is much contingency in these relationships, as was clearly observed in the studies noted above. Some species respond positively, and some negatively, to habitat loss and isolation, and the net impact on plants will depend on the roles played by various animals and their responses to habitat alteration.

POSITIVE–POSITIVE INTERACTIONS
Pollination

A critically important ecological interaction for plants is the mutualistic relationship between plants and their pollinators, which may also be affected by a variety of mechanisms that stem from changes in habitat spatial characteristics (Jules and Shahani 2003). Because many plants are self-incompatible, and thus require pollen from other individuals to successfully reproduce, the movement of pollen among plants is essential. And most pollen movement occurs via animal pollinators, including mammals, birds, and insects. There has been much concern recently over declines in pollinators and their services (e.g., Buchmann and Nabhan 1997; Kremen et al. 2002; Biesmeijer et al. 2006). In addition to the use of pesticides, one likely contributing factor in their decline is the loss and fragmentation of suitable habitat for pollinators.

Two early observational studies showed both that pollinators may be less abundant in habitat fragments than in continuous habitat, and that this may have negative impacts on plant reproduction in the fragments. Jennersten (1988) studied pollination and seed set of the butterfly-pollinated plant, *Dianthus deltoides* (Caryophyllaceae), within an agricultural landscape in southwest Sweden. Through both pollinator observations and careful studies of plant reproduction, Jennersten found that plants growing in habitat fragments were much less likely to be visited by pollinators, and hence set fewer seeds, than those in a continuous, "mainland" habitat. Aizen and Feinsinger (1994) observed similar patterns by studying pollination in a completely different landscape, the dry subtropical forests of northwestern Argentina. Their study was broader in scope, involving 16 different plant species (from eight different families) that were visited by a wide variety of pollinators, including butterflies, moths, bees, wasps, beetles, and hummingbirds. Their goal was to assess "community-level health" in response to habitat loss and fragmentation by examining a suite of plant-pollinator systems within the same landscape. Pollen loads, fruit set, and seed set were observed for subsets of plants growing in small (<1 ha) fragments, larger (>2 ha) fragments, and continuous forest. For the plant community as a whole, pollination was substantially higher in continuous forest than in habitat fragments, and this reduction in pollination was, to a lesser extent, reflected in lower overall seed set in the forest fragments.

Additional observational studies in fragmented habitats have revealed significant declines in pollination and seed set. For example, pollination and seed set of tropical forest trees in both Thailand and Costa Rica (Ghazoul and McLeish 2001) were lower in fragmented than in continuous forest. The authors suggested that the mechanism for these declines was that forest fragmentation affected pollinator foraging. In particular, because trees were scarce in small habitat fragments, pollinators spent more time foraging elsewhere in these fragments, and so did not move as frequently among the trees, thus lowering seed set for these trees relative to those in continuous forest. Lennartsson (2002) observed reductions in viability of the grassland herb *Gentianella campestris* in fragmented versus continuous grassland in Sweden, due to declines in pollinator visitation and seed set. Demographic data were incorporated into a stochastic population model, which revealed likely extinction thresholds for small populations in grassland fragments.

A crucial phenomenon for small populations is the *Allee effect* (Allee 1951), the reduction in reproduction that occurs due to the decreased ability to locate mates. In the case of many plants, mate-finding occurs via the presence and efficacy of pollinators. Martha Groom (1998) performed experiments with an annual beefly-pollinated plant, *Clarkia concinna* (Onagraceae), in California, where she manipulated both plant patch size and isolation to examine effects on plant reproductive success. Groom trimmed naturally-occurring *Clarkia* patches to three sizes: *tiny* (1–10 individuals per patch), *small* (11–50 individuals per patch) and *large* (>50 individuals). For each patch size category, she also created isolation distances ranging from 1 m to 2.2 km, and then measured pollen receipt and seed set for focal plants in each patch size/isolation treatment combination. Plants in small, isolated populations were much less likely to receive pollen and set seed than their counterparts in large populations close to neighbors. Beyond these straightforward results, which were consistent with theoretical expectations, Groom showed an intriguing statistical interaction between patch size and isolation. Plants in larger patches could essentially withstand being isolated, but plants in small populations were particularly vulnerable to isolation. Moreover, there was a threshold isolation distance beyond which plant reproduction in small populations declined to zero, but such a threshold was not observed in large populations. This reduction in reproductive success was a likely explanation for the vulnerability of the tiny populations—over 75% of the populations in that size category went extinct during the study, compared to no extinctions in the large patches. Because landscape change due to human activities is likely to lead to smaller, more isolated plant populations, the findings from this study suggest that plant populations may be particularly susceptible to extinction unless restoration and planning can ameliorate those effects.

Is this selection of studies on pollination and fragmentation representative of broader patterns? To address that question, we can look to a recent meta-analysis of pollination and fragmentation studies (Aguilar et al. 2006). These authors quantitatively reviewed the results from studies of pollination in relation to habitat loss and isolation published during the period from 1987 to 2006. They observed an overall significant, negative effect of habitat loss and isolation on pollination and plant reproductive output, but the effect was more acute for self-incompatible plant species than for self-compatible ones. This result is compelling, but not too surprising, since self-incompatible plants are obviously much more reliant on pollinator visitation for reproduction than are self-compatible species. No other plant reproductive trait helped to explain the variation in effect sizes among species' responses, however. Two previous reviews of pollination and fragmentation failed to reveal such strong patterns (Aizen, Ashworth, and Galetto 2002; Ghazoul 2005), but Aguilar et al. (2006) conducted the first formal meta-analysis of these studies, and this approach is much more powerful than simply tallying studies that show one response versus another.

The primary focus of pollination research in relation to habitat loss and isolation has been to evaluate effects on native plant species in habitat remnants. But the pollination of species that are planted for human consumption may also be influenced by habitat spatial structure, and this has clear and profound implications for our well-being. The best example of research so far on the provision of pollinator services in modified landscapes is the work of Claire Kremen and colleagues, who have been studying the pollination of several food crops in northern California as a function of the composition and configuration of native habitats in the landscape surrounding crop fields (Kremen, Williams, and Thorp 2002; Kremen et al. 2002; Kremen et al. 2004; N.M. Williams and Kremen 2007; see fig. 7.4). For example, their research has shown that the abundance and diversity of native pollinators, and their contribution to the pollination of watermelon, is strongly influenced by landscape context (Kremen et al. 2002; see chapter 5). Specifically, farms that are surrounded by a relatively high proportion of natural habitat (30% or more cover of native habitat within 1 km of the farm) enjoy higher pollen deposition on watermelon flowers than farms surrounded by a low cover of native habitat (less than 1% within 1 km of the farm). The patches of native habitat provide critical nesting and foraging habitat for native bees (Kremen et al. 2002), so the abundance and diversity of bees is higher at these farms. Moreover, survival and reproduction of the native solitary bee *Osmia lignaria* appeared to be positively affected by the amount of native habitat present in this largely agricultural landscape (N.M. Williams and Kremen 2007).

Seed Dispersal

Many plants also depend on animals to disperse their seeds to suitable germination sites at some distance away from the parent plants, in order to avoid inbreeding. Seed-dispersing animals include mammals of many sizes and shapes (primates, bats, elephants, rhinoceros, rabbits, rodents, and bears), many bird species, and insects (including ants and dung beetles). In some instances, human activities, such as hunting, have reduced the abundance of these animal

Figure 7.4. Top: An organic farm in northern California, showing the landscape context with native habitat nearby. Photo courtesy of Claire Kremen, University of California–Berkeley. *Bottom:* Bees pollinating a sunflower—*Apis mellifera* (honeybee, left) and *Bombus vosnesenskii* (bumblebee, right). Photo © by Sarah S. Greenleaf, University of California–Davis.

mutualists, and their associated plants are likely to suffer the consequence of reduced population viability (e.g., Wang et al. 2007). Similar outcomes would be predicted for changes in seed disperser abundance and distribution caused by habitat loss and isolation.

The dynamics of seed dispersal, such as the proportion of seeds dispersed or the destiny of the seeds, may be disrupted or altered in fragmented landscapes. For instance, several ant species in northeastern Georgia disperse seeds of bloodroot (*Sanguinaria canadensis*), a forest understory herb. Ness (2004) studied ant seed dispersal in forest edges and interiors, and found that native ants tended to avoid forest edges, choosing instead to disperse seeds away from these edges and into the forest interior. Hence ants appeared to deliver bloodroot seeds to appropriate germination sites, and they may help to alleviate the negative impact of forest edges on bloodroot persistence. In Atlantic coastal forest fragments in Brazil, lower rates of scatter-hoarding by rodents in small fragments reduced seed dispersal for the endemic palm *Astrocaryum aculeatissimum* (Galetti et al. 2006). Rates of seed dispersal were higher in large fragments and sharply reduced in small fragments, which resulted in lower densities of palm seedlings and juveniles in the small fragments. In contrast, seed dispersal of *Prunus africana* by a suite of 36 fruit-eating bird species in western Kenya was slightly greater in five forest fragments than in a continuous forest site (Farwig, Böhning-Gaese, and Bleher 2006), despite lower overall frugivore species richness in the fragments than in the main forest sites. It is unclear from this study, however, how seed dispersal translated into plant recruitment in fragments versus continuous forest.

Mutualisms Involving Ants

Most studies of mutualistic interactions in fragmented habitats have centered on pollination and seed dispersal, two highly conspicuous and critically important phenomena for the viability of many plant species. Yet very little attention has been given to some other, perhaps more obscure, but no less essential interactions among species. Ant-plant mutualisms are common in temperate and tropical ecosystems, including the intriguing ant-acacia relationships found in both Central America and Africa (e.g., Janzen 1966; Stanton et al. 1999). Ants associated with plants typically defend the plant against herbivore attack, and in return the plant provides rewards in the form of food and shelter. Ants are also involved in mutualistic relationships with other insects, especially aphids (Homoptera) and larval butterflies in a couple of families (Lepidoptera: Lycaenidae and Riodinidae). In these associations, ants protect the herbivores from predators, and aphids and caterpillars secrete a sugary liquid that is eaten by the ants. However, even less is known about these types of interactions and how they may be influenced by habitat loss and isolation (Styrsky and Eubanks 2007).

Within the context of the Amazonian forest fragmentation experiment (BDFF) discussed previously, Bruna, Vasconcelos, and Heredia (2005) examined whether ant-plant mutualisms appeared to be affected by habitat loss and isolation. They surveyed a subset of the BDFF long-term study sites, including four 1 ha fragments and four continuous forest sites. They inspected over 1000 myrmecophytes (plants with associated ants) across the sites and recorded the identity, abundance, and richness of their attendant ants. The results were complex, and, contrary to their expectations, the researchers detected no differences in either plant or ant species richness or the total ant-plant density among the forest fragments and the continuous forest sites. They did note, however, that eight of the plant species occurred at higher average densities in continuous forest than in the 1 ha fragments. This observation led them to suggest that perhaps, due to low population sizes in the fragments, both ants and plants in these isolates may be more susceptible to stochastic processes that could drive them toward local extinction.

Mycorrhizal Associations

Mutually beneficial relationships among plants and fungi (*mycorrhizal associations*) are ubiquitous in nature and fundamental to functioning ecosystems. Underground fungi colonize the external or internal surfaces of plant roots and facilitate the uptake of key nutrients, such as nitrogen and phosphorus, from the surrounding soil. The fungi benefit by carbohydrates that are exuded from the plant roots. Two major groups of mycorrhizal fungi infect plant roots: arbuscular mycorrhizae that penetrate plant roots, and ectomycorrhizal fungi, whose hyphae form sheaths on the external surfaces of plant roots. There are few examples specifically showing how mycorrhizal associations vary in the context of habitat loss and fragmentation, but we can make some inferences by piecing together the results of a handful of relevant studies. The construction of the Panama Canal and its associated Lake Gatun created a series of different-sized tropical forest islands (former hilltops) in the midst of the lake. Researchers have compared many taxa on these islands to the adjacent mainland, in order to discern the effects of habitat area and isolation on species richness. Similarly, Mangan et al. (2004) surveyed arbuscular mycorrhizal fungi on several of these islands by collecting soils and screening for fungal spores. They observed that island sites had distinct arbuscular mycorrhizal fungal communities when compared to mainland sites, so habitat fragmentation did affect community composition significantly. There were no systematic differences, however, in species richness or the diversity of fungal spores with decreasing forest fragment size. The authors suggested that the different composition of fungal inoculum in island sites may cause subsequent shifts in the patterns of seedling growth and survival in these sites, since mycorrhizal fungi positively affect both of these characteristics of plant performance. The authors suggest that disruption of this mutualism could ultimately shift above-ground species composition.

In a quite different study system involving temperate forests, habitat area did have a significant effect on fungal species richness. Peay et al. (2007) studied ectomycorrhizal fungi on habitat islands of the host tree, *Pinus muricata*, in northern California. They found a distinct and substantial increase (over five orders of magnitude) in the species richness of fungal spores as tree-island area increased.

Although these are naturally-occurring islands, and not the direct result of human disturbance, the implication is that habitat loss would diminish ectomycorrhizal fungal species richness.

Experimental studies of mycorrhizal fungi in two different contexts further suggest that dispersal may limit the colonization of hosts by fungal spores. Dickie and Reich (2005) found that oak seedlings (*Quercus macrocarpa*) in abandoned agricultural fields in Minnesota were more likely to be infected by mycorrhizal fungi if they were growing close to the forest edge (0–10 m) rather than farther from it (15–20 m away). Similarly, the prevalence of ecotomycorrhizal infections in two oak species (*Q. ellipsoidalis* and *Q. macrocarpa*) was higher in seedlings growing near adult trees than in seedlings more distant from adult trees (Dickie et al. 2007). These results suggest that if adult trees became more scattered due to landscape change, the rates of natural colonization of seedlings by ectomycorrhizal fungi would be curtailed. Limited dispersal also appeared to affect arbuscular fungi that colonized corn plants in fallow agricultural fields in Zimbabwe (Lekberg et al. 2007). Arbuscular mycorrhizal communities were more similar in study sites that were close together, but differed greatly in widely-spaced study sites. Thus the dispersal of fungal spores may strongly affect community composition, with the inference that if dispersal is disrupted by habitat loss and fragmentation, then colonization may be similarly reduced.

SOME UNKNOWNS

There are at least two other interactions that have been described by ecologists, but we know little about them in the context of changes in habitat spatial structure. A fascinating but relatively little-studied interaction is *floral larceny* (Irwin, Brody, and Waser 2001), whereby animals steal nectar from flowers but get away without rewarding the plant with successful pollination. Although this interaction can clearly play a role in plant reproductive output and, by extension, is likely to affect population persistence in altered landscapes, there appear to be no published studies yet that focus explicitly on how habitat spatial structure influences the rates and patterns of nectar robbery. It seems reasonable to expect that nectar robbers will be attracted to flowers in a similar way as pollinators, so at least the part about "finding the interacting partner" would seem to be affected by habitat loss and fragmentation in ways that resemble plant-pollinator interactions. One interesting unanswered question is how persistent an antagonistic interaction may be, relative to an interaction that is mutualistic. Metacommunity theory is just beginning to devise a formal framework to deal with this (e.g., see Hoopes,

Holt, and Holyoak 2005 for the development of spatial mathematical models of competition, predation, and mutualism). Mutualistic interactions among plants (*facilitation*) are more common than was once thought (Bertness and Callaway 1994), and they may be affected by changes in habitat spatial structure in different ways than plant competition is. Although there is recognition that plant-plant interactions may shift between competition and facilitation, depending on the environmental context (e.g., Graff, Aguiar, and Chaneton 2007), there appears to be a lack of formal studies that examine the effects of habitat loss and fragmentation on facilitation among plants.

FOUNDATION SPECIES

Some species have disproportionately large impacts on ecosystems and thus make major contributions to ecosystem function. These interactions often do not fall neatly into a single category, since the species may ultimately enhance the richness of the entire community but might have negative effects on a particular species, since they may be competitors or even predators of that species within the ecosystem. These high-impact species are the *foundation species* of Soulé et al. (2003), and include "ecosystem engineers" like prairie dogs, beavers, and termites, as well as key food resources, such as fig trees in tropical forests or the American chestnut (the latter is now functionally extinct in eastern United States forests). The function of these foundation species may vary in a linear fashion with density (as density declines, function declines monotonically), or it may vary non-linearly (there may be critical threshold densities below which function is effectively zero; see fig. 7.1). Because these foundation species are highly interactive species, Soulé et al. argue that they should be prioritized in conservation efforts.

Prairie dogs in grasslands in the western United States provide a clear example of this issue. One of the most critically endangered mammals in North America is the black-footed ferret, which is an obligate predator of prairie dogs. Ferrets have large home ranges; in fact, about 40–60 ha of prairie dog colonies are required for a single ferret. So when grassland habitat is reduced and fragmented, prairie dog colonies become smaller and more isolated from one another, and ferrets cannot persist in these altered landscapes. In this case, the function of prairie dogs as foundation species (for ferrets, at least) declines in highly fragmented landscapes.

MULTIPLE INTERACTIONS IN THE SAME SYSTEM

The Savannah River experiment in South Carolina provides a unique example where multiple species interactions have been studied within the same experimental system (Tewksbury et al. 2002). Various researchers at this site have studied how habitat corridors affect species occurrence and movement patterns (see chapter 4), predator-prey relationships, seed predation, seed dispersal, and pollination. The main message that emerged from this comprehensive study is that corridors affected each of these interactions significantly, probably due to their effects on individual movement patterns. For example, Brinkerhoff, Haddad, and Orrock (2005) tested whether small mammal (prey) foraging behavior varied in response to habitat corridors and perceived predation risk (simulated by adding predator urine to half of the experimental blocks). Corridors synchronized small mammal foraging activity among connected patches, and they also provided a conduit for movement away from areas of high perceived predation risk. The implication of this study is that the observed changes in prey foraging behavior may results in shifts in overall small mammal assemblages in connected versus isolated habitat patches.

Corridors affected invertebrate, rodent, and avian seed predators on pokeweed (*Phytolacca americana*), but the effects were complementary (Orrock et al. 2003). In other words, invertebrates removed fewer seeds in patches that were connected by corridors than in unconnected patches, but rodents removed more seeds in connected patches. Seed removal by birds was similar in the connected and unconnected patches. In sum, total seed removal by all seed predators was not affected by corridors, because invertebrates removed more seeds in those patches where rodents removed fewer seeds, and vice versa. Because these seed predator taxa forage differently, and because they preferentially consume seeds of different species, corridors may ultimately cause shifts in the overall composition of plant and seed predator communities.

Seed dispersal by birds was a slightly different story. The dispersal of fruits and seeds of two shrub species, *Myrica cerifera* and *Ilex vomitoria*, was higher in connected than in unconnected patches (Tewksbury et al. 2002; Levey et al. 2005). Researchers followed the fates of seeds dispersed by birds using seed traps situated beneath artificial perch sites. Because one of the plant species, *M. cerifera*, was so common throughout the study area, both sets of researchers used a clever method to distinguish seeds from fruits collected specifically from their experimental plants. They sprayed the experimental fruits with a fluorescent powder, then examined the dispersed seeds under a fluorescent microscope to confirm which seeds had come from fruits in particular study plots. Levey et al. (2005) also tracked the movements of the major seed disperser, the eastern blue-bird (*Sialia sialis*), and incorporated movement and seed dispersal data into an individually-based simulation model. Taken together, the results from both studies showed strong positive effects of habitat corridors on seed dispersal, and both experiments demonstrated that observations of small-scale movements by birds can be accurately scaled up to predict broad-scale patterns of seed dispersal by these birds.

Because isolation may negatively affect pollinators and pollination, habitat corridors may positively contribute to pollination success. Pollen movement via pollinators appeared to be affected significantly by habitat corridors within the broad-scale Savannah River fragmentation experiment. To examine pollen movement in relation to corridors, Tewksbury et al. (2002) used holly (*Ilex verticillata*). They planted male individuals (pollen donors) in the central patches of their experimental array and planted female holly (pollen recipients) in each of the peripheral patches, including both patches connected by a corridor and unconnected patches. Because they used cultivars of holly, and not the native holly at their study site, they could attribute pollen flow directly to their experimental plants, not to plants in the surrounding landscape. These holly cultivars were visited by a wide variety of potential pollinators, including flies, wasps, bees, and butterflies. Consistent with other results showing significant positive effects of corridors (see chapters 5 and 6), the proportion of flowers that were successfully pollinated and produced fruit was higher in connected patches than in unconnected ones, presumably because the corridors facilitated the movement of insect pollinators in this landscape.

INTERACTIONS BETWEEN INTERACTIONS

Ecological science has a tendency to start with the simple and progress to the more complex. The studies discussed above generally focused on one species interaction at a time, although we know that every species interacts in different ways with many other species within ecological communities. Several studies have grappled with this complexity by simultaneously evaluating the effects of more than one type of interaction on a particular species. Groom's (1998) study of pollination of *Clarkia concinna* in relation to patch size and isolation was discussed earlier in this chapter, but she also conducted experiments on this plant to examine both pollination and herbivory in relation to patch isolation (Groom 2001). Groom's results from this latter study showed that isolated plant patches enjoy lower levels of herbivory, but they also suffer lower rates of pollination than plants in patches close to other patches. Thus it appears that insects have a more difficult time locating isolated plants, which has both costs and benefits for plants in isolated patches.

In particular, the consideration of multiple species interactions involving plants helps us to understand the net effect of habitat loss and fragmentation for plant regeneration. Erik Jules and colleagues conducted a comprehensive study of several processes and interactions related to plant reproduction in the long-lived forest understory herb *Trillium ovatum* (Liliaceae) in fragmented forests in the Pacific Northwest (Jules 1998; Jules and Rathcke 1999; Tallmon et al. 2003; Kahmen and Jules 2005). The study began with the observation that *Trillium* recruitment was close to zero near (within about 65 m of) forest edges, but was higher in forest interior habitat (Jules 1998). Subsequent experiments revealed that near edges, seed production was lower, possibly due to decreased pollinator activity there, and seed predation by mice was higher (Jules and Rathcke 1999, fig. 7.5). Seed dispersal and seed germination did not differ between the forest edge and interior sites, so declining plant recruitment was likely to have been caused by decreased seed production and enhanced seed predation. Deer mice, major seed predators, appeared to thrive in clear-cuts and attained densities that were 3–4 times higher at forest edges than in interiors (Tallmon et al. 2003). Given that *Trillium* is a long-lived plant that occurs within forests that can be several hundred or even thousands of years old, what do these results tell us about the patterns of long-term recovery of *Trillium* in disturbed forests? Kahmen and Jules (2005) surveyed recovered forests that differed in age and confirmed that population recovery may take as long as centuries unless certain individuals that survive periodic disturbances are able to serve as sources for seed germination within sites.

Plants lend themselves well to studies of multiple species interactions. García and Chacoff (2007) adopted a similar approach and concurrently considered three interactions—pollination, seed dispersal, and seed predation—for hawthorn trees (Rosaceae: *Crataegus monogyna*) in northern Spain. In this case, the different interactions were accomplished by different species groups that had disparate ranges of movement. Frugivorous birds (seed dispersers) were highly mobile, insects (pollinators) were a bit less mobile, and woodmice (*Apodemus sylvaticus*, seed predators) were the least mobile of the three groups. The key results from this study were that isolated hawthorn trees were less likely to be pollinated, less likely to have their seeds dispersed, yet more likely to have their seeds eaten by mice than were trees surrounded by extensive forest cover. Thus the multiple

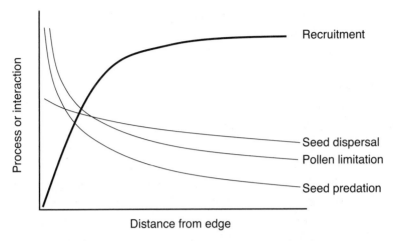

Figure 7.5. Recruitment, seed dispersal, pollen limitation, and seed predation of *Trillium ovatum* in relation to the distance from a forest/clear-cut edge. *Trillium* recruitment (heavy line) was the net effect of multiple interacting processes, including those shown and others not shown but cited in Jules and Rathcke (1999). Recruitment data were generalized from Jules (1998), and seed dispersal, pollen limitation, and seed predation were generalized from data presented in Jules and Rathcke (1999).

effects related to plant regeneration in this system were additive, rather than compensatory. The effects of different interactions varied with spatial scale, consistent with the movement patterns of the different animal species involved. The implication is that isolated hawthorn trees are in big trouble. They appear to be able to produce fewer seeds, and many of the seeds that they do produce are consumed by mice. The next step would be for ecologists to integrate the effects of these and other multiple-species interactions into models or experimental studies of plant performance in different spatial contexts.

SYNTHESIS

Several messages emerge as a result of this review of studies of species interactions in fragmented landscapes. First, it is clear that because species are often tightly linked within food webs or via other functional relationships, declines in the abundance of one species may cause similar declines in a linked species. What Harrison and Bruna (1996) call "chains of ecological interactions" can result in positive feedback loops that cause species to swirl down the extinction vortex. Species interactions may thus play a huge role in these processes.

Second, the study of species interactions is an excellent opportunity to link animal behavior with population and community dynamics. Because animals use behavioral cues to locate and move toward or away from each other, the functional connectivity of landscapes will largely influence the outcome of these behavioral decisions.

Third, it is clear that a single type of interaction (pollination, seed dispersal, competition) can be strongly influenced by habitat spatial structure, usually because it affects the ability of one or both of the interacting partners to encounter each other within patchy landscapes. But when additional interactions are considered, the net effect may be that the positive and negative impacts cancel each other out, such as in Groom's (2001) example of *Clarkia concinna*, which experienced lower herbivory but also lower pollination in isolated patches. Or the negative impacts may stack up to create not just one, but several reasons for species declines in fragmented landscapes, such as the hawthorn trees in Spain that experienced reduced pollination and seed dispersal *and* higher seed predation in isolated settings (García and Chacoff 2007). In addition to their importance in fragmented landscapes, the outcomes of multiple species interactions are becoming more widely recognized for their broad relevance to ecology (Morris et al. 2007).

Finally, it is apparent from the discussion on multiple species interactions that there is some work to do to develop formal conceptual models to predict the outcome of different types of species interactions in fragmented landscapes. In terms of maintaining species diversity in patchy landscapes, metacommunity theory highlights the importance of both the strength of species interactions and the amount of dispersal. But unanswered questions remain. For example, is patchiness likely to affect antagonistic interactions more strongly than mutualistic ones? Or will there be an impact on interactions that involve trophic connections versus those that do not? Our ability to refine predictions of species responses and the outcomes of species interactions in fragmented landscapes should enhance our success at planning and managing landscapes that support the entire suite of biological interactions, so that our forests (and grasslands, scrublands, savannas, and oceans) may truly remain full.

CHAPTER EIGHT

Parasites, Pathogens, and Disease Emergence

> Human-induced land use changes are the primary drivers of
> a range of infectious disease outbreaks and emergence
> events.
> —*Jonathan A. Patz, Peter Daszak, Gary M. Tabor, A. Alonso
> Aguirre, Mary Pearl, Jon Epstein, Nathan D. Wolfe, A. Marm
> Kilpatrick, Johannes Foufopoulos, David Molyneux, David J.
> Bradley, and members of the Working Group on Land Use
> Change and Disease Emergence (2004)*

In recent years, media coverage has startled readers with tales of outbreaks of mysterious maladies in humans, wild animals, trees, and agricultural crops— West Nile virus, Lyme disease, chronic wasting disease, amphibian limb deformities, sudden oak death, and soybean rust. Reports of the resurgence of more familiar human diseases—plague, rabies, and malaria—have appeared as well. The likely causes of this recent disease emergence and resurgence include climate change (Harvell et al. 2002; P. Anderson et al. 2004), the invasion of exotic species (J.K. Brown and Hovmøller 2002; Derraik and Slaney 2007) and water pollution (Rejmánková et al. 2006; Hrudey and Hrudey 2007). But we are also discovering compelling cases where parasite and pathogen prevalence are closely related to changes in landscape spatial configuration (e.g., Kitron 1998; Ostfeld 2005). It turns out that anthropogenic landscape change is a pivotal factor in the dynamics of many diseases that pose risks to humans, wildlife, forests, and agricultural crops (Patz et al. 2004).

Several notorious diseases appear to emerge as the result of complex interactions among hosts, vectors, and pathogens, and landscape change may shift the dynamics of these interactions, thereby increasing the risk of exposure to disease. There are several ways in which parasite and pathogen prevalence may be altered in response to landscape change, and these are not necessarily mutually exclusive. First, we know that land transformation via urbanization, deforestation, and agricultural intensification can cause habitat loss and fragmentation, introduce novel landscape contexts, and increase the amount of edge, all of which may alter the relative abundance and richness of hosts and vectors. Second, in human-modified landscapes, changes in functional connectivity may alter dispersal rates and transmission routes for hosts, vectors, or pathogens and parasites. Third, native habitats surrounded by unfamiliar landscape contexts may experience declines in habitat quality, such as the eutrophication of lakes, streams, or wetlands adjacent to heavily fertilized agricultural fields.

From a conservation standpoint, disease increasingly threatens the persistence of native species and ecosystems that are of conservation concern. Species that are already rare or declining in fragmented habitats may show an enhanced susceptibility to disease. In particular, landscape change or habitat degradation may result in a reduction in the hosts' capacity to defend themselves against parasites and pathogens (*immunocompetence*), due to stress or food limitation within small or degraded habitat fragments. Habitat loss and isolation may also reduce genetic diversity in small populations, diminishing the capacity for species to evolve resistance to new parasites and pathogens. Species may also be exposed to novel diseases for which they have little or no built-in resistance. For example, many different coral species in the Caribbean are now threatened by a suite of previously unknown pathogens, which collectively pose a serious threat to the integrity of coral reef ecosystems (Porter et al. 2001). A bit of promising news is that some corals have begun to show increased resistance to some pathogens, either through induced defenses or evolved resistance (Kim and Harvell 2004).

In 1998, Wilcove and colleagues published an often-cited analysis of the five major threats to imperiled plants and animals in the United States. Habitat destruction and degradation ranked the highest, with over 80% of these species in jeopardy for this reason. Disease came in last, with only 2% of the species reportedly threatened by this factor; most of these cases were of Hawaiian birds that have declined due to avian malaria. More recently, however, several authors have highlighted the increasing threat of wildlife disease for species of conservation concern (Daszak, Cunningham, and Hyatt 2000; Lafferty and Gerber 2002). Schloegel et al. (2006) reported what may be the first *documented* case of a species extinction in the wild due to an infectious disease, the recent decline to extinction of the sharp-snouted day frog (*Taudactylus acutirostris*) in northern Queensland, Australia, due to infection by the pathogen *Batrachochytrium dendrobatidis* (chytrid fungus). It is probably the case that, given the increasing number of reports of rare species negatively affected by these maladies, such as upper respiratory tract disease in desert tortoises or the recent rampant spread of Tasmanian devil facial tumor disease (McCallum and Jones 2006), the percentage of species at risk of extinction from disease has increased both in the United States and worldwide in just the past 10 years.

Due in part to these observed increases in the impacts of disease on native species and ecosystems, there are now interdisciplinary research efforts that link human, wildlife, and plant diseases to anthropogenically caused changes in landscape spatial configuration or habitat quality. These endeavors fall under various research umbrellas, referred to as disease ecology, conservation medicine, landscape pathology, landscape epidemiology, and spatial epidemiology (table 8.1). Investigations typically focus on how shifts in species composition, driven by landscape change, may dramatically influence the emergence and spread of infectious diseases. These "new" fields of inquiry build upon a rich history of theoretical and empirical research in epidemiology that integrates spatial patterns and the processes of pathogen transmission and spread.

Because of the burgeoning literature on these topics, this book features an entire chapter devoted to parasite and pathogen ecology in fragmented landscapes. Many of the studies reviewed here describe the dynamics of zoonotic diseases— those shared between wildlife and human hosts—but this chapter also highlights examples of pathogens and parasites known only to non-human species, as well as a few plant pathogens. First there is a brief review of ecological theory relevant to disease ecology, and then the rest of the chapter is organized around each of the major mechanisms for changes in parasite and pathogen prevalence in human-modified landscapes.

SPATIAL THEORY AND MODELING

The transmission and spread of pathogens reveals what is essentially a spatial record of interactions among individual hosts and vectors. Therefore, detailed analyses of observed patterns in disease incidence can help clarify the factors that affect pathogen transmission and spread. Comprehensive spatial and temporal

TABLE 8.1

Glossary of key terms related to parasites, pathogens, and disease occurrence, starting with the simplest terms and progressing to others that build upon and refer back to them. Terms are cited from Schmidt and Ostfeld (2001); Haydon et al. (2002); Tabor (2002); Holdenrieder et al. (2004); McCallum et al. (2004); Fenton and Pedersen (2005); and Ostfeld, Glass, and Keesing (2005).

Term	Definition
Pathogen	A disease-causing microorganism that multiplies within its host, generally controlled by the host response.
Parasite	An organism that lives in an intimate association with one host individual of another species per life-history stage, with a detrimental effect on the host.
Macroparasite	Typically, multi-cellular metazoans—such as helminths—that cannot complete an entire life cycle within one individual host.
Microparasite	Typically, unicellular microorganisms—such as viruses, bacteria, and protozoans—that can multiply rapidly within a host.
Disease	A condition that impairs the normal function for a plant or animal. An *infectious disease* is caused by a transmissible biological agent.
Zoonosis	Diseases that can be shared among wild or domestic animals and humans.
Transmission	The passing of a disease from an infected individual or group to a previously uninfected individual or group. A *directly transmitted* disease does not involve a vector, but is passed by direct contact of one individual with another.
Vector	A mobile organism that transmits a parasite from one host to another.
Reservoir host	One or more epidemiologically connected populations or environments in which a pathogen can be permanently maintained and from which infection is transmitted to the defined target population.
Reservoir competence	The ability of a host to transmit a specific pathogen to a feeding vector.
Vector competence	The ability of a vector, once it has fed on an infected host, to infect a naïve host.
Dilution effect	Host communities with high species richness or evenness are likely to contain a high proportion of hosts that are inefficient in transmitting the disease agent to a feeding vector, thereby reducing disease risk.
Conservation medicine	An integrative field of study that examines the health of both individuals and groups of individuals, and the landscapes in which they live, by simultaneously considering animal health, human health, and ecosystem health.
Landscape pathology	The incorporation of landscape ecological concepts and methods into the science of forest pathology.
Landscape epidemiology	The identification of geographical areas where disease is transmitted, and the analysis of risk patterns and environmental risk factors.
Community epidemiology	The consideration of multiple hosts, vectors, and pathogens that interact in ecological communities.
Spatial epidemiology	The study of spatial heterogeneity in the risk or incidence of infectious disease.

data sets on human diseases are especially valuable for understanding disease dynamics in spatial contexts. For example, Brian Grenfell and collaborators have used decades of observations of childhood measles in the United Kingdom to build mathematical models that can reconstruct the observed spatial and temporal patterns of epidemics. By retroactively modeling the data to fit observed patterns, these authors have revealed the importance of key variables (such as spatial isolation and town size) on the persistence and periodicity of measles (Bolker and Grenfell 1995; Grenfell, Bjørnstad, and Kappey 2001; Grenfell, Bjørnstad, and Finkenstadt 2002). What is most relevant to the present discussion is the inference that the spatial structure of populations and communities may strongly influence the transmission and spread of pathogens. For example, given what we know about animal movement, pathogen dispersal, and landscape connectivity, we might expect disease transmission to be more likely in connected versus fragmented landscapes.

Recall that chapters 3 and 6 contain lengthy discussions of the potential benefits of habitat corridors for species persistence. A few dissenting voices (Simberloff and Cox 1987; Simberloff et al. 1992; Hess 1994, 1996a, b) cautioned that corridors may be ultimately detrimental to populations and communities, because they may facilitate the spread of disturbance and disease among otherwise isolated populations. Their argument was that pathogens may easily disperse along corridors among otherwise isolated and disease-free populations. So perhaps there is an advantage to habitat isolation, if pathogen movement is restricted relative to continuous landscapes.

In three closely related papers, Hess (1994, 1996a, b) synthesized concepts from metapopulation theory and spatial epidemiology to explore both the likelihood of pathogen spread in several different landscape configurations and its implication for metapopulation persistence. Hess (1996a) simulated different spatial arrangements of populations and found that spatial configuration differentially affected the probability of metapopulation extinction due to disease, but only under certain conditions. For example, when migration between populations was relatively low, the *necklace* configuration (a set of linearly arranged patches, where individuals could only move to the adjacent patch during each of the model's time intervals) had the lowest probability of extinction when disease reduced the life span of organisms by 50%–60%. Based on the results of simulation models, Hess concluded that "too much movement" among local populations could be hazardous, as it could facilitate the spread of disease. Movements that occurred with relatively low frequency, such as those involved in migration or dispersal, were not as detrimental in terms of disease transmission as higher-frequency movements, for example, forays by animals within their home range. The negative effects of connectivity were most acute for highly contagious diseases, and Hess thus recommended several management options for such situations, including quarantine, vaccination, or even limiting the movement of individuals by blocking movement pathways.

But under what conditions might the benefits of corridors for population persistence outweigh the costs of disease transmission? Subsequent expansion and modification of Hess's original formulations helped to clarify this issue. Gog, Woodroffe, and Swinton (2002) argued that Hess's models may not have broad applicability, because of research demonstrating that many pathogens have multiple hosts (Haydon et al. 2002). When the presence of alternative hosts was incorporated into models of disease spread, the results showed an overall net positive effect of inter-population movement on population persistence, rather than a negative effect (Gog, Woodroffe, and Swinton 2002). This finding was largely due to the incorporation of *background infection* into the models, which means that the pathogen was already widespread in the system because it was present in multiple species, and increasing the movement rates among populations did not exacerbate the negative effects of disease spread.

McCallum and Dobson (2002) extended Hess's models further and broadened this discussion to consider the effects of habitat fragmentation on disease persistence and transmission more generally. They also added two key components: they modeled metapopulation dynamics in disease systems where local populations were allowed to recover from infection, and they included the presence of a second, reservoir host species. Recovery was incorporated into their models, since in many disease systems, pathogens *fade out* to local extinction when the abundance or density of susceptible hosts becomes very low. The local population is thus disease-free until the number of susceptible individuals increases to some level where the pathogen can invade them. When recovery was included, the target host species rapidly recolonized connected patches, and the benefits of recolonization for population persistence outweighed the cost of increased pathogen movement among connected patches. When models included a reservoir host (Haydon et al. 2002; see table 8.1), the outcomes depended on specific dispersal abilities of the target species in relation to the pathogen and reservoir host. In these models, the target host species could essentially escape the pathogen in connected patches if the target host could disperse either to patches where the reservoir host had gone extinct, or to patches that had resistant reservoir populations. McCallum and Dobson (2002) concluded from their models that, in situations where a novel pathogen invaded a system and the target host species was unable to disperse to other patches, it may be necessary to quarantine or to manage the movement of individuals among populations to prevent disease spread.

Taken together, the various spatial models of metapopulation dynamics of hosts and pathogens suggest that conservation biologists should carefully consider the potential impact of pathogen transmission among isolated populations. Connecting otherwise isolated populations may be costly or might be beneficial, depending on these conditions: (1) whether the pathogen is specialized on a single species or has multiple hosts, (2) whether recovery of populations from infection is possible, and (3) the relative dispersal abilities of hosts and pathogens (e.g., Thrall and Burdon 1999). In many of the cases explored in modeling studies, however, the positive impact of corridors on population exchange and species persistence far outweighed the cost of enhanced disease spread.

SHIFTS IN COMMUNITY COMPOSITION

A common theme that emerges from studies of landscape change and disease risk is that habitat modification typically simplifies ecological communities, causing shifts in the abundance of hosts and vectors that may then result in higher pathogen prevalence. So if species are lost because of habitat loss and fragmentation—either through edge effects or changes in patch size, isolation, or landscape context—there are likely to be profound effects on pathogen-vector-host relationships that may result in increased human exposure to pathogenic organisms (Ostfeld and Keesing 2000a, b). The reduction in disease risk in species-rich communities, known as the *dilution effect* (table 8.1), may be a primary means by which landscape change influences disease risk. Diverse communities may promote the dilution effect via several mechanisms: (1) reducing the probability of an encounter between host and pathogen, (2) reducing the probability of pathogen transmission between hosts, (3) limiting the abundance of particular host species via competition or predation, (4) causing increased mortality in infected species via competition, or (5) facilitating the recovery of infected species via mutualistic interactions (Keesing, Holt, and Ostfeld 2006). The study of various mechanisms responsible for the dilution effect may ultimately facilitate intervention or management strategies to ameliorate the risk of disease exposure for humans.

Despite a limited understanding of the ways in which species diversity affects disease risk, examples from a handful of disease systems provide some evidence for the dilution effect. Most examples involve vector-borne diseases with multiple hosts, and include Lyme disease, West Nile virus, tick-borne encephalitis, bartonellosis, Chaga's disease, and ehrlichiosis (summarized in Ostfeld and Keesing 2000b; Dobson et al. 2006; Keesing, Holt, and Ostfeld 2006). There are also a few examples from directly transmitted animal pathogens (e.g., hantaviruses) and plant pathogens. Dobson et al. (2006) succinctly summarized the benefits of bio-diversity for reducing the risk of several well-known, vector-borne zoonotic dis-eases—in these cases, habitat loss and fragmentation have reduced the abun-dance and richness of predators and competitors from ecological communities, causing an increased abundance of a competent host species and thus increased pathogen prevalence. These authors warn that, as we reduce biodiversity, we are compromising the ability of natural systems to dilute pathogens, as well as in-creasing our own exposure to zoonotic diseases. The following sections review three well-studied systems where species richness reduces disease risk: Lyme disease, West Nile virus, and foliar plant pathogens in perennial grasslands.

Lyme Disease

The observation that host species richness and pathogen prevalence are nega-tively correlated has been most extensively explored in studies of Lyme disease by Richard Ostfeld and colleagues (Van Buskirk and Ostfeld 1998; Ostfeld and Keesing 2000a, b; Schmidt and Ostfeld 2001; LoGiudice et al. 2003; Ostfeld and LoGiudice 2003; Ostfeld, Keesing, and LoGiudice 2006). Lyme disease in the northeastern United States is a zoonosis involving several bird, mammal, and reptile hosts; it has a single vector, the black-legged tick (*Ixodes scapularis*), and a single pathogen, the spirochete *Borrelia burgdorferi*. The disease manifests itself in humans by a skin rash, joint swelling, and flulike symptoms, and it affects thousands of humans per year. Ticks have a two-year life cycle and take one blood meal in each of three life stages: larva, nymph, and adult. The main hosts of the larval and nymphal ticks are small mammals, birds, and sometimes lizards, while the main hosts for adult ticks are white-tailed deer (*Odocoileus virginianus*). Both modeling and field observations have shown that the number of human cases of Lyme disease, and the prevalence of the Lyme spirochete, are reduced in species-rich host assemblages relative to species-poor assemblages (Ostfeld and Keesing 2000a; Schmidt and Ostfeld 2001; LoGiudice et al. 2003).

The particularly relevant piece of the Lyme disease story for this chapter is that the loss and fragmentation of northeastern deciduous forests has caused shifts in the species composition of mammal assemblages, which has implications for the risk of Lyme disease for humans. Specifically, B. Allan, Keesing, and Ostfeld (2003) observed that in the fragmented forest landscape of Dutchess County, New

York, smaller forest fragments (<2 ha) supported higher densities of tick nymphs (as well as a higher prevalence of infected nymphs) than did larger forest fragments (4–7 ha). The net result was a higher density of infected nymphs in small versus large fragments (fig. 8.1a). Because forest loss and fragmentation tend to reduce vertebrate diversity, the authors suggested that the increase in the prevalence of nymphal infection that they observed was likely due to a higher abundance of the generalist species, the white-footed mouse (*Peromyscus leucopus*), in small fragments. This mouse has been observed to dominate small mammal communities in forest fragments relative to continuous forest (Nupp and Swihart 1998)—and because the white-footed mouse is the most competent host for the Lyme disease pathogen, this means that in forest fragments there are likely to be more tick meals taken from white-footed mice and, ultimately, a higher proportion of nymphal ticks infected with the pathogen. Smaller forest patches probably support larger populations of white-footed mice because of this species' tolerance for declines in patch area, and because predators and competitors tend to be less common in small forest patches. The implication is that humans who live near small forest fragments may have an enhanced risk of exposure to Lyme disease relative to those near large forest fragments (B. Allan, Keesing, and Ostfeld 2003).

A similar study performed in Connecticut by a different research group compared tick density, tick infection prevalence, and the incidence of Lyme disease in humans in relation to deciduous forest fragmentation (Brownstein et al. 2005). Forest fragmentation was described using patch size and isolation measures obtained from satellite imagery. The researchers also sampled 30 field sites for ticks and summarized human cases for each town in Connecticut. They observed positive relationships between tick density and forest fragmentation, as well as between tick infection prevalence and fragmentation. But in contrast to the results from the previous study (B. Allan, Keesing, and Ostfeld 2003), Brownstein et al. (2005) found that the number of human cases was negatively, rather than positively affected by fragmentation. In other words, more human cases were recorded in the towns with more continuous forest. Because both studies showed higher tick densities and infection rates in fragmented forests, why did they differ in their observed patterns of human cases? One reason may be that the measurements were not identical, so the disparity could be due to variation in the sampling methods. It could also be that patterns of residential development or human behavior differ in subtle ways between these two landscapes, and that is sufficient to cause differences in the incidence of Lyme disease. However, it seems safe to conclude from both sets of studies that tick density and tick infec-

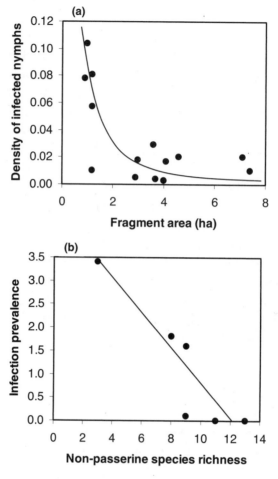

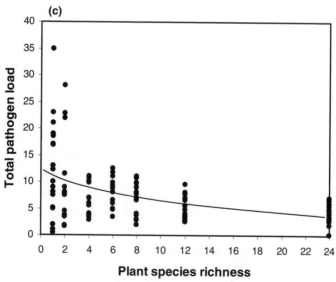

tion prevalence are higher in fragmented landscapes, having at least the potential to enhance the human risk of exposure to Lyme disease significantly.

West Nile Virus

West Nile virus (WNV) is an arthropod-borne virus (*arbovirus*) that causes encephalitis in humans, potentially resulting in permanent neurological damage or death. Like Lyme disease, WNV is a vector-borne disease that resides in multiple host species, so it is reasonable to expect that its prevalence may be affected by host diversity via the dilution effect. However, at least three ecological differences between the two disease systems are worth noting: (1) WNV is transmitted by mosquitoes rather than ticks, (2) the pathogen is a virus rather than a bacterium, and (3) the main hosts are birds rather than mammals.

Ezenwa et al. (2006) conducted field studies in southern Louisiana in a landscape comprised of small forest fragments of loblolly pine (*Pinus taeda*) and wetlands. They gathered data from two different spatial scales to examine links between host diversity and the human risk of WNV. First, they sampled six field sites to measure both mosquito infection with WNV and avian species richness. Second, they gathered data on human WNV cases and avian species richness for all the counties in Louisiana that had reported human cases of this illness. Their results strongly conformed to the dilution effect: mosquito WNV infection rates were negatively associated with the species richness of non-passerine birds (fig. 8.1b), and the number of human cases per county was similarly negatively associated with the richness of non-passerine species. The significant relationship with non-passerine species was expected in this study, since these birds are thought to be relatively incompetent reservoirs for the pathogen, which is a prerequisite for the dilution effect (Ostfeld and Keesing 2000b). Although Ezenwa et al. did not directly explore links between landscape change and avian species richness, the implication is that habitat loss and fragmentation would likely reduce species richness and thereby increase WNV risk. Another study of WNV

Figure 8.1. (opposite) Evidence for a dilution effect for three different disease systems. (a) The density of tick nymphs infected with *Borrelia burgdorferi* versus forest fragment area for 14 fragments in Dutchess County, New York. Figure redrawn from B. Allan, Keesing, and Ostfeld (2003). (b) The prevalence of mosquitoes infected with West Nile virus versus non-passerine bird species richness for six study sites in southern Louisiana. Figure redrawn from Ezenwa et al. (2006). (c) The total foliar fungal pathogen load across the plant community versus plant species richness for experimental grassland plots in Minnesota. Figure redrawn from Mitchell, Tilman, and Groth (2002).

along an urban gradient in Georgia found a weak but positive correlation between WNV prevalence in wild birds and the amount of urban/suburban land
cover (S. Gibbs et al. 2006). These results similarly suggest that WNV risk may be
greater in areas of lower bird diversity, since bird diversity typically declines in relation to urbanization.

In addition to the risk imposed by WNV for humans, the virus is also fatal for
many bird species. LaDeau, Kilpatrick, and Marra (2007) analyzed 26-year population trajectories of 20 North American bird species and found that seven of the
20 species (including the eastern bluebird, American robin, and American crow)
have declined significantly since the first reported WNV outbreak in New York in
1999. They suggested that, similar to the mass mortalities experienced by Hawaiian birds due to avian malaria or by amphibians that suffer from chytridiomyco-
sis, WNV may pose a significant impact to diverse groups of birds throughout
North America. Moreover, the shifts in bird community composition driven by
the disease have the potential to engender positive feedbacks to WNV prevalence
and subsequently enhance the risk to humans. Specifically, if primarily incompetent hosts are killed by the virus, then the relative abundance of competent
hosts may increase, further increasing the prevalence of WNV.

Foliar Pathogens

Changes in plant community structure may affect pathogen prevalence in plants,
but a vital factor in determining the direction of the relationship between plant
species richness and pathogen prevalence may be whether the pathogens are specialized to a few species or are generalist pathogens that can infect a wide variety
of plants (Mitchell and Power 2006). Because of variation among plant species in
their susceptibility to pathogens, higher-diversity plant communities may reduce
overall pathogen loads. For example, Mitchell, Tilman, and Groth (2002) studied
specialist foliar fungal pathogens in experimentally created, perennial grassland
communities in Minnesota and found evidence for the dilution effect. In species-
poor communities (1–4 species), pathogen loads across the entire plant community were almost three times higher than in species-rich (12 or 24 species)
communities (fig. 8.1c).

The proposed mechanism for this result was as follows: because specific
pathogens tended to affect only a few plant species within the community, an increase in the relative abundance of a particular host plant species in species-poor
communities meant that there were more host individuals available to be in-

fected by the pathogens that occur on that species. Simply stated, highly diverse communities had lower abundances of each host species, thereby reducing the spread of specialist pathogens that depend on the density of their respective hosts. In this manner, the observed effects were driven by changes in host density, rather than in diversity per se. The dynamics of the dilution effect differ slightly in this system from that observed for Lyme disease, since these plant pathogens were relatively specialized on particular host species, while the Lyme spirochete is a generalist pathogen. However, because the white-footed mouse is the most competent reservoir host for Lyme disease, the increase in relative abundance of this host in species-poor communities is analogous to the increase in relative abundance in species-poor plant communities of a particular host plant for a pathogen that is specialized on that species.

CHANGES IN LANDSCAPE CONNECTIVITY

Despite previously discussed concerns regarding the potential for increases in disease transmission via conservation corridors, I know of no field studies in which such effects have been documented directly. There are, however, examples where populations may escape disease outbreaks by virtue of their isolation, or where management decisions may bring novel pathogens in closer contact with susceptible plant or animal populations. My research group has observed the first phenomenon, in which isolated populations are relatively safe from disease outbreaks, in our collaborative studies of plague (*Yersinia pestis*), which is transmitted among mammalian hosts by fleas in grasslands in the western United States. Black-tailed prairie dogs (*Cynomys ludovicianus*) are particularly susceptible to plague and typically suffer greater than 90% mortality in affected colonies. We analyzed two long-term data sets of prairie dog colony die-offs due to plague, one from Boulder County, Colorado, and the other from Phillips County, Montana (Collinge et al. 2005). In both study areas, plague occurrence over the past 20 years was more likely on spatially-clustered colonies than in more isolated colonies (fig. 8.2). Colonies that were farther from plague-positive colonies were less likely to be exposed to plague than colonies closer to plague-positive ones. However, we do not know whether, in these isolated prairie dog populations, the benefit from escaping disease outweighs the costs of reduced immigration. Further, our analyses showed that colonies near landscape features, including roads, streams, and reservoirs, were less likely to experience plague outbreaks than colonies that were not surrounded by these features. One hypothesis to explain

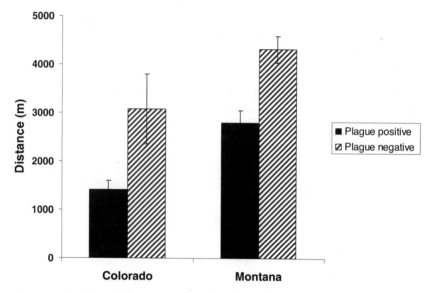

Figure 8.2. The occurrence of plague caused by *Yersinia pestis* in black-tailed prairie dog (*Cynomys ludovicianus*) colonies in Boulder County, Colorado, and Phillips County, Montana. Isolation was measured as the distance (in meters) to the closest plague-positive colony. Plague-negative colonies were significantly farther from plague-positive colonies than were other plague-positive colonies in both study areas, suggesting that plague outbreaks are spatially clustered and that isolated colonies are less likely to be exposed to plague. Redrawn from Collinge et al. (2005).

this result is that roads, streams, and reservoirs may serve as barriers to the movement of prairie dogs or other hosts, thereby limiting pathogen spread.

Isolated populations may escape their natural enemies, including parasites, since successful parasitism requires parasites or parasitoids to locate their hosts, which may be more difficult in fragmented landscapes. Kruess and Tscharntke (1994) investigated the interaction between parasitoids and hosts in fragmented habitat by establishing a field experiment with planted patches of red clover (*Tri-folium pratense*) in an agricultural landscape. The researchers placed five 1.2 m² clover patches within naturally occurring meadows that contained red clover, and 13 equally-sized "clover islands" at locations ranging from 100 to 500 m from naturally occurring meadows. They examined the abundance of endophagous insects (herbivorous insects that feed within plant tissue, including seed-feeders and stem borers), as well as species richness and the rates of parasitism of the herbivores by small parasitic wasps. The more isolated clover patches (those located farther from the naturally occurring meadows) had a lower abundance of herbivorous insects than did clover patches near meadows, suggesting that herbivore migration to distant patches was limited. The diversity of parasitoids and the rates of parasitism on herbivores declined even more precipitously with isolation. The net result was that herbivores were essentially able to escape their natural enemies in isolated habitats, which implies that biological control of pest species may be less effective in this type of habitat. The authors suggested that secondary consumers, such as parasitoids, may be more vulnerable to habitat isolation than primary consumers, such as herbivores, because parasitoid populations depend on the successful establishment of herbivore (host) populations. They argue that this has broader implications for ecology: "Food chains should be shorter in small islands than in larger habitats" (p. 1584).

Actions taken in the context of forestry, fisheries, livestock management, or wildlife management may alter landscape connectivity in ways that may affect the spread of pathogens or parasites. For example, the movement of infected sheep and cattle among distant farms, markets, and slaughterhouses was largely responsible for the spread of foot-and-mouth disease in the United Kingdom in 2001. The epidemic ultimately reached most of the country and resulted in the culling of hundreds of thousands of sheep, cattle, goats, pigs, and deer, with an economic impact of over eight billion U.S. dollars (Kao 2002). The outbreak would have been much more localized if humans had not moved these animals so rapidly. Further, many new problematic plant pathogens, such as sudden oak death in California, have emerged as a result of the global transport of forest products, which is essentially an issue of increasing connectivity among otherwise isolated geographic realms.

The continued spread of novel pathogens, once established in a new place, may depend significantly on landscape features such as roads and river corridors, as well as spatial characteristics such as habitat area, isolation, and landscape context. Those who study the epidemiology of plant diseases, following the lead of animal disease ecologists, are now turning to broader spatial scales and using tools from landscape ecology and metapopulation biology to address patterns of the spatial spread of pathogens across landscapes (Holdenrieder et al. 2004). As is the case for many animal diseases, Holdenrieder et al. suggest that there are two outstanding questions for broad-scale plant pathology. How does landscape connectivity affect pathogen spread? And can the advantages of connectivity for maintaining populations and communities be outweighed by its potentially disadvantageous pathological effects?

Two examples of the spread of plant pathogens in heterogeneous landscapes relate directly to functional landscape connectivity. Perkins and Matlack (2002) analyzed the spatial distribution of forest types in and among pine plantations within a matrix of agricultural land in the southeastern United States. Loblolly (*Pinus taeda*) and slash pine (*P. elliottii*) were historically restricted to watercourses, so these species were connected within watersheds but were relatively isolated between them. Plantation forestry using these two species has expanded their range into upland areas, so that stands are now in much closer proximity (an average distance of 35 m) than they were historically (at 600–900 m). Loblolly pine in particular is susceptible to fusiform rust (*Cronartium quercuum*), and oaks (*Quercus* spp.) are the alternate host for *C. quercuum*. This alternate host also occurs within most pine plantations. Hence modern forestry practices have changed the spatial juxtaposition of pine and oak stands—and reducing the distance between the two kinds of stands has increased the functional connectivity and the ease of spread for this pathogen.

In northern California and Oregon, another more problematic, non-native pathogen colonizes the roots of Port Orford cedar (*Chamaecyparis lawsoniana*), which typically grows in riparian areas. The root rot, caused by *Phytophthora lateralis*, is fatal to these trees (Jules et al. 2002). The authors reconstructed the spatial and temporal spread of this pathogen using tree cores, and they found that most of the infection events originated from dirt roads that crossed creeks. Vehicle traffic apparently facilitated the spread of this disease, since the large vehicles associated with timber harvests along these frequently muddy roads carry organic material, and probably fungal spores, from place to place. Foot traffic also appeared to enhance the spread of the pathogen at local scales. Thus the changes in landscape connectivity created by the road network in this landscape served to increase the spread of a non-native pathogen.

Similarly, in the animal realm, an unanticipated change in connectivity among salmon farms and marine fjords enhanced the probability of parasite infection in migrating wild salmon (McVicar 2004; Krkošek, Lewis, and Volpe 2005; Krkošek et al. 2006). Sea lice (*Lepeophtheirus salmonis*) are copepod (small crustacean) ectoparasites of salmonids that are directly transmitted among individuals; their life cycle has a free-living larval stage and parasitic juvenile and adult stages. The establishment and expansion of salmon farms (one form of marine aquaculture), particularly in British Columbia, Canada, has enhanced the spread of sea lice from farmed to wild salmon. This has occurred because the migrating juveniles must swim through narrow fjords on the first 80 km or so of their outbound journey to the Pacific Ocean, and, en route, they pass in rather close proximity to salmon farms. These juvenile wild salmon (who are free of sea lice under natural conditions) are thus exposed to adult farmed salmon that are infected

Figure 8.3. A seafood shop on Salt Spring Island, one of the Gulf Islands near Vancouver, British Columbia, Canada. Photo by S. K. Collinge.

with the lice (Krkošek, Lewis, and Volpe 2005; Krkošek et al. 2006). These salmon farms are *point sources* of sea lice, which greatly increases the probability that the migrating juveniles will become infected. The overall population impacts of sea lice infestations of wild salmon have not yet been quantified, but given the high rates of infection and mortality in wild salmon, there is likely to be some impact on populations (Hilborn 2006). It's just another reason to say no to farmed salmon (fig. 8.3)!

Wildlife populations living in protected areas may be exposed to pathogens and parasites through their spatial proximity to infected domestic animal hosts living on reserve boundaries. For example, several species of carnivores in the Serengeti ecosystem of northern Tanzania have been severely affected by outbreaks of diseases that originated with domestic dogs along the borders of the protected area. Domestic dogs serve as reservoirs for rabies virus, canine distemper virus, and canine parvovirus, and these pathogens have been especially harmful to wild dogs (*Lycaon pictus*) and lions (*Panthera leo*) (Packer et al. 1999; Cleaveland et al. 2000, 2007; Woodroffe 2001; Vial et al. 2006). The borders between protected areas and human settlements are relatively fluid in this region—

there are few fences, and wildlife often move beyond reserve boundaries—so the most effective solution appears to be vaccinating domestic dogs for rabies and canine distemper (e.g., Vial et al. 2006) to prevent the spillover of pathogens to wild hosts.

If this is a topic of interest, there are several additional examples of disease spread related to landscape spatial configuration and functional connectivity that would reward a more detailed exploration. These cases pose particular challenges to conservation biologists and land managers and include Sin Nombre virus in Canada (Langlois et al. 2001), brucellosis in bison and elk in Yellowstone National Park (Dobson and Meagher 1996), bovine tuberculosis in Britain (Woodroffe et al. 2005), the spread of rabies in the eastern United States (Real and Childs 2006), the emergence of Nipah virus in Indonesia (Daszak et al. 2006), and infectious diseases transported by the wildlife trade (Fèvre et al. 2006).

HABITAT DEGRADATION

Thus far, this chapter has examined how shifts in community composition and changes in landscape connectivity may influence the prevalence of parasites and pathogens. It now turns to a third mechanism by which landscape change may influence pathogen and parasite occurrence—through shifts in habitat characteristics as a result of human activities. As discussed at length previously (see chapter 5), differing landscape contexts may profoundly affect species and communities living within native habitat remnants. For example, wetlands, estuaries, forests, or grasslands that are situated in landscapes with high-intensity agriculture, forestry, or urbanization will undoubtedly experience different inputs and disturbances compared to those same habitat types in a relatively undisturbed landscape. The effects of landscape context may be especially acute for aquatic systems, since both the physical and biological characteristics of these systems are largely affected by the composition of land cover and land uses in the surrounding watershed. This section presents several examples where habitat degradation due to changes in land use or landscape spatial configuration affects the occurrence of pathogens or parasites.

One of the most comprehensive studies of land use change and disease risk comes from northern Belize, where Eliska Rejmánková and colleagues have studied the links between agricultural fertilization and the risk of exposure to malaria (Rejmánková et al. 1996, 1998, 2006). To divulge the punch line first, the fertilization of sugarcane fields leads to changes in wetland vegetation, which increases the amount of suitable habitat for the mosquito species that is the superior vector for malaria, thereby increasing the risk of human exposure to malaria. Each of the links in this chain of events has been carefully studied, and here's how it works. Two major types of herbaceous wetlands exist in northern Belize: on low-nutrient soils, wetlands are occupied by short, sparse vegetation, dominated by the rush *Eleocharis* spp. (Juncaceae), with floating cyanobacterial mats interspersed with the herbaceous vegetation; and on high-nutrient soils, wetlands are characterized by tall, dense macrophytes, dominated by the cattail *Typha domingensis* (Typhaceae). Adult mosquitoes in the genus *Anopheles* are primary malaria vectors worldwide; in Belize, *A. albimanus* and *A. vestitipennis* are two key species in this disease system, but *A. vestitipennis* is the more competent vector. Larval *A. albimanus* occur primarily in *Eleocharis* (short, sparse) marshes, and larval *A. vestitipennis* occur primarily in *Typha* (tall, dense) marshes and swamp forests. Sugarcane agriculture involves the input of nitrogen- and phosphorus-containing fertilizers, which enrich the surface waters of wetlands. This causes a shift in plant species composition from *Eleocharis*-dominated wetlands to *Typha*-dominated wetlands. Rejmánková et al. (2006) used GIS analyses to show that *Typha* cover in wetlands was positively related to the amount of agricultural land in the surrounding uplands. With more *Typha*-dominated wetlands, there are more *A. vestitipennis* mosquitoes. Hence the close proximity of wetlands to agricultural fields ultimately increases the risk of human exposure to malaria. Although malaria is not a major cause of human mortality in Belize, these results are likely to apply to other regions where malaria remains problematic.

The intensive nutrient fertilization associated with agricultural practices may ultimately affect vertebrate populations as well, mediated by complex interactions among other members of aquatic food webs. Recent evidence suggests that the eutrophication of ponds and wetlands due to agricultural or livestock activities may be responsible for limb malformations in amphibians. Amphibian biologists have been particularly perplexed by the dramatic increase in observations of severe amphibian limb malformations (extra limbs, missing limbs, or malformed limbs) in North America since the 1990s (P. Johnson et al. 2002; P. Johnson and Chase 2004). Deformed animals rarely survive to sexual maturity, and high levels of such malformations are expected to threaten population viability (Blaustein and Johnson 2003). As with the malaria example above, this phenomenon involves a rather complex chain of events. P. Johnson and Chase (2004) argued that evidence from across North America supports the following scenario: nutrient-rich runoff from agricultural fields or livestock pastures has caused the eutrophication of ponds and wetlands, with subsequent increases in planorbid (Planorbidae) snails, which function as intermediate hosts for the trematode parasite *Ribeiroia ondatre*. The increased abundance of *Ribeiroia* then results in higher parasite abundance in amphibians, and higher infection levels lead to more limb malformations (fig. 8.4). Although the cause of limb malformations has not been definitively determined in every case, mounting evidence supports the *Ribeiroia* hypothesis. P. Johnson and Chase (2004) built a convincing case for *Ribeiroia* as the culprit by showing a significant positive relationship between malformation frequency and *Ribeiroia* infection levels for 11 species of frogs and toads within 56 populations from California, Oregon, Washington, Minnesota, and Wisconsin. The causal link between nutrient inputs to freshwater and *Ribeiroia* infections of amphibians was persuasively demonstrated in recent experimental studies with field mesocosms (P. Johnson et al. 2007). Both the density of intermediate hosts (planorbid snails) and the per-snail production of *Ribeiroia* infective stages (cercariae) increased in high-nutrient mesocosms compared to low-nutrient ones, and *Ribeiroia* infection in frogs (*Rana clamitans*) was correspondingly higher in high- versus low-nutrient mesocosms. Moreover, amphibian infection is simultaneously influenced by the activity of frog-eating birds, which act as the parasite's definitive host, further highlighting the complex relationships among environmental conditions, host interactions, and parasite abundance.

Activities associated with urbanization, as well as those involving agriculture, may import higher nutrient loads into surface waters, with consequences for aquatic parasites and pathogens. In two related studies, amphibians in wetlands along urban gradients in the northeastern United States were screened for limb malformations (B. Taylor et al. 2005) and for parasites (Skelly et al. 2006). B. Taylor et al. (2005) surveyed frogs in 42 wetlands in northern Vermont for evidence of limb malformations in relation to wetland landscape context. In this study, higher rates of amphibian malformation were associated with the close proximity of wetlands to agricultural fields and lawns. Despite this clear pattern and logical causal link to nutrient runoff and eutrophication, these researchers found no evidence for infection by *Ribeiroia ondatrae*. They speculated that an agricultural pollutant, rather than *Ribeiroia*, may have been the direct cause of developmental malfunctions within the amphibians. In a second study, this time of two groups of parasitic trematodes (echinostomes and *Megalodiscus*) that infected two species of amphibian hosts, Skelly et al. (2006) detected no significant relationship between the amount of urbanization surrounding 60 sampled wetlands in Connecticut and parasite prevalence. Three of the wetlands that were sampled, however, had extremely high infection intensities of one of the parasites (echinostomes) on one of the hosts (green frog, *Rana clamitans*); these three wetlands

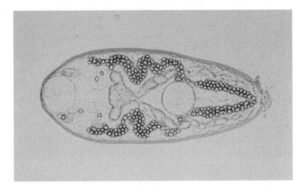

Figure 8.4. Top: Limb malformation in *Rana pipiens. Bottom:* Infective stage (meta-cercariae) of the suspected causative agent, *Ribieroia ondatrae.* The anterior end is to the left in the photo. Photos courtesy of Pieter T. J. Johnson.

were also surrounded by highly urbanized uplands and had the highest snail densities of any of the wetlands sampled. As with *Ribeiroia*, snails are intermediate hosts for the echinostome parasite, so it makes sense that snail abundance was positively correlated with frog infection.

A final example of habitat degradation and parasitism in amphibians comes from a field study of wetlands in forests and pastures in the lowlands of Costa Rica. McKenzie (2007) surveyed three species of frogs and their parasites, both in water bodies situated within intact forests and in those in nearby cattle pastures that had been cleared of forest. The rates of parasitism by the majority of parasites, as well as parasite species richness in one of the frog species, were higher in pasture wetlands than in forest wetlands. As noted above, the likely explanation is that changes in water quality, in particular nutrient enrichment, within cattle pastures relative to forests positively influenced the rates of amphibian parasitism.

In marine as well as freshwater systems, water quality is strongly influenced

by land uses and by the land cover of watersheds that drain into rivers that ultimately reach the ocean. Southern sea otters (*Enhydra lutris nereis*) in California
were undergoing a promising population recovery during the past several decades after being overharvested in the 19th and early 20th centuries. In the late
1990s, however, this recovery stalled, with actual population declines between
1996 and 2002. Researchers autopsied dead sea otters and found that the proto-
zoan parasite *Toxoplasma gondii* occurred in 52% of dead sea otters (M. Miller et
al. 2002, 2004; Jessup et al. 2004; Conrad et al. 2005). *Toxoplasma gondii*'s definitive hosts are felids, and it is suspected that domestic cats may contribute to the
high prevalence of infection in sea otters, due to issues of landscape context. Sea
otters sampled near areas of high freshwater runoff exhibited a three-fold increase in the presence of antibodies (*seropositivity*) to *T. gondii*, as compared to otters sampled in areas that were more distant from these possible sources of infection (M. Miller et al. 2002). Cat feces may be entering marine systems at these
maximal runoff areas and concentrating these sources of infection for the sea otters. It is not yet clear whether *T. gondii* or other parasites that affect sea otters,
such as *Sarcocystis neurona* (whose definitive host is the opossum), are severe
enough to regulate populations of sea otters, but the potential exists for their impacts to reverberate throughout these nearshore marine communities. Because
sea otters play a keystone role in these ecosystems by controlling the abundance
of kelp-eating sea urchins, population declines due to infectious disease and parasites are likely to affect the integrity of kelp forests in these systems (Collinge,
Ray, and Cully 2008).

Other nearshore marine communities may suffer from unanticipated consequences of nutrient-rich runoff. Several coral disease outbreaks have been reported in the Florida Keys, Mexico, and other parts of the Caribbean, starting in
the 1990s (Porter et al. 2001); this has the potential for enormous impacts on
these valuable ecosystems. One of these pathogens, *Aspergillus sydowii*, is a soil
fungus that has caused severe die-offs of the sea fan *Gorgonia ventalina* in the
Florida Keys (Harvell et al. 1999). It has not yet been possible to forge direct links
between the source(s) of *A. sydowii* and the sites of sea fan infection, but hypotheses consider either runoff from terrestrial systems or the deposition of ae-
olian dust as possibilities (Kim and Harvell 2004). Moreover, nutrient enrichment may play a role in the local severity of pathogen infection on sea fans. In a
field experiment in the Yucatan Peninsula of Mexico, Bruno et al. (2003) showed
that fertilized sea fans experienced increases in the severity of *A. sydowii* infection compared to unfertilized sea fans. The authors suggested that, although nutrient enrichment is not the likely cause of the overall epidemic of aspergillosis,

localized nutrient enrichment may explain some of the local spatial variation in sea fan infections. They also postulated that nutrient input from human sources may exacerbate the spread of this now-established disease. Through both field and laboratory experiments, Voss and Richardson (2006) observed similar links between nutrient enrichment and the severity and spread of black band disease in massive starlet coral (*Siderastrea siderea*).

To give a quick example from a terrestrial system, Mackelprang, Dearing, and St. Jeor (2001) observed a higher prevalence of Sin Nombre virus (SNV, a hantavirus) infection in deer mice (*Peromyscus maniculatus*) in shrublands in central Utah that had been disturbed and degraded by off-road vehicles. In their surveys, the authors found that SNV infection in deer mice, the primary host, was three times greater than that found in other studies of deer mice in more intact landscapes. They hypothesized that off-road vehicle use has compacted soils, removed vegetation, and produced more dirt roads, all of which favor the more generalist deer mouse. With higher densities of deer mice, the authors speculate that higher transmission rates of SNV have led to overall higher levels of SNV prevalence.

INCREASED HOST SUSCEPTIBILITY
Stress-Related Responses

A fourth way in which habitat loss, isolation, and degradation may influence parasites and pathogens is by promoting stress and food limitation in host populations, thus rendering them more susceptible to parasitism or pathogens. For instance, several studies have documented the increased prevalence of a collection of gastrointestinal parasites in primates occupying forest fragments and selectively logged forests, as compared to those in undisturbed forest, in western Uganda (Gillespie, Chapman, and Greiner 2005; Chapman, Speirs, et al. 2006; Chapman, Wasserman, et al. 2006; Gillespie and Chapman 2006; Salzer et al. 2007). Surveys for parasites in fecal samples of red-tailed guenons (*Cercopithecus ascanius*), red colobus monkeys (*Pilocolobus tephrosceles*), and black-and-white colobus monkeys (*Colobus guereza*) revealed that parasite prevalence was higher in logged versus undisturbed forests for red-tailed guenons, but not for the two colobine species (Gillespie, Chapman, and Greiner 2005). For both the red and the black-and-white colobus, however, the proportion of individuals with multiple parasite infections was greater in forest-edge than forest-interior primate groups (Chapman, Speirs, et al. 2006). There were no differences in nematode parasite prevalence among the endangered red colobus monkey in nine forest fragments that ranged in size from 1.2 to 8.7 ha (Gillespie and Chapman 2006). In that study, tree stump density, which was an index of human use and therefore habitat disturbance or degradation, explained the greatest amount of variance in parasite prevalence. Recent population declines of red colobus in forest fragments appear to be due primarily to lower food availability, but poor nutrition also enhanced the rates of parasite infection (Chapman, Wasserman, et al. 2006). Two additional intestinal parasites, *Cryptosporidium* spp. and *Giardia* spp. (both protozoa), infected red-tailed guenons and red colobus monkeys in three highly disturbed fragments, but these parasites were absent from undisturbed forests. Protozoal infection was not detected in black-and-white colobus monkeys from these sites (Salzer et al. 2007). Taken together, these comprehensive studies of parasites within the same study system convincingly demonstrate that habitat disturbance and degradation influence parasite prevalence in two of the three monkey species studied. These results for the red colobus monkey are of particular concern, since it is a rare species undergoing population declines.

Increased animal densities and heightened parasitic infections have also been observed in response to intentional wildlife management practices. Ezenwa (2004) reported the results of a survey of gastrointestinal parasites of impala (*Aepyceros melampus*) in five protected game reserves in Kenya. Parasitism by generalist, strongyle nematodes of impala increased in small fenced reserves that also had a high diversity of native bovids, such as buffalo, waterbuck, eland, and bushbuck. Both reserve size and bovid diversity may have affected impala parasitism by enhancing transmission as well as susceptibility. Although behavioral and physiological responses of impala were not measured in this study, Ezenwa (2004) speculated that reserve size could indirectly affect host infection risk by restricting movement, reducing resource availability, or disrupting impala social structure, all of which may induce stress and compromise the immune responses of impala.

Increased disease transmission may occur within populations whose densities have been artificially increased as a result of human activities. For example, the infection rates of chronic wasting disease (CWD) in mule deer (*Odocoileus hemionus*) in north-central Colorado were twice as high in herds near human settlements as in herds that were farther away from them (Farnsworth et al. 2005). Chronic wasting disease is caused by an infectious protein (a prion, similar to "mad cow disease") that has received increasing attention in North America over the past 20 years, with particular concerns for the risk of transmission to humans via the consumption of infected meat. Although the exact mechanism for the increase in CWD prevalence near developed areas is not known, the authors suggested that human development may have concentrated the deer into smaller areas, resulting in increased pathogen transmission in two ways. Food provisioning near settlements could have increased local deer densities, and development could have eliminated suitable deer habitat, forcing the same number of deer into smaller and smaller areas. Further, hunting pressure is typically reduced immediately adjacent to human settlements in this region, which could result in a longer adult lifespan of deer and a lengthier infectious period.

Loss of Genetic Variation

Habitat loss and isolation may cause declines in population size, ultimately leading to reduced genetic variation. Because genetic variability may enable the evolution of resistance to parasites and pathogens, limited genetic variation may limit the adaptability of populations. To test this hypothesis, Field et al. (2007) surveyed parasites of genetically-characterized earthworm (*Lumbricus terrestris*) populations occupying urban habitat fragments in northwestern Germany, within the city of Münster. They hypothesized that habitat loss and fragmentation in this urban setting would lead to small, genetically depauperate populations that would exhibit a lower resistance to parasite attack. They measured infection by sporocysts of the gregarine parasite *Monocystis* spp. (Apicomplexa or Sporozoa) in earthworms collected from habitat patches ranging in size from 0.29 to 2.14 ha (large patches) and from 0.018 to 0.078 ha (small patches). These protistan parasites (close relatives of *Toxoplasma* and *Plasmodium*) are about 0.5 mm long and infect the internal organs of earthworms. Contrary to expectations, earthworm genetic variability did not vary systematically with fragment size, nor did earthworm parasite loads. The authors suggested that perhaps the sampled areas had not been isolated long enough for genetic change to occur and convey resistance. It may also be that the range of fragment sizes was too small to detect changes in genetic variation and parasite infection rates among earthworm populations.

SYNTHESIS

Collectively, the studies reviewed here show that changes in landscape spatial configuration can and do influence the prevalence of parasites and pathogens through modifications in species composition, shifts in landscape connectivity, the degradation of habitats (especially aquatic ones), and increased host susceptibility. More than one of these mechanisms can act simultaneously to produce an undesired outcome—for example, habitat degradation appears to alter aquatic food webs and increase the infection of frogs by *Ribeiroia ondatrae*.

There is a clear message that for many organisms that humans find creepy and crawly, abundance increases either as an indirect outcome of landscape fragmentation, or as a direct result of practices associated with land conversion, such as agriculture or forestry. Bradley and Altizer (2007) concluded that the effects of urbanization on wildlife disease emergence are likely to differ from the effects of forest fragmentation or agricultural intensification, because "the extreme changes that accompany urbanization probably cause declines or losses of most wildlife species and their associated parasites" (p. 101). Although there is some evidence presented in this chapter that urbanization may influence pathogen and parasite prevalence, most of these studies were conducted in relatively low-density, suburban-style developments rather than in high-density urban centers such as Shanghai or Nairobi. The relationships between biodiversity and disease prevalence in urban centers await further study.

A related message that emerges from these studies is that there is a strong but under-appreciated link between biological diversity and human health. Observed relationships between species richness and pathogen prevalence in several important disease systems reveal another vital ecosystem service for the well-being of humans—that healthy human populations are critically dependent on the lowered disease risk provided by species-rich biological communities.

The emergence of interdisciplinary fields of study, such as conservation medicine, that has been stimulated by the increased prevalence of diseases worldwide offers promising opportunities for collaborative research into the causes and consequences of disease emergence. These interdependent research efforts promote an understanding of disease-related processes across such fields as medicine, ecology, epidemiology, and parasitology, and offer the hope that they can provide creative strategies for the management of and intervention for emerging diseases in humans, wildlife, forests, and agricultural crops.

CHAPTER NINE

Modeling

All models are wrong, but some models are useful.
—*George E.P. Box and Norman R. Draper (1987)*

Models can be somewhat intimidating to the uninitiated ecologist, because they often involve some complex mathematical equations. But a model in its simplest definition is just an abstraction of reality. We all know that a toy car is a model of a real car, but because children are not big enough to drive real cars, we give them less dangerous forms of vehicles that don't have accelerators. And a map is a model or representation of a place on the Earth, but obviously it doesn't contain all of the components of the real place. Models in general are central to scientific inquiry; in particular, ecological models craft abstractions of ecological reality. Because the natural world is infinitely complex, we can begin to understand it by creating models that pare down its overwhelming complexity into pieces and parts that we attempt to manipulate and comprehend.

In the case of understanding fragmented landscapes, there is an enormous variety of spatial and temporal patterns of landscape change. They occur at multiple scales and strongly affect the ecological systems that we study. We basically have three choices if we want to learn more about these systems: (1) we can conduct observational studies of existing conditions, (2) we can perform manipulative experiments that attempt to isolate the causes of particular responses, and (3) we can construct models in which we manipulate variables under a variety of conditions and examine the outcomes. Observational studies are a necessary starting point, but they often fail to reveal the mechanisms for responses (see chapter 3). Fragmentation experiments may uncover mechanisms, but they are often logistically difficult and time consuming (see chapter 4), and it may even be unethical and detrimental to perform them. Hence many ecological studies have turned to mathematical models to estimate or simulate the effects of spatial variation in landscape structure and examine its consequences. For example, many metapopulation models reveal how different rates of migration affect metapopulation persistence, and null landscape models provide researchers with the opportunity to vary the amount and spatial arrangement of habitats in cyberspace and examine the movement and persistence of model organisms. These modeling efforts often provide insights that are useful in guiding empirical studies and in suggesting potentially important variables that would only be discovered after years or decades of intensive field study.

Ecological modeling allows us to test and refine theory and see what is possible—what *can* happen under certain conditions, but not necessarily what *does* happen in the world outside. Models are meaningful heuristic tools, in that they can help to further guide investigation by revealing critical data gaps or potential mechanisms underlying responses. Most relevant to the current topic is that we are often interested in ecological responses to landscape change, which typically occur at a broad spatial scale and over long periods of time. Thus exploring the effects of landscape configuration often requires the use of models, because it is impossible to perform manipulations in the field at the temporal and spatial scales relevant to studying these effects.

Ecological models are often misunderstood by empirical ecologists. Many hard-core field ecologists are often skeptical of models because they do not report the results of "real" observations of "real" organisms. And many simplifying assumptions must often be made in order to keep the models from becoming unmanageable. However, models are not built in a vacuum. They are based on field observations, but they allow the examination of a variety of conditions and possibilities for interactions among factors. As field ecologists, perhaps we are envious that modelers don't have to struggle to collect data from real organisms, so they can perform analyses and publish papers without enduring the pelting rainstorms or searing heat or foul smells sometimes encountered during field work. That is probably partly true, but in my collaborations with modeling experts, I have been surprised to learn that it isn't as easy as empiricists think to build mathematical models that actually work. Creating a model that simulates reality in a reasonable way is often challenging. Levins (1966) suggested that models in population biology must make tradeoffs between realism, precision, and generality; no single model could maximize all three goals simultaneously. So there are many decisions that must be made in constructing a model to achieve the optimal combination of those three corners of the triangle. "All models are wrong, but some models are useful" (Box and Draper 1987, p. 424), so as long as we accept that models are a constructed abstraction of reality, we can learn a lot from their use in ecology.

This chapter reviews the use of models in studies of fragmented landscapes by providing a glimpse of the categories of questions that have been addressed and the major types of models that have been used. Prior to the detailed discussion of models, it introduces how ecologists have described landscape spatial patterns through the development and use of landscape pattern indices. These indices provide the necessary background for understanding models of landscape change, which constitutes the second section of this chapter. The landscape change models featured here include examples of landscapes modified by anthropogenic disturbances such as forestry, agriculture, and urbanization. Third, it examines a suite of modeling approaches and results that project population and community responses to landscape change. For example, these models include projections of metapopulation dynamics and spatially explicit population viability analyses. For each category, there is at least one and sometimes a few detailed examples to illustrate the types of data required, the range of approaches used, and the variety of outcomes observed. Many of these modeling efforts are explicitly geared toward conservation management and planning goals, which are explored further in chapter 11.

DEVELOPMENT OF LANDSCAPE INDICES

A major goal of landscape ecology is to decipher and describe spatial heterogeneity in ecological systems and to understand the effects of this complexity on ecological processes (e.g., Forman and Godron 1986; M. Turner 1989). Toward this end, O'Neill and colleagues (1988) published a foundation paper in landscape ecology; their intent was to develop a collection of simple indices that could describe and discriminate among landscape spatial patterns. They examined digitized maps of land cover from 94 landscapes across the eastern United States, and they computed three indices for each landscape: dominance, contagion, and fractal dimension. *Dominance* was derived from information theory (as is the familiar Shannon-Wiener index of species diversity), and its value reflects the relative proportions of different land cover types in a landscape. By analogy to species evenness, dominance is the inverse—high dominance means one land cover type tends to dominate the landscape, whereas low dominance values mean that there are several land cover types that are relatively equally distributed across that landscape. *Contagion* refers to the degree to which land cover types are clumped together in the landscape; a high value for contagion means that land cover types tend to be clustered spatially, whereas a low value means that land cover types are relatively dispersed across the landscape. The third index, *fractal dimension*, derived from fractal analysis (see chapter 2), refers to the complexity of shapes or patterns in the landscape. A high value for this index reflects complex shapes, generated, for example, by highly variable topography, whereas a low value indicates relatively simple geometric shapes and boundaries.

Calculation of these three simple indices for the 94 landscapes showed that they ranged widely and thus were able to discriminate amongst a large variety of landscapes. High values of dominance were observed in agricultural landscapes, such as in southern Illinois, as well as in forested landscapes, such as in West Virginia (O'Neill et al. 1988). Moreover, fractal dimension was highly negatively correlated with the proportion of the landscape in agricultural or urban land cover, and highly positively correlated with the proportion of the landscape in native forest cover, suggesting that fractal dimension could be used as an accurate index of the amount of human activity in a landscape. This paper achieved its goal of finding indices that succinctly described landscape pattern, and it laid the foundation for observational and modeling studies of how landscape patterns change over time, as well as how populations and communities respond to landscape change. The marriage of these indices with data on the abundance and distribution of populations and communities in varied landscapes came later, and potentially provides a powerful tool for understanding the ecological consequences of landscape change.

Soon after these indices were published, researchers began testing their applicability with sample data sets and, in some cases, developed additional metrics. For some situations, these indices were quite useful in describing ecologically relevant spatial patterns, but in other instances, the indices were criticized for not doing so. For example, Groom and Schumaker (1993) summarized patterns of landscape change on a global scale, with the goals of characterizing estimates of habitat loss and evaluating spatial measures of landscape change in forests. Most relevant to the current discussion, they examined changes in the spatial distribution of forest cover in the Olympic National Forest, Washington, by comparing frequency distributions of forest patch sizes between 1940 and 1988. The mean patch size decreased substantially over this time period, as did variance in patch size. In other words, logging in this region shifted the distribution of patch sizes toward many small forest patches with few large unbroken areas. They evaluated the use of three landscape metrics—perimeter-to-area ratio, shape index, and fractal dimension—and concluded that these did not effectively represent the differences in spatial pattern among forest landscapes in this region, at least in describing the degree of fragmentation. They suggested that an additional index, the direct measurement of interior area, accurately captured the forest loss and fragmentation that was observed in this managed forest. Further, they called for the development and testing of more "biologically meaningful indices of the isolating effect of fragmentation" (p. 44).

The charge to invent ecologically relevant metrics of landscape pattern was taken up by McGarigal and Marks (1995), who developed a software program called FRAGSTATS. Their goal was to provide an easy-to-use tool to calculate many different landscape indices, and, since its invention, FRAGSTATS has been extensively used in studies of species' responses to landscape patterns. The landscape metrics in FRAGSTATS were designed to capture features of both landscape composition (how much of what cover type?) and landscape configuration (how are land cover types arranged?). A key challenge inherent in blending these indices with ecological data is that the FRAGSTATS user can compute many different indices of landscape pattern, yet may still be uncertain about how most of these indices affect individual behavioral decisions, movement patterns, or, ultimately, population and community persistence for their particular organism or study system. In their original documentation (in the first line of the first page), McGarigal and Marks (1995) cautioned users to be VERY careful (they did in fact use capital letters!) to select appropriate indices and to apply them in appropriate contexts when using this program. They also gave explicit warnings about putting garbage into the model and getting garbage out. So the authors of the software were keenly aware of the potential to misuse these indices.

Not surprisingly, users did not always follow the directions, and so papers were published that reported lists of correlations of landscape indices with ecological data (such as species richness or population density) without a clear or meaningful ecological interpretation of the patterns. And criticism followed. For example, Gustafson (1998) reviewed the use of spatial pattern indices and urged users to (1) analyze pattern at the appropriate scale, (2) use an analysis method that is relevant to the goals of the study, (3) choose metrics that are relevant to the ecological process being studied, and (4) develop a priori hypotheses, based on ecological theory, that relate a particular index to an ecological process. Despite such clear guidelines, several years later H. Li and Wu (2004) again reviewed the use of landscape indices and argued that many studies used these indices without a clear knowledge of the ecological relevance of landscape patterns to ecological processes. They also noted that landscape indices may not indicate changes in habitat quality (usually they just indicate cover type), a finding which may have major implications for species persistence.

This is reminiscent of the old adage about using the right tool for the job. If used in the intended way, calculating landscape metrics can reveal meaningful structural differences among landscapes, changes in landscape pattern over time, and relevant implications for ecological processes (including species persistence, the spread of disturbance, or invasions of exotic species). As with any tool, the improper use of spatial pattern indices may, at best, obfuscate ecological relationships, or, at worst, convey interpretations that are simply wrong.

PROJECTING LANDSCAPE CHANGE

Once the appropriate descriptors of landscape spatial pattern have been thoughtfully identified, these indices can be used to project future landscape change. The basic approach for these projections is to quantify landscape patterns from aerial photos or satellite images taken in two or more time periods and then compare patterns across time. Other existing landscape patterns may be used as initial conditions in simulation models to project future landscape change, based on current or alternative trajectories. There are several useful syntheses of the construction and utilization of these types of landscape models (Baker 1989; M. Turner and Gardner 1991; Baker and Mladenoff 1999), and there is an excellent guide for students just learning these analytical tools (Gergel and Turner 2002).

Forestry

The old-growth conifer forests of the Pacific Northwest are remarkably valuable for their timber, and also for their unusually rich biological communities. The struggle between economic and ecological interests in this region became particularly acute in the 1980s, when accumulated forest harvests from the previous four decades began to have noticeable effects on species persistence and water quality. Because of its economic and ecological value, landscape patterns and projections of future landscape change in this region have been comprehensively studied. Harris (1984) published an ecological critique of the *dispersed patch* cutting method that had been used since the 1940s, because this harvest regime resulted in highly fragmented forests and the accompanying losses of biological diversity. Harris proposed an alternative cutting strategy of circular *long-rotation islands*, which were based on the conceptual foundation of island biogeography theory and designed with an old-growth patch in the center that maximized patch size and minimized patch isolation. Franklin and Forman (1987) compared several alternative cutting strategies for this region and measured the mean patch size and amount of edge for each alternative. They suggested that future forest-cutting practices should aggregate, rather than disperse cutting units, in order to reduce the amount of forest fragmentation and its accompanying edge effects. Several studies subsequently analyzed patterns of land use change for this region and compared alternative forest cutting patterns by using simulation models to project future patterns of landscape change and infer their ecological impacts (H. Li et al. 1993; Spies, Ripple, and Bradshaw 1994; Wallin, Swanson, and Marks 1994; Ripple, Hershey, and Anthony 2000).

One particularly striking result from this collection of studies was that spatial patterns of past and current land use can leave a long-term spatial legacy. For these forests—comprised of trees that live for hundreds of years and for which forest-cutting rotations are on the order of 40 to 80 years—the spatial pattern of cutting established decades ago may take a very long time to erase. Wallin, Swanson, and Marks (1994) simulated the development of landscape spatial patterns 300 years into the future, based on aggregated versus dispersed-patch forestry schemes. They compared the spatial metrics of edge density and mean interior forest patch size for these two scenarios and found, not surprisingly, that the aggregated pattern resulted in lower edge density and greater interior patch size than the dispersed cutting strategies.

Most interestingly, though, the authors projected edge density and mean interior forest patch size for forests that had initially been harvested using the dispersed-patch model, but were then switched after 20, 40, or 60 years of cutting to the aggregated scheme. In none of these scenarios did the forests that were initially harvested using the dispersed-patch model exhibit the favorable edge densities and patch sizes produced by the aggregated cutting model, even for simulations that were switched after only 20 years. The rather sobering take-home message from this exercise was that once a pattern of landscape change is established, it may take decades or centuries to nullify (at least in long-lived forest ecosystems), even if policies dictate a shift in spatial patterns of change relatively early in the process.

Although these examples have highlighted forests in the western United States, there are many published studies that use similar approaches to analyze spatial patterns of deforestation and project future landscape change. This method has been used to evaluate landscape change due to deforestation at sites all over the world, including China, Finland, Russia, Honduras, India, and Turkey, to name just a few.

Agriculture

Clearing native vegetation for agricultural production is the most significant change in the structure of ecosystems at the global scale. In particular, nearly one-fourth (24%) of Earth's terrestrial surface has been converted to cultivation, and more land was converted to cropland in the 30 years after 1950 than in the 150 years between 1700 and 1850 (Millenium Ecosystem Assessment 2005).

Tropical deforestation for slash-and-burn agriculture is one of the most highly publicized forms of landscape change, and it has resulted in rampant forest clearing in the past few decades. More recently, the burgeoning soybean and beef industries in places like Brazil have similarly converted vast areas of rainforest to agricultural lands (Nepstad, Stickler, and Almeida 2006). Dale et al. (1994) studied patterns of widespread deforestation in the state of Rondônia, Brazil, by farmers clearing forests to support slash-and-burn agriculture. Because these farms are typically only productive for six to eight years, tracts of land are abandoned relatively quickly and new forest areas are cleared. To provide guidance for future land use change, these researchers developed three scenarios of forest clearing and calculated landscape metrics (contagion, dominance, and fractal dimension) for each scenario. The scenarios differed primarily in the amount and timing of forest clearing and in whether plots were abandoned and then recolonized by subsequent farmers. Dominance and contagion indices followed similar trajectories over time, since the simulations involved shifts from one dominant and continuous cover type (forest) to another (cropland). Contagion was lowest for the best-case scenario, which had the lowest overall amount of forest clearing, and was higher for the other two scenarios, which involved clearing most of the forest after about 20 years. The authors suggested that the low contagion for the best-case scenario reflected a more heterogeneous landscape that would be likely to have higher habitat diversity and perhaps provide movement corridors for forest species. Fractal dimension was highest for the best-case scenario; in the other two scenarios it dropped abruptly, and then recovered, as land was converted from forest to cropland. These alternative scenarios of projected land use change revealed significant shifts in the spatial patterns (both of habitat area and configuration) of forest and cropland and provided guidelines for desirable trajectories for both biological diversity and human well-being.

Another biologically diverse landscape is the Cape Floristic Region (CFR) of South Africa, which hosts remarkably high levels of plant species richness and endemism and is a key target for conservation efforts. Rouget et al. (2003) described the extent of habitat transformation in this 88,000 km² region, based on current land use and projected future trends in land conversion due to agriculture, invasion by exotic species and urbanization. Agriculture (particularly plantation forestry) is the most likely form of land conversion in this region, and it is of concern, since it would destroy habitat for the region's endemic plants. The authors used satellite imagery to estimate the amount of land conversion that has taken place, based on the distribution and abundance of current land cover types. They projected future changes in land use for the next 20 years by employing a rule-based model that incorporated decision-making processes, as well as a statistical model that projected land use based on the distribution of existing land uses and correlated environmental variables. These projections were done for 16 categories of vegetation that comprise the flora of the Cape Region, so the analyses revealed which land cover types were likely to be most susceptible to conversion over the next 20 years. Although the authors did not specifically present indices of landscape spatial pattern in this paper, this spatially explicit, multiple threat approach would be quite useful for clarifying which vegetation types were likely to undergo conversion to agriculture and, hence, for targeting specific conservation actions (for more discussion of this topic, see chapter 11).

Similar analyses of land conversion from native habitat to agriculture have been conducted in Saskatchewan (Hobson, Bayne, and Van Wilgenburg 2002); in the United States for the period from 1950 to 2000 (D. Brown et al. 2005); for deforestation due to agricultural intensification in Oaxaca, Mexico (Gómez-Mendoza et al. 2006); and for forest loss and fragmentation in the eastern lowlands of Colombia (Madriñán et al. 2007). These latter authors measured fragmentation indices by comparing aerial photographs from 1939 to 1997.

Urbanization

Land conversion for urbanization is another dominant land use transition (Wear, Turner, and Naiman 1998; Duncan, Larson, and Schmalzer 2004), especially given continued shifts in human population from rural to urban areas (fig. 9.1). The world's urban population increased 10-fold from 1900 to 2000, and the number of large cities (those in excess of 1 million people) increased from 17 in 1900 to 388 in 2000 (Millenium Ecosystem Assessment 2005). In many locations around the globe, recent transitions to urban land cover follow an initial shift

Figure 9.1. Urban development near Fairfield, California, in valley grassland habitat. Photo by S. K. Collinge.

decades or even centuries ago from native habitat to agricultural cultivation. For example, in China, recent land conversion near large cities is transforming agricultural fields into urban development (e.g., Weng 2002; W. Li et al. 2005).

N.S. Williams, McDonnell, and Seager (2005) studied patterns of land use change due to urban development near Melbourne, Australia, looking specifically at the conversion of native grassland remnants to urban development. Using an approach similar to those of the other studies, they compared the distribution and spatial characteristics of grassland patches in a large study area (188,000 ha) using field data and existing GIS data sets on these grassland sites from 1985 and 2000. In addition to grassland conversion, this study had the further objective of determining the degradation of grassland patches due to invasions by exotic species. Their analysis showed that 44% of the grasslands present in 1985 had either been converted to housing or industry by 2000 (23%) or were degraded by the invasion of exotic species (>60% cover of non-native species over 21% of the grassland area). Not surprisingly, given the levels of habitat loss, the remaining grassland patches were fewer in number and were located farther away from other grassland patches in 2000 than they had been in 1985. Grassland patches were more likely to be converted to urban development if they were closer to the central business district, close to major roads, and far from streams. Interestingly, the predictor variables for grassland degradation were different from those for destruction. The perimeter-area ratio of grassland patches was significantly positively associated with the probability of degradation, suggesting that invasive species were facilitated along habitat edges (see chapter 5). In this system, land conversion for urban development resulted in characteristic spatial patterns of habitat loss and isolation that would be likely to increase threats of extinction for the native species that inhabit these endangered ecosystems. Despite further development pressure in this region, the results of this study may prove to be useful in guiding future development and avoiding particularly biologically valuable native grasslands.

Patterns of urban growth may be restricted or shaped by particular policies that limit building in certain areas, which may ultimately have positive effects on biological diversity. In Southern California, urban development has historically occurred primarily in the valleys and flatter areas, but as building sites have become scarcer, development occurs on steeper slopes. Syphard, Clark, and Franklin (2005) explored the impacts of slope restrictions on urban growth adjacent to the Santa Monica Mountains northwest of Los Angeles. They used a cellular automaton model (described in more detail below) to simulate landscape change and quantify its effects on the spatial characteristics of native vegetation patches. The authors computed general landscape metrics—such as patch area, the amount of edge, and the *largest-patch index* (the percentage of the landscape occupied by the largest patch, which is a measure of connectivity)—because, they argued, these were likely to be the most relevant for the persistence of native species. In these simulations, landscape metrics changed in somewhat erratic and non-intuitive ways over time, so it was crucial for the researchers to interpret these indices in the context of the particular spatial characteristics of this landscape. Key results from their simulations were that restrictions on the development of steep slopes (those >60%) reduced overall habitat loss, and also prevented fragmentation of the largest patch in the landscape. The scenario that allowed land with slopes greater than 60% to be developed fragmented the largest patch; it also reduced landscape connectivity for the mountain lion (*Felis concolor*), a large, wide-ranging carnivore in this region, by blocking a key movement corridor—a canyon that connected the Santa Monica Mountain Range with a large national forest to the north.

Projections of spatial patterns of land conversion to support urban and exurban growth have likewise been developed for other regions in the world. These models typically include analyses of both the spatial pattern and the implications of such land conversion patterns for ecological processes. For example, Theobald, Miller, and Hobbs (2005) developed such projections for the entire continental United States. Interestingly, a large number of recent models concern urban areas in China, including Beijing (Qi et al. 2004), Jinan (Kong and Nakagoshi 2006), Guangzhou (Yu and Ng 2007), and Haikou city (Tian, Yang, and Xie 2007).

In summary, spatial pattern indices have been developed and widely employed to characterize patterns of landscape change. Some are useful in some contexts, while others have been criticized for having little biological relevance. In the end, the choice of landscape indices depends on the goals of the study, and those must be clearly stated and rationalized. Moreover, if the goal of the analysis is to project the ecological responses of particular landscape patterns, then there should be justifiable, empirical links between spatial patterns and ecological processes. But in defense of spatial pattern indices, they are quite useful in distilling and describing broad-scale patterns of complex landscapes into a language that ecologists can understand. The next section, on modeling studies of ecological responses to landscape spatial pattern, is the flip side of this process.

POPULATION AND COMMUNITY RESPONSES

Models of ecological responses to landscape change have typically addressed three main issues relevant to conservation and management: (1) the spread of natural disturbances, such as fires, across fragmented landscapes (M. Turner et al. 1989; M. Turner and Romme 1994), (2) invasions of exotic species (D. Peters 2004; With 2004), and (3) the persistence of native species or communities (Hanski 1999; Beissinger and McCullough 2002). Given the focus of this book on population and community responses to landscape change, this chapter emphasizes models that feature population and community persistence. These models have been collectively referred to as *metapopulation models* (Hanski 1999; Akçakaya et al. 2004), although many now include multiple interacting species in metacommunities (Holyoak, Leibold, and Holt 2005), as well as in meta-ecosystems and metalandscapes (Loreau, Mouquet, and Holt 2003; With, Schrott, and King 2006; see also chapter 2). Further, modeling approaches to confront issues of species persistence treat spatial variation in different ways (fig. 9.2): there are (1) *spatially implicit* models, which include spatially structured data, but not particular spatial locations of habitat patches or populations, for example, two-population models in which two separate populations are connected by migration, but where the populations are not explicitly located in space (Hanski and Simberloff 1997; Hanski 1999); (2) *spatially explicit* models, in which individuals or habitat patches are

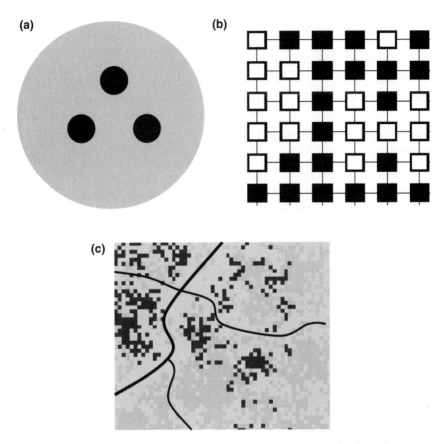

Figure 9.2. Diagrams of ecological models that represent spatial variation in different ways. (a) Spatially-implicit models, where ecological entities are patchily distributed, but the particular spatial location of the patches is not considered. (b) Spatially-explicit models, where individuals or habitat patches are situated in particular spatial locations. (c) Spatially-realistic models, in which the modeled landscape simulates the particular spatial arrangement of a real, spatially heterogeneous landscape. Figures modified from diagrams prepared by Chris Ray.

situated in particular spatial locations (e.g., Gilpin 1996; D. Peters 2004); and (3) *spatially realistic* models, which are spatially explicit models in which the modeled landscape represents the particular spatial arrangement of a real, spatially heterogeneous landscape (Hanski 1999). The models of landscape change described in the previous section were all spatially realistic models, in that they modeled the dynamics of real landscapes.

Spatially Implicit Models

Spatially implicit models of population or community persistence are typically designed to evaluate how migration between populations or communities influences the dynamics of the system. A *two-population model* is an extension of a model for the dynamics of a single population, in which the single population is now connected to another population via migration (e.g., Gilpin 1996; Hanski 1999). This type of metapopulation model is thus quite relevant in the study of habitat loss and fragmentation, since migration is often disrupted in fragmented landscapes. Some source-sink models of population dynamics (e.g., Pulliam 1988; see also chapter 2) are an example of spatially implicit models, since they evaluate the dynamics of two or more populations that exchange individuals via migration. With two populations, these models are usually tractable analytically, that is, they can be computed mathematically. If more populations are added to the system (*n*-population models), computer simulation is sometimes required. For example, the models of disease transmission among multiple populations in fragmented landscapes (e.g., Hess 1996a; McCallum and Dobson 2002), that were discussed in the previous chapter follow this spatially implicit, *n*-population model structure.

One of the first spatially implicit models applied to a single species in fragmented habitat was constructed by Lande (1987) for a generic territorial species, and then applied to the northern spotted owl (*Strix occidentalis caurina*) (Lande 1988), which occurs in old-growth forests of the western United States. In the 1987 generic model, Lande likened the patches of Levins's metapopulation model to the individual territories of a territorial species. He described local extinction as the death of an individual in a territory, and colonization of the patches as the settling of an individual in a suitable unoccupied territory. This analytical model was designed to determine the proportion of suitable habitat occupied by this territorial population at equilibrium. In conservation terms, the model could be used to project the minimum proportion of suitable habitat in a region that would be necessary to ensure population persistence or, conversely, the impact of habitat destruction or fragmentation on population size. This model was useful as a starting point for revealing how habitat loss may influence the probability of extinction for a territorial species in a patchy habitat. Lande's 1987 model admittedly did not incorporate several features that would have made it more realistic, but he encouraged spatially explicit elaborations of this basic model structure using information from real landscapes on territory spatial distribution and abundance. These spatially explicit (With and King 1999) and spatially realistic models were later constructed for the spotted owl (e.g., McKelvey, Noon, and Lamberson 1993; Lamberson et al. 1994) and for other species (see below).

Single-species metapopulation models have been expanded to two-species metacommunity models in order to test ideas about the persistence and dynamics of a species interaction (Hoopes, Holt, and Holyoak 2005). For example, Amarasekare (2004) devised a spatially implicit model of mutualistic interactions involving two species—a relatively mobile partner and a relatively sedentary partner—based on attributes of plant-pollinator mutualisms. Interactions among the mutualists were modeled in a fragmented environment in which patches were linked by the dispersal of the mobile mutualist (i.e., the pollinator) (Groom 1998, 2001; see also chapter 7). Dispersal of the mobile mutualist from source communities facilitated the rescue (sensu J.H. Brown and Kodric-Brown 1997) of sink communities in these simulations, but this effect depended on how dispersal affected the source community. If dispersal was by "surplus" individuals that would not have contributed to the growth rate of the source community, then regional persistence occurred. But if dispersal involved emigrants that would have contributed to the reproductive output of the source, then emigration could not exceed a particular threshold, or the entire metacommunity would collapse.

In the context of managing species interactions in fragmented habitats, Amarasekare (2004) concluded that the impact of dispersal on source communities is the key variable to consider. For instance, if the dispersal of a species from a source community has negligible impacts on the local reproduction of the source community, then it may be desirable to increase connectivity among habitat fragments via corridors, stepping stones, or matrix management, because the source-sink dynamics will enhance both local and regional persistence. But if the dispersal of a species from a source community causes reduced local reproduction in that community, then it may be desirable to regulate dispersal via management, so that it is low enough that the growth rate in the source community does not suffer. This modeling study highlights the importance of both the amount of dispersal from source communities and the consequences of dispersal for these communities when determining the regional persistence of communities. This emphasis on the amount of dispersal with regard to the persistence of interactions was discussed in experimental studies of predator-prey dynamics (Holyoak and Lawler 1996; see also chapter 4).

Spatially Explicit Models

Spatially explicit models specify the spatial locations of biological entities (e.g., individuals, populations, communities, and ecosystems) and evaluate how particular spatial configurations affect ecological processes. The application of cellular automata (CA) models to ecology has facilitated an extensive exploration of the role of landscape structure on the propagation of disturbance, species invasions, and the persistence of populations. This section explores the particular application of CA models (also called *grid-based* or *lattice* models) to project the probabilities of species persistence in fragmented landscapes. Our understanding of the consequences of habitat fragmentation via the use of CA models has been greatly advanced by the contributions of Kimberly With (With and Crist 1995; With 1997, 2002; With and King 1999, 2001; A. King and With 2002; With, Schrott, and King 2006) and Lenore Fahrig (Fahrig 1997, 2001, 2003; Bender, Tischendorf, and Fahrig 2003; Tischendorf, Bender, and Fahrig 2003; Bender and Fahrig 2005; Tischendorf et al. 2005). I illustrate this spatially explicit modeling approach by focusing on studies that have examined extinction thresholds, which is a critical issue in conservation biology. Because this work is only briefly summarized here, I encourage readers to explore this body of research further.

Landscapes can be modeled in a spatially explicit manner by representing different land uses or land cover types as black or white squares within a binary array of grid cells, i.e., as two types of land use (fig. 9.2b). Employing these grid-based, cellular automata models (see also chapter 2) became a key component of landscape ecological research through the pioneering work of investigators at the Oak Ridge National Laboratory in the 1980s (e.g., Gardner et al. 1987; M. Turner 1987; M. Turner and Gardner 1991). A special subset of cellular automata models is a type of neutral model in which maps are generated via random processes or spatially correlated processes (using algorithms from fractal geometry) but are not based on landscape patterns derived from known ecological processes. These *neutral* or *null models* are thus analogous to null models in other areas of ecology (e.g., Connor and Simberloff 1979). They can be compared to real landscapes in which patterns have been generated by ecological processes to evaluate key disparities among random versus ecologically generated patterns. Gardner and Urban (2007) recently provided a historical review of neutral models and their current status, and they describe the development of a new analytical tool that is more realistic than neutral models, but that is not as data-demanding as the spatially explicit, individual-based models discussed below.

The use of neutral models has profoundly influenced our understanding of the effects of habitat loss and fragmentation (With and Crist 1995; With 1997; Fahrig 2001). In neutral models, grid-based maps typically represent cells as either suitable habitat or unsuitable habitat. Each cell has a state, which could, for example, be the number of individuals of a particular target species in that cell. *Individual-based models* place individuals within cells, and these individuals are followed spatially over the course of their lives, behaving and reproducing according to rules specified by the modeler (Gilpin 1996). Most importantly, the amount and spatial configuration of habitat can be controlled and varied independently within these simulation models. One area of particular interest is the presence of *critical thresholds*. Percolation theory predicts that there are thresholds where small changes in the number of grid cells that are occupied result in a transition from the grid being disconnected to its being connected (the *percolating cluster* discussed in chapter 2). Analogous to fragmented landscapes, there may be critical thresholds where a further small loss of habitat abruptly reduces landscape connectivity. Neutral models have explored a key, two-part question related to habitat loss and fragmentation—whether there are extinction thresholds where small changes in habitat amount or configuration result in sharp declines in species abundances, and what characteristics of species or landscapes may influence those thresholds.

With and King (1999) explored extinction thresholds by combining the spatially implicit metapopulation model of Lande (1987), discussed above, with neutral landscape models in which the amount and spatial configuration of habitat could be varied independently. They modified Lande's model by varying the spatial distribution of habitat (from randomly distributed, as in Lande's model, to landscapes differing in their degree of clumping, or contagion), shifting the searching behavior of the territorial species (from random searching to searching only adjacent cells in the grid-based model), and exploring a range of reproductive outputs. The goal was to determine whether these modifications would influence the extinction threshold, which is the minimum amount of habitat in the landscape at which the population becomes extinct. Model results showed that, in general, populations in the clumped landscapes were able to persist over a wider range of values of habitat loss than was predicted by Lande's (1987) model of randomly distributed habitat. Thus the extinction threshold for most permutations of the model either occurred at a lower overall amount of habitat than in Lande's model or did not occur at all. By varying the degree of spatial aggregation of the habitat but keeping the habitat amount constant, With and King (1999) showed that a more clumped distribution of habitats was generally more favorable for species persistence than more fragmented habitats.

Given this observance of extinction thresholds, Fahrig (2001) explored the conditions under which extinction thresholds are most likely to occur. She constructed a neutral landscape model and incorporated four factors thought to influence the extinction threshold into it—namely, the reproductive rate of the organism, the rate of emigration from suitable habitat, the spatial arrangement of habitat (fragmentation), and the quality of the matrix (i.e., the probability of species' survival in unsuitable habitat)—to examine the relative importance of each of these factors on the extinction threshold. This was an individual-based model in which reproduction, movement, and the survival of individuals was tracked across different landscapes. The results of these simulations showed that the extinction threshold varied widely, from 1% to 99% of suitable habitat, depending on the relative values of the four factors that were modeled. Reproductive rate and the rate of emigration had the highest relative impact on the extinction threshold, followed by a moderate effect of matrix quality, and very small effect of habitat configuration. Because reproductive and emigration rates are characteristic of organisms, they cannot be easily manipulated by management actions. Matrix quality, however, could be improved by management strategies that increase the survival of organisms moving through these areas, and this would be likely to decrease the extinction threshold (fig. 9.3). Fahrig concluded that this may be the most promising means of enhancing population persistence in human-modified landscapes. Finding that habitat spatial arrangement has relatively little impact on the extinction threshold is in contrast to With and King's (1999) results, but it is consistent with other work by Fahrig which suggests that habitat amount, rather than spatial configuration, is the strongest influence on species persistence (Fahrig 1997, 2003).

The discrepancy between model results regarding habitat spatial configuration (fragmentation per se)—whether it strongly (With and King 1999) or weakly (Fahrig 1997, 2001, 2003) influences the extinction threshold—may depend on subtle differences in model assumptions and approaches (Fahrig 2002). Based on a comparison of these methods and outcomes, Fahrig concluded that habitat fragmentation is likely to influence the extinction threshold when habitat loss does not result in an increase in the movement of individuals into the matrix. For example, if an organism is constrained strictly to one type of habitat in the landscape, then as habitat is lost and fragmented, the patches of habitat become smaller, and the probability of local extinction increases. With no movement across the matrix, the overall extinction probability depends on the size of the largest patch. Thus reducing fragmentation in this landscape would increase the size of the largest patch and thereby increase persistence time.

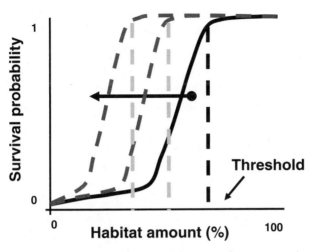

Figure 9.3. A diagrammatic representation of a reduction in the extinction threshold for a species. The dark black line represents the original relationship between the amount of habitat and the survival probability for a modeled species. The extinction threshold is the amount of habitat (*x*-axis) required to ensure survivability, so it occurs where the survival probability reaches 1.0 (*y*-axis). Management activities, such as the improvement of matrix quality, may shift this curve to the left, as shown by the two sets of dashed dark gray lines. For these shifted curves, the extinction thresholds are reduced (the dashed light gray vertical lines). Based on the modeling results of Fahrig (2001).

Spatially Realistic Models

Spatially realistic models are similar to spatially explicit approaches in that particular spatial locations are specified. Spatially realistic models go a step further, however, and use habitat configurations from real landscapes (fig. 9.2c). The goal of these efforts is typically to project probabilities of species persistence in particular landscapes, as well as to explore the effects of alternative strategies of habitat removal or configuration on these probabilities. Such models typically use data on the presence-absence or abundance of a species in particular spatial locations, and they are variously referred to as incidence function models (Hanski 1999), state transition models (Sjögren-Gulve and Ray 1996; S. Harrison and Ray 2002), individual-based models (Gustafson and Gardner 1996; Zollner and Lima 2005), and spatially explicit population models (Pulliam, Dunning, and Liu 1992; Akçakaya, McCarthy, and Pearce 1995; Lindenmayer and Possingham 1995; Haines et al. 2006).

The use of *incidence-function models* for projecting metapopulation dynamics was developed and has been widely used by Hanski (1994, 1999), especially in the context of long-term, spatially extensive studies of the dynamics of the Glanville fritillary butterfly (*Melitaea cinxia*) and its natural enemies, which inhabit dry meadows of the Åland Islands, Finland. Suitable habitat for this butterfly species consists of patchily distributed meadows that contain larval host plants. Incidence-function models are relatively easy to parameterize, because they require a map of the spatial distribution of the habitat patches, as well as observations of site occupancy by the target species during one or more years. So, with a snapshot of species occurrence patterns, it is possible to estimate model parameters and use these to simulate dynamics over time, as well as to vary the spatial configuration, amount, or quality of habitat and observe their effects on metapopulation persistence. Results from the incidence-function modeling exercises using data from the Åland Island study system have revealed several important aspects of butterfly metapopulation dynamics. For example, patch occupancy could be accurately predicted in relation to patch size and isolation for some, but not all, parts of the study area (Hanski et al. 1996), prompting researchers to investigate additional environmental variables that may influence patch occupancy. A subsequent analysis showed that incorporating patch quality and more complex measures of patch isolation did not improve model predictions (Moilanen and Hanski 1998; but see Fleishman et al. 2002 for an example of significant effects of patch quality on butterfly metapopulation dynamics).

One of the Glanville fritillary's natural enemies is a parasitoid (Hymenoptera: Braconidae: *Cotesia melitaearum*) that, in turn, is infected by its own hyperparasitoid (Hymenoptera: Ichneumonidae: *Gelis agilis*) (Lei and Hanski 1997). Field observations showed that the distribution of the parasitoid was positively associated with host population size and patch area and negatively associated with the distance from other parasitoid populations. Using incidence-function models to model parasitoid population dynamics revealed that the parasitoid metapopulation was not at equilibrium, probably because the high turnover of hosts in the patches resulted in a lag in parasitoid occurrence as they tracked the spatially dynamic host metapopulation.

Hanski's research group has employed Bayesian approaches to account for the uncertainty in estimating the parameters for metapopulation dynamics in this system (O'Hara et al. 2002) and to rank management scenarios that would minimize the risk of metapopulation extinction (Dreschler et al. 2003). The combination of a spatially rich data set and a straightforward modeling approach has contributed greatly to clarifying the factors influencing metapopulation dynamics of single and multiple species in a patchy landscape.

In cases where a patchy landscape has been surveyed repeatedly over time, *state-transition models* can be used to identify factors that may be responsible for colonization and extinction events. For example, Sjögren-Gulve and Ray (1996) used this approach to model the dynamics of metapopulations of the pool frog, *Rana lessonae*, which occurs in permanent ponds in forests along the east coast of Sweden. In this case, a state transition refers to colonization or extinction events that occur in pools between surveys. For instance, an extinction would include a pool that was occupied in one census and then was unoccupied in the next. The authors used logistic regression to model environmental factors that may explain four possible state transitions, that is, comparing continually occupied ponds, ponds that were colonized, ponds that went extinct, and ponds that were continually unoccupied. Models of this system revealed that forestry practices affect pool frog distribution; in particular, large ditches that have been constructed to drain clear-cut areas and increase accessibility to wet areas have hampered the ability of frogs to move across the landscape. This is a clear example of Fahrig's (2001) conclusion regarding the importance of matrix quality as a determinant of population persistence in patchy landscapes. This state-transition approach was similarly used by S. Harrison and Ray (2002) for five plant species in California serpentine outcrops. Their simulation results showed that long-term metapopulation persistence was highly likely, and that persistence was only weakly dependent on patch connectivity. This result may be because these plant species maintain a relatively long-lived seed bank, which may buffer populations against stochastic variation in environmental conditions that affect annual patch occupancy.

In *spatially explicit population models* (SEPMs in Dunning et al. 1995; Turner et al. 1995; also called *n*-population simulation models in Hanski 1999), each local population has its own dynamics, and migration among populations is modeled explicitly, often including some details about specific behavioral decisions that affect movement. These models are also spatially *realistic*, since they usually aim to make projections about the dynamics of a specific metapopulation within the context of habitat spatial distribution in a particular landscape. Although they can be powerful tools for the management of a particular species (and especially for endangered species), relatively few of these models have been constructed, because they require intensive and extensive data about the system, information which is often not available. For example, SEPMs have been parameterized for the Florida scrub jay (Root 1998), Eurasian badger (*Meles meles*) in the Netherlands (Apeldoorn et al. 1998), Caribbean spiny lobster (*Panulirus argus*) in the Florida Keys (Butler 2003), Siberian tiger (*Panthera tigris*) in the Russian Far East

(Carroll, Miquell, and Dale 2006), and the herb *Boltonia decurrens*, an endemic plant that occurs within the Illinois River floodplain (Mettler-Cherry, Smith, and Keevin 2006).

Spatially explicit models for particular species have been useful in directing conservation efforts and allowing the informed selection of alternative management scenarios. For example, Cox and Engstrom (2001) evaluated alternative scenarios of the spatial configuration of conserved lands for the endangered red-cockaded woodpecker (*Picoides borealis*) in a 4000 km² region of old-growth long-leaf pine forests in northern Florida and southern Georgia. The authors combined a stochastic demographic model with geographically based information on habitat distribution to weigh the costs and benefits of various strategies for land acquisition or protection through conservation easements. The analysis provided guidance for choosing among strategies that were likely to achieve long-term population persistence.

Despite their utility in particular situations, spatially explicit population models have also been criticized for requiring too much information and being quite sensitive to uncertainty in parameter estimation (Ruckelshaus, Hartway, and Kareiva 1997; Beissinger and Westphal 1998). Ruckelshaus, Hartway, and Kareiva (1997) suggested that *less* detailed models might provide a better fit between the complexity of the model and the quality of data available to incorporate into the models. In their reflection on the status of research on fragmented populations, Melbourne et al. (2004) questioned whether more complex models are needed in all instances, and urged researchers to confront the same data using both complex and simple models to determine if the more complex models necessarily provide better answers. To address the second criticism, more recent models have incorporated estimates of uncertainty in model parameters by using Bayesian approaches (O'Hara et al. 2002; Snäll et al. 2005).

Combining Landscapes and Metapopulations

The beginning of this chapter introduced two general categories of models related to habitat loss and fragmentation: models that project patterns of future landscape changes, and models that project population and community responses to landscape change. These have generally been pursued as separate efforts, since they are sufficiently complicated in themselves. When combined, they may provide a particularly powerful tool for estimating the consequences of landscape change for population and community persistence. The issues of data-intensity and uncertainty still need to be addressed, but some progress has been made in this realm (e.g., see MacKenzie et al. 2006 for a treatment of estimating detectability).

Efforts to combine landscape change with population models are a promising tool for understanding species persistence in changing landscapes. For example, Akçakaya et al. (2004) combined data on forest dynamics and population dynamics to assess the viability of the sharp-tailed grouse (*Tympanuchus phasianellus*) in the Pine Barrens region of northwestern Wisconsin. This approach allowed the incorporation of landscape dynamics, such as forest succession, disturbances (e.g., fire), and silviculture. Wilcox, Cairns, and Possingham (2006) brought together information on habitat disturbance and recovery, as well as patch colonization and extinction parameters, into a spatially explicit metapopulation model. Hilty, Lidicker, and Merenlender (2006) combined models of vineyard expansion in northern California with models of wildlife-habitat relationships in order to estimate the probable effects on particular species of wildlife. And studies of the brown creeper in old-growth forests of Ontario, Canada, assessed population viability in the context of alternative forest harvest schemes (Wintle et al. 2005; Chisolm and Wintle 2007).

SYNTHESIS

Models that simulate fragmented landscapes to assess the ecological consequences of habitat loss and fragmentation have been exceedingly useful in informing and guiding research and conservation efforts. Models have described patterns of landscape change and have projected the responses of species or communities to varying habitat spatial configurations. Although all models must make simplifying assumptions, they provide insights into what can happen in nature in situations where it is impossible to conduct systematic observations or controlled experiments. One of the most important findings from modeling studies, identifying the factors that influence the extinction threshold for species in fragmented landscapes, is one that we most likely would not have discovered through empirical studies. Because the modeling environment allows the examination of multiple factors that are critical to species persistence, such as matrix quality, the results from these studies provide novel interpretations of existing field data and can guide future management efforts.

The future of this field promises to include ever-more-sophisticated modeling efforts. Increasingly, modelers explicitly incorporate uncertainty in parameter estimation into their simulation models. One aspect of uncertainty that is being recognized more and more is in our limited human ability to detect species when they are actually present in the landscape. Also, more modeling studies are beginning to blend population and landscape approaches, which is an exciting development. There are hints that moving from metapopulation models to metacommunity models may produce non-intuitive outcomes. As we simultaneously consider the effects of multiple species, it may help us to further define data gaps and factors that we need to study more.

As Beissinger et al. (2006) concluded in their review of integrating the efforts of field biologists with modeling, "the best conservation decisions will occur where cooperative interaction enables field biologists, modelers, statisticians, and managers to contribute effectively" (p. 1).

CHAPTER TEN

Restoration

Habitat restoration is primarily the initiation and
coalescence of *growing habitat fragments.*
— *Daniel H. Janzen (1988)*

A few years ago I had the good fortune to accompany the illustrious ecologist
Dan Janzen on a field trip near his home in the tropical dry forests of Guanacaste, Costa Rica. As we drove toward Santa Rosa National Park through a landscape of scattered forest fragments and cattle pastures with the occasional solitary tree, Janzen spoke passionately about the rampant destruction of the dry
forests and the thousands of species, including his favored 3000 or so species of
moths and butterflies, which depended on these forests. He bemoaned the fact
that the alluring tropical wet forests seemed to get all the attention from the public and from conservationists, when in reality the tropical dry forests were in
more immediate danger of disappearing. Since the arrival of European colonists
to Costa Rica in the 1500s, thousands of hectares of this towering forest have
been cleared and planted with jaragua (*Hyparrhenia rufa*), an African pasture
grass that provides nutritious forage to support the booming cattle industry in
this region of northwest Costa Rica.

The bus stopped, and we walked along the edge of a pasture dense with
jaragua to a small tree that bordered the pasture. That's when Janzen launched
into the subject of how to reverse the loss of these splendid forests. In addition to
protecting remaining tracts of dry forest, Janzen mentioned the prospects for

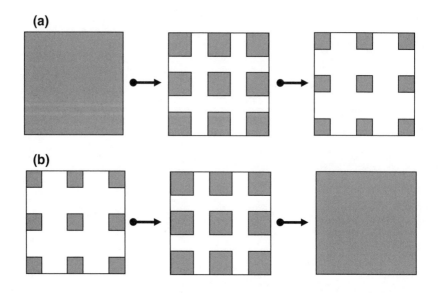

Figure 10.1. Conceptual diagrams of (a) habitat loss and fragmentation and (b) the reverse process, habitat restoration. The gray areas are native habitat, and the white areas are transformed habitat.

restoration. He noted that the tropical dry forest did not appear to be recovering on its own in southern Guanacaste, since most of the forest had been cleared, and there were only small, scattered remnants that remained. In the central part of Guanacaste, however, where we stood, more forest was left and the landscape was less fragmented. In this landscape with relatively higher, more continuous forest cover, there were ample sources of seeds from forest trees to colonize abandoned pastures and grow up into forest. But in the landscapes further south that had little forest cover, there were few sources of propagules, and natural reforestation was stalled. I was struck at the time by this anecdote, since it clearly revealed a critical aspect of habitat recovery and restoration that had not yet been widely appreciated by ecologists. But as Janzen (1988) had already pointed out, just as we consider the effects of habitat loss and fragmentation on species declines, we should also consider these spatial effects when the habitat fragments are growing via recovery or restoration (fig. 10.1).

Because habitat destruction and degradation are the leading causes of declines in biodiversity (Wilcove et al. 1998), it logically follows that efforts to restore native ecosystems to historic conditions should enhance native diversity. Mark Vellend (2003), who has studied the recovery of deciduous forests in the northeastern United States, emphasized that "theoretical and empirical research on habitat loss and fragmentation has focused almost entirely on species dynamics in remnant habitat patches . . . If remnant habitat patches provide the source of colonists for restored patches, it follows that the extent of habitat loss prior to abandonment should influence reestablishment of populations and communities via the reduction of potential sources of colonists. *This suggests a potentially important link between two major themes in ecology and conservation biology: habitat loss and habitat restoration*" (p. 1158; italics mine).

Identifying the ecological consequences of habitat loss and fragmentation has been a major theme of conservation biology for the past 40 years, and much of this literature has been covered in the past several chapters. Studies of habitat loss and fragmentation are essential to conservation biology, where the goal is to protect, preserve, and maintain biological diversity and ecosystem services. But the restoration of fragmented landscapes is an increasingly important strategy for conservation biologists, because it may enlarge the available habitat for native species, as well as bring back vital ecosystem services or *natural capital* (A.P. Dobson, Bradshaw, and Baker 1997; Clewell 2000; Daily and Ellison 2002; Clewell and Aronson 2006). Theoretical and empirical studies of habitat loss and fragmentation can guide both restoration and conservation activities, because the spatial position of restored habitats may strongly influence the success of restoration actions for native species and communities. Indeed, the spatial considerations discussed throughout this book, including patch size, isolation, and landscape context, are likely to be highly relevant to species persistence in restored landscapes (Fry and Main 1993; Saunders, Hobbs, and Ehrlich 1993; MacEachern, Bowles, and Pavlovic 1994; P. White 1996; Maschinski 2006). For example, because dispersal success is reduced in fragmented landscapes, the colonization of restored habitats is often hindered, so restoration efforts must carefully consider the spatial position of restored areas within the landscape.

There are literally hundreds of restoration projects throughout the world that have been initiated to recover endangered species, restore habitats degraded by human activities, reduce the cover of invasive species, prevent soil erosion, enhance water quality or quantity, provide resources for local communities, or promote carbon sequestration. There is much to be learned from these efforts regarding how to successfully introduce ecological processes or species to degraded sites. Given the emphasis of this book, the focus in this chapter will be narrowed to include restoration efforts that explicitly focus on the spatial aspects of restoration, that is, how the amount and spatial arrangement of habitat patches influences ecological processes. Several excellent works are available for those who would like to learn more about theory and practice relevant to other key topics in the field of restoration ecology, including the use of successional theory and community assembly theory in designing restoration efforts: Weiher and Keddy (1999); T. Young, Chase, and Huddleston (2001); Daily and Ellison (2002); Temperton et al. (2004); van Andel and Aronson (2005); T. Young, Petersen, and Clary (2005); and Falk, Palmer, and Zedler (2006).

This chapter considers the conceptual frameworks relevant to spatial considerations in restoration and reviews models that project how the amount and configuration of habitats might affect the success of restoration for populations and communities, based largely on colonization and extinction dynamics. Next, it reviews case studies from several different landscapes where spatial aspects in restoration have been considered. These examples include observations and experiments designed to understand the spatial factors that influence the regeneration of understory herbs in north-temperate deciduous forests, the colonization of closed landfills in the New York metropolitan area, tropical forest restoration in southern Costa Rica, the restoration of riparian vegetation in California, wetland community restoration, and the restoration of oak savannas and prairies for butterflies and moths.

CONCEPTUAL BACKGROUND

In very simple terms, the process of restoration may be viewed as the reverse of the processes of habitat loss and fragmentation—simply rewind the tape and voilà!—restoration is accomplished. I say this somewhat facetiously, however, since it is far from clear if restoration is truly just the flip side of habitat loss and fragmentation. For example, there may be significant lags in the colonization of restored habitats (e.g., Tilman, Lehman, and Kareiva 1997), especially if there are critical thresholds in the amount of available restored habitat in the landscape that are required for colonization. Or there may simply be few (or no) propagules available for dispersal to restored sites, particularly if source sites are located far away, which would require that humans intervene to introduce appropriate species to restored sites. This type of intervention is done routinely, which suggests that in many cases, unaided dispersal to restored sites is not feasible. Hilderbrand, Watts, and Randle (2005) referred to this phenomenon—the lack of natural dispersal—as the *Field of Dreams* myth of restoration ecology. This moniker is a reference to the nostalgic 1989 movie starring Kevin Costner, who played an Iowa farmer driven to build a baseball diamond in his cornfield when he hears a mysterious voice say "If you build it, they will come." The myth regarding restoration is that if the physical structure of a habitat is restored, species will colonize on their own. In reality, there may either be no suitable sources, or there may be significant barriers to dispersal that will prevent the colonization of even highly suitable sites. Further, populations may be limited not by available habitat, but by demographic constraints, such as reproductive potential (Schrott, With, and King 2005), in which case habitat restoration may not be that effective in ensuring population recovery.

Given all of those caveats, if dispersal among remnant habitat patches limits species persistence, based on the amount or spacing of habitat patches, then the theories that have been discussed, especially island biogeography theory and metapopulation theory (see chapter 2), should be applicable to restoration. Because Levins's (1969) original formulation of metapopulation theory derived from the problem of how to establish persistent populations of biological control agents in agricultural fields, it provides an analog to the restoration of declining or extinct populations. Movement among habitat patches is a key component of both theories, as are colonization and extinction events. In situations where the probability of dispersal can be enhanced through restoration efforts, theory predicts that species persistence should also be enhanced. In addition to island biogeography theory and metapopulation theory, percolation theory may be particularly relevant to restoration when the objective is to restore connectivity among otherwise isolated patches (J. Williams and Snyder 2005; see also chapter 2).

RELEVANT SPATIAL MODELS

In drawing conclusions from their mathematical model of habitat destruction and competitive coexistence (a variation on the metapopulation theme), Nee and May (1992) noted its pertinence to restoration: "The model can be looked at from viewpoints other than those adopted here. The consequences of patch addition may be of more relevance to European ecology, if the farmland that is being removed from agriculture over the next decades is not simply paved over" (p. 39).

Despite their recognizable relevance, the application of spatial population and community models to problems of habitat restoration has a much shorter history than their use in projecting ecological changes in response to habitat loss and fragmentation. Tilman, Lehman, and Kareiva (1997) provided a useful starting point in their consideration of restoring habitat in fragmented landscapes. They used a spatially explicit cellular automata model (see chapter 9) to simulate habitat destruction and restoration. Their model specified (1) the random selection of habitat patches to be restored, and (2) no reintroduction of target species to re-

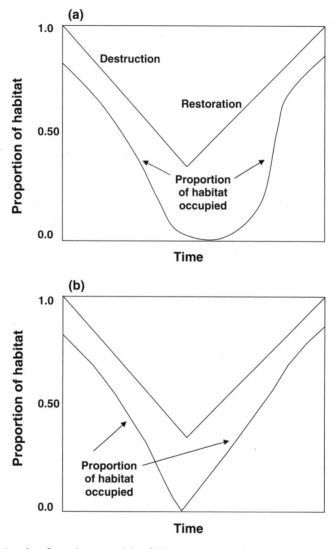

Figure 10.2. Results of simulation models of habitat restoration following destruction. The proportion of habitat, plotted on the y-axis, is the amount of habitat available in the landscape. In both scenarios, the top line indicates the decline of available habitat during a period of habitat destruction, and then its increase during a period of habitat restoration. The bottom line indicates the proportion of habitat occupied by a species during the destruction and restoration phases. (a) The random restoration of sites. Based on results from Tilman, Lehman, and Kareiva (1997). (b) The restoration of sites adjacent to currently occupied sites. Based on results from Huxel and Hastings (1999).

stored habitat. Even though random selection may seem unreasonable at first glance, restoration in the real world is a rather opportunistic enterprise. Sites selected for restoration are often chosen because they are available for purchase, are inexpensive, or are accessible, but rarely are they chosen strictly on the basis of their biological potential. The results of Tilman, Lehman, and Kareiva's (1997) simulations showed a lag in the proportion of habitat occupied by the target species during restoration (fig. 10.2a). In other words, the model showed that a significant number of patches had to be restored before a positive effect on habitat occupancy (population abundance) occurred. Huxel and Hastings (1999) further developed this idea and showed, via similar simulation models, that the spatial location of restoration sites critically influenced the proportion of habitat occupied. In the simulated scenario that specified *non-random* placement of restored sites adjacent to occupied patches (the *adjacent scenario*), there was no lag in occupancy during restoration (fig. 10.2b). When restored sites were adjacent to occupied sites, the species recovery was rapid, and a greater proportion of the sites were occupied early in the process of restoration. This obviously suggests a less risky scenario to ensure population persistence.

Habitat corridors have been widely discussed as a means of connecting fragmented habitats (see chapters 3 and 6), and they may also be constructed to enhance connectivity in conjunction with restoration projects. J. Williams and Snyder (2005) used a neutral model in the context of percolation theory to identify optimal corridor placement in fragmented landscapes. Recall that neutral models are comprised of a rectangular lattice in which cells are designated as either habitat or non-habitat, and a percolating cluster is defined as a collection of connected habitat cells that extends from one side of the lattice to the other. The authors varied the amount of habitat (p) in the simulated landscape and used either the 4-, 8-, or 12-cell neighbor rule (see chapter 2) to define connectivity in each simulation. For each combination, they estimated two types of corridors: the *geometric-shortest path*, which was defined as the shortest continuous corridor across the lattice, and the *least restoration path*, which was the path that involved the restoration of the fewest cells. Simulations showed that either of these shortest-path connections resulted in far fewer habitat cells that required restoration than would be the case for randomly located restoration sites, which is analogous to the adjacent-scenario results of Huxel and Hastings (1999) for habitat patches. Additionally, the length of the shortest paths required for connectivity generally declined with the amount of habitat in the landscape; with more habitat, the number of restored cells required was lower, but the least restoration path required fewer restored cells to achieve connectivity than did the geometric-shortest path. Although this modeling framework imposed many assumptions on the system that may be unrealistic (based on empirical studies of animal and plant use of corridors), this method of identifying corridors may be useful for deciding among alternative corridor configurations when planners are faced with restoring landscape connectivity in real landscapes.

The models discussed so far adopt the logical assumption that habitat restoration should reverse declines in species that have suffered from the effects of habitat loss and fragmentation. But what if that assumption is not always valid? Before restoration actions are initiated, it would probably be wise to discern whether populations that have declined due to habitat loss would be likely to be recovered by habitat restoration. To test this essential assumption, Schrott, With, and King (2005) used a spatially structured demographic model in combination with a neutral landscape model to ask whether habitat restoration would benefit populations with varying sensitivities to habitat loss and fragmentation. The model was based on the features of three generic species of migratory songbirds that differed in their sensitivity to habitat edges. Parameters for the highly edge-sensitive species dictated that they would have the largest decline in reproductive success with increasing edge-area ratio (analogous to the perimeter-area ratio) of habitat fragments, while medium-sensitivity birds were modeled to exhibit intermediate declines, and low-edge-sensitivity birds to have the lowest declines. The model projections indicated that habitat restoration was most effective for (1) species with low-to-moderate edge sensitivities and (2) in landscapes that were not highly fragmented. For populations of species that were highly edge-sensitive, or any species in heavily fragmented landscapes, in order to be successful, restoration needed to be imposed well before the population was deemed to be vulnerable to extinction. In most real-life situations, habitat restoration does not occur until a population is at risk of extinction, but the model results suggest that some populations cannot be recovered at this point solely through habitat restoration. This is because habitat loss and fragmentation, and the accompanying negative edge effects, have compromised the species' reproductive inertia. The authors argued that in these situations, reversing declining populations would need to rely more on improving the demographic potential—via management actions that enhance survival, fecundity, or dispersal success—than on habitat restoration. This is an intriguing result and it suggests the need for further exploration of the conditions under which restoration is likely to positively influence population recovery.

The simulation models just discussed were constructed to explore generalizations for species with generic life-history traits; they were general models that could apply to many different species or situations. But in some cases, comparable models have been designed and evaluated to address restoration needs for distinct species in particular settings. For example, the notion that the spatial position of restoration sites is likely to influence restoration success may be especially true for rare species that are habitat specialists and have relatively limited dispersal. One such species is the valley elderberry longhorn beetle, or VELB (Huxel and Hastings 1999; Collinge et al. 2001; Huxel et al. 2003). The VELB is a threatened species that inhabits elderberry bushes that grow in riparian woodlands along rivers and streams in California's Central Valley. Repeat surveys of VELB populations between 1991 and 1997 (Collinge et al. 2001) suggested that the colonization of suitable habitat patches occurred within, but not between, separate drainages. Mitigation or habitat restoration is routinely used as an aspect of VELB recovery throughout the Central Valley (U.S. Fish and Wildlife Service 1984; Talley, Wright, and Holyoak 2006), and it involves planting elderberry bushes along rivers and streams. Because most habitat loss occurs near Sacramento as a result of urban development, most mitigation sites have been located relatively near that city.

To investigate the role of mitigation site placement on population persistence, Gary Huxel developed an individual-based, spatially explicit simulation model to examine demographic and stochastic factors that may influence the viability of VELB populations. Similar to the approach used in Huxel and Hastings (1999), he also explored the role of restoration site location on overall population persistence (Huxel and Collinge, unpublished data). In addition to significant effects of varying demographic parameters on population persistence, the model revealed that mitigation site placement also had a significant effect on VELB site occupancy (table 10.1). In particular, when sites were located adjacent to occupied sites throughout the 250 km long Sacramento Valley, regional population persistence was greater than that for the present configuration of mitigation sites clustered near Sacramento. The results from these as-yet unpublished data support the generic findings of Huxel and Hastings (1999), namely, that the placement of mitigation sites adjacent to occupied sites may influence regional population persistence. Consistent with these model projections, Talley, Wright, and Holyoak (2006) reported that surveys of mitigation sites for VELB along the Sacramento River showed that a greater percentage of VELB exit holes were found at sites in close proximity to existing riparian vegetation, which is where extant VELB populations would occur.

Models such as the one described for the VELB can provide recommendations for a choice of restoration sites, which may greatly enhance conservation effi-

TABLE 10.1

ANOVA results of a spatially explicit, individually based simulation model for the valley elderberry longhorn beetle (VELB) that varied the parameters associated with dispersal distance, juvenile survival, and mitigation site placement to project the proportion of sites occupied by the VELB. Mitigation site placement significantly influenced site occupancy ($P < 0.05$), and the effects of dispersal distance on site occupancy depended significantly on juvenile survival (dispersal distance × juvenile survival, $P < 0.001$). SYSTAT 7.0 for Windows (SPSS Inc., Chicago, IL) was used for the data analyses.

Source of variation		Site occupancy		
	df	MS	F-ratio	P
Dispersal distance	1	6055.2	111.7	0.001
Juvenile survival	1	9073.8	167.4	0.001
Mitigation site placement	1	231.2	4.2	0.043
Dispersal distance × Juvenile survival	1	2101.2	38.8	0.001
Dispersal distance × Mitigation site placement	1	1.2	0.02	0.880
Juvenile survival × Mitigation site placement	1	1.2	0.02	0.880
Dispersal distance × Juvenile survival × Mitigation site placement	1	45.0	0.83	0.365
Error	72	54.2		

ciency in a world of limited financial resources. Schultz and Crone (2005) constructed a model for another rare species, the Fender's blue butterfly (*Icaricia icarioides fenderi*), which occupies scattered patches of upland prairie habitat in the Willamette Valley of Oregon. Because most of the prairie habitat has been converted to agriculture, the long-term recovery of this species will be likely to involve considerable habitat restoration (Schultz 2001). Field studies of this butterfly over several years revealed particular aspects of its behavior and demography that led the authors to recommend a couple of simple, straightforward rules for prioritizing restoration sites. These rules were that sites chosen for restoration should be within 1 km of an occupied site (based on butterfly movement behavior) and at least 2 ha in size (based on butterfly demography). Schultz and Crone (2005) compared this set of simple rules to the results obtained from two simulation models to assess whether the rules made sense in terms of long-term butterfly persistence. They constructed two models—an incidence function model and a spatially explicit individual-based model—to simulate population dynamics in the existing network of habitat patches. The incidence function model (see chapter 9) was a useful approach for modeling the long-term steady-state occupancy of habitat patches in the network. The individual-based model was useful for identifying the short-term population dynamics of butterflies within the network of patches. To identify high-priority sites for restoration, they used the individual-based model to simulate population dynamics in 146 separate iterations, with the addition of one potential restoration site (out of 146 possibilities) per iteration.

The simulation results showed consistency between the simple rules and the more complex models—all converged on the finding that the restoration of large connected patches would provide the greatest benefit to the Fender's blue butterfly population. More specific results from the individual-based model revealed that small connected patches would have a higher restoration value than large isolated patches. Output from the incidence-function model, however, suggested that patch size and connectivity were equally important criteria for choosing restoration sites. The authors reasoned that these different results emerged from assumptions made with the two different model types, where the incidence-function model predicted long-term, stochastic, steady-state conditions and the individual-based model projected colonization dynamics over a shorter period (25 years were modeled in the simulations). Schultz and Crone concluded that restoration sites should be prioritized based on the results of shorter-term simulations, which meant that their recommendations were to select sites that were less isolated, regardless of size, since butterflies would be more likely to colonize nearby sites. More generally, they concluded that restoration should prioritize less isolated sites in situations where colonization dynamics are likely to be a key aspect of population persistence. As is always the case in any study, other features of butterfly habitat could have been included in models designed to set priorities for restoration, but this analysis provided a pragmatic, simple, and highly repeatable approach to selecting restoration sites that was firmly based in species biology and confirmed with simulation models. This method certainly holds promise for other systems in which these key features of species behavior and demography have been well characterized.

Chinook salmon are another declining species for which habitat restoration is likely to be a critical component of species recovery. Isaak et al. (2007) constructed statistical models to determine associations between habitat quality, size, and connectivity and the occurrence of chinook salmon (*Oncorhynchus tshawytscha*) nests. Data included annual censuses of salmon nests and field measurements of habitat characteristics at 43 sites across a network of streams in central Idaho. The results showed that the most likely models for nest occurrence included habitat size and connectivity, with habitat quality being of little impor-

tance in the models. Habitat connectivity was strongly associated with nest occurrence, but connectivity interacted with habitat size, in that connectivity was
relatively more important when habitat size (i.e., a patch that contained salmon
nesting sites) was small. The authors concluded that habitat restoration should
either target the expansion of existing nesting areas or create new habitats in sites
located in places where they would increase habitat connectivity. Although the
authors did not specifically evaluate particular restoration sites based on their
size and connectivity, as in the butterfly example above, the rigorous analysis of
census data with the spatial features of salmon habitat provided defensible guidelines for restoration activities. Mindful of these findings from modeling studies
about what spatial features may influence restoration success, this chapter now
turns to empirical studies that have explored habitat spatial features and their influence on the success of population and community recovery.

CASE STUDIES FROM THE FIELD
Understory Herbs in Deciduous Forests

It is estimated that between 3000 BC and AD 1086, forest cover in the United
Kingdom plummeted from 85% to 15% of total land cover (Peterken 1996). Forest cover reached a low point of 4% of total land cover in 1895, but by 1992, forest cover was estimated at 11%, indicating regrowth. A similar pattern of land
conversion occurred in northeastern North America, but it started much later—
widespread forest clearing began in the early 1700s, and by about 1850 deforestation reached its peak, with 80% of the forests having been cleared for agricultural activities (Foster 1992). By 1970, however, New England was 80%–90%
forested. This natural recovery of forests on both sides of the Atlantic occurred
without active restoration efforts, so forest species relied on opportunistic recolonization of abandoned agricultural fields from old-growth remnant forests
nearby.

Vellend (2003) wondered whether, in particular, the recolonization of forest
understory herbs in recovered forests might have been influenced by landscape
context. In Europe, the designation *ancient forest* refers to forests that are either
primary forests that were never cut, or regrowth forests that are so old that they
pre-date maps of the region. In North America, ancient forest generally refers to
primary old-growth forests. Vellend compiled data sets on ancient forest herb diversity (referring to herbs that had occurred with higher frequency in ancient versus recent forests) and landscape composition from 10 sites in Europe and North
America—including Germany, Denmark, the United Kingdom, Belgium, the

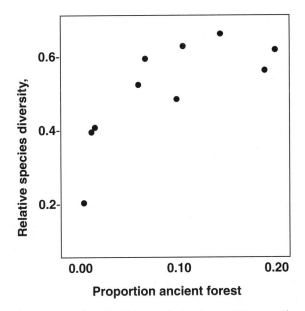

Figure 10.3. The proportion of ancient forest in the landscape (P_A) versus the relative diversity of understory herbs in recent forest (S_R) to ancient forest (S_A) for 10 sites in the United Kingdom, Europe, and North America. Regression of log P_A on S_R/S_A, partial r^2 = 0.76, with $P < 0.001$. Redrawn from Vellend (2003).

Netherlands, Poland, and three sites in northeastern North America—to assess associations between forest cover and forest herb diversity. Data on forest herb diversity were analyzed in relation to the proportion of ancient forest in the surrounding landscape, the proportion of recent forest, and the time since agricultural abandonment. Further, Vellend developed a metapopulation model to project the patch occupancy of ancient forest herbs in recent forest patches, based on the relative proportion of ancient and recent forest patches in the landscape and on their relative dispersal ability (i.e., *slow colonizers* versus *fast colonizers*).

Despite differences in forest types and geography, Vellend's results showed very clearly that the proportion of ancient forest in a landscape accounted for over 65% of the variation in species diversity of forest herbs (fig. 10.3). Put simply, the proportion of old-growth forest in the surrounding landscape was strongly positively associated with the diversity of forest herbs in recent, post-agricultural forest patches. When the ratio of forest herbs in recent forests versus ancient forests was high, this indicated that recent forests had understory herb diversity similar to that of ancient forests (fig. 10.3). So with a higher cover of ancient forest in the surrounding landscape, the recovery of species diversity in secondary forest patches appeared to proceed more readily. Interestingly, Vellend's modeling results also showed that the preservation of ancient forests in a landscape is far more important for the recovery of forest herbs than the area covered by recent forests, which argues more urgently for the conservation of old-growth forests wherever they occur. Vellend concluded that "severe habitat loss greatly delays the process of natural restoration" (p. 1163). This fits closely with the observations of tropical dry forest regrowth with which this chapter began—in areas of low forest cover, recolonization is often stalled, perhaps for decades or even centuries.

Additional studies of plant recolonization patterns in recovered forests have confirmed and expanded upon these initial results from Europe and North America. For example, Vellend (2004) showed that both forest herb diversity and the genetic diversity of *Trillium grandiflorum*, a representative forest herb, were higher in primary forests than in secondary forests in central New York State. The assumption was that lower forest herb diversity in secondary forests was due to the fact that forest herbs were still in the process of recolonization following forest clearing from 70 to 100 years ago. The genetic diversity of *T. grandiflorum* proved to be lower in secondary forests, probably due to lower population sizes relative to those in primary forests. So the effects of forest clearing on species and genetic diversity in this region appear to have persisted for decades, probably because of the limited dispersal of forest herbs into cleared sites. This has obvious implications for restoration in fragmented landscapes. The results suggest that leaving sites alone to recover on their own (the "let nature take its course" routine) may require waiting a very long time, or forever, until native diversity is reestablished.

Differences in forest recovery across regions may reflect differences in species dispersal related to landscape spatial configuration. The recovery of understory herbs in recent forests across three regions of Flanders, Belgium, showed a strong positive association between the degree of recovery and patch connectivity and age (Verheyen et al. 2006). Environmental conditions appeared to play a lesser role than landscape context in the recovery of these forests. Forests that were further along the trajectory of recovery tended to have a higher proportion of both vertebrate-dispersed species and species with short-distance dispersal, suggesting the importance of spatial factors for the colonization of dispersal-limited species.

The story on lower forest herb diversity in recent forests appears to hinge not only on limited dispersal from ancient forests, but also on sub-optimal environmental conditions in secondary forests. Vellend (2005) observed that for *Trillium grandiflorum*, adult plants had higher population densities, grew larger, and were more likely to flower in primary versus secondary forests in central New York State, suggesting that restoration activities for these species must consider dispersal limitation as well as appropriate environmental conditions. Key environmental factors supporting higher *T. grandiflorum* growth were not specifically identified in this study, but Vellend posited that populations may be limited by herbivory, pathogen infection of seedlings, lack of mycorrhizal associations, or competition with other plant species.

Collectively, these studies of the natural colonization of secondary forest patches by understory herb species suggest a strong role for landscape context in influencing processes and patterns of recovery. Even though forest recovery in these examples did not involve active human intervention, the observational data and modeling efforts clearly reveal the importance of ancient forest patches for forest recovery and provide tangible guidance for north-temperate forest conservation and restoration efforts.

Woody Plants on Closed Landfills

Although much of the northeastern deciduous forest in the United States has recovered since the 1850s, there remain many severely degraded sites. For example, native forests in certain locations were cleared to create sanitary landfills (sites where garbage is dumped and buried over several decades). Once these landfills are full, they are usually capped with soil, planted with grass, and then left alone. Steven Handel and colleagues began a research program in the 1990s to investigate delayed natural succession on closed landfills (G. Robinson, Handel, and Mattei 2002). They visited a number of such sites in the New York metropolitan area and observed that many had not developed natural plant communities, even a couple of decades after closure. Both observations and experiments ultimately demonstrated that woody plant succession on capped landfills was limited by the lack of seed dispersal to these sites. Many trees and shrubs of the eastern North American flora are bird dispersed, so the authors suspected that the absence of succession was due to the lack of avian seed dispersal of woody species to these sites.

Two sets of observations suggested that, in fact, dispersal may be limiting the recruitment of woody species on closed landfills. G. Robinson, Handel, and Schmalhofer (1992) studied the natural woody plant colonization of a landfill in New Jersey, which was located within a small established plantation that had been used to test tree species suitable for planting on landfills. The authors observed that many new plants had recruited to the site, and most were bird-dispersed species. Given the close proximity of the recruiting plants to the planted trees, these authors reasoned that the small tree plantation had offered perching sites for avian frugivores, which dispersed seeds in the plantation and facilitated recruitment. G. Robinson and Handel (1993) studied another site, the Fresh Kills landfill on Staten Island, New York, where clusters of trees and shrubs had been planted on the capped landfill one year prior to their observations. They found that over 1000 woody plants had colonized the site naturally and that the number of colonizing seedlings was positively related to the number of plants that were originally transplanted into the plots. Thus local plant density was positively associated with seedling recruitment. These results confirmed the notion of *nucleation* for colonizing species, which is the idea that patches of vegetation can serve as focal points for the rapid spread of invading species. Nucleation has generally been discussed in the context of unwanted, invasive exotic species (e.g., Moody and Mack 1988). However, the strategic placement of focal plantings may also be quite effective in promoting the colonization of desirable species as well. At a broader spatial scale, the number of seedlings of bird-dispersed species was associated significantly with the distance to a natural source of colonists (a nearby woodland remnant); more seedlings occurred in plots that were closer to nearby woodlands (G. Robinson and Handel 1993).

Subsequent experiments by Handel's research group confirmed that limited seed dispersal to closed landfill sites may be largely responsible for delayed succession. In one experiment at Hackensack Meadowlands, New Jersey, G. Robinson and Handel (2000) installed habitat islands to see whether clusters of scattered trees would attract avian frugivores, and whether the size of the trees that were planted would influence visitation by avian frugivores. This was essentially a test of the conditions required to restore this vital plant-animal interaction in a fragmented landscape. They also measured the woody plant colonization of adjacent empty plots (i.e., ones that were not planted). The planted plots had 2–4 times the densities of plant recruits (primarily represented by seedlings of avian-dispersed species) as did the unplanted plots, which demonstrated that plantings did attract avian frugivores and confirmed the nucleation concept. Seedling recruitment was concentrated in close proximity to experimentally planted plots, and recruitment was also associated with the distance from a natural source of colonists. As noted in the observational studies above, spatial effects on woody plant colonization occurred at both local and broad spatial scales. Notably, the woody flora of the woodland-remnant source habitat was impoverished compared to intact deciduous forests in this region, and the relatively few species that colonized the landfill site reflected that low diversity.

In a second experiment at the Fresh Kills landfill on Staten Island, researchers varied the size of the plantings, ranging from 7 to 70 individuals per cluster (G. Robinson, Handel, and Mattei 2002). The dispersal of seeds into the planted clusters was high, as was observed in the previous experiment, but this experiment showed that cluster size did not influence seed dispersal into the clusters significantly. Clusters that were in close proximity to a woodland remnant, which was a natural seed source, received significantly more seeds than those farther away, demonstrating a strong spatial effect of the seed source (fig. 10.4). Equally interesting was the observation that all clusters, even those at the far end of the site that were 600 m from the woodland remnant, received many seeds, suggesting that all clusters functioned to some extent in attracting avian frugivores. A serendipitous but revealing observation from this experiment was that during the second year of the study, about half of the area of the source woodland remnant was cleared, and seed rain into the site also declined substantially. This accidental part of the experiment helped to confirm the significance of the woodland remnant as a source of seeds for the colonization of the landfill.

Spatial effects were demonstrated at two spatial scales in these studies of landfill succession. First, seedling recruitment was higher around the focal plantings, which probably offered perching sites for birds, thereby enhancing dispersal and recruitment. Second, recruitment declined with increasing distance from the woodland remnants, which served as a broad-scale source (hundreds of meters) of fruiting trees and shrubs. In their summary of observational and experimental studies of landfills, G. Robinson, Handel, and Mattei (2002) emphasized the general notion that native seed sources were essential for promoting ecological succession on degraded sites, and thus that habitat restoration is absolutely dependent on the conservation of remnant habitats. Although Handel and colleagues have focused on woody plant succession, their findings echo those of Vellend (2003) for understory herbs, in that the presence and composition of source habitats is critical for the colonization of secondary or restored habitats.

Tropical Montane Forests

In tropical forests, as in temperate forests, many seeds are dispersed by animals, so the recovery of forests is likely to depend on the behavior, abundance, and spatial distribution of these species. In southern Costa Rica, as in the dry forests of

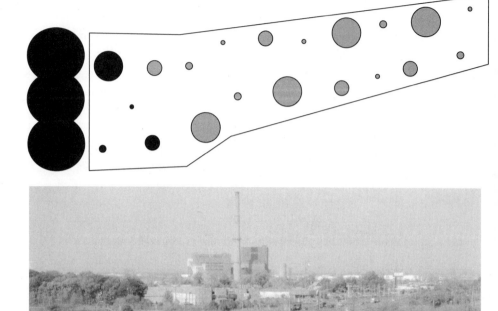

Figure 10.4. Top: The experimental design of plantings to evaluate the effect of cluster size on seedling recruitment at the Fresh Kills landfill, Staten Island, New York. The size of the circles is scaled to the number of plants per cluster: small = 7 plants, medium = 21 plants, large = 42 plants, and extra-large = 70 plants. A native woodland remnant is located on the left edge of the experimental planting area. The four darker gray circles are those four clusters that received the most seeds. The distance from one end of the site to the other is 570 m. *Bottom:* Woody plant colonization (foreground) of the Fresh Kills landfill, Staten Island. Diagram and photo courtesy of Steven N. Handel, Rutgers University.

Guanacaste, much forest has been cleared for cattle pasture. Karen Holl studied forest regeneration at a previously forested site that was cleared for a coffee plantation, and then used as a cattle pasture until it was abandoned in 1995. In a summary of several years of research, Holl et al. (2000) concluded that the lack of dispersal of forest seeds to the site, as well as competition from pasture grasses, imposed the greatest obstacles to forest regeneration in this abandoned pasture.

Several observations and experiments revealed the relative importance of bird-perching structures, planted trees and shrubs, and remnant pasture trees for promoting seed dispersal and seedling establishment. Bird perches have been used in other systems to facilitate avian seed dispersal (e.g., McClanahan and Wolfe 1993), and in this tropical setting perches did attract birds to deposit seeds, but seedling establishment was still low under these structures, due to competition from pasture grasses. Planting native trees and seeding shrubs showed potential in accelerating forest succession, but the results were variable. Some of the planted trees were attacked by herbivores; many of the shrubs germinated, but their seedlings did not survive due to competition from grasses. However, the trees and shrubs that did survive facilitated avian seed dispersal and enhanced the establishment and survival of forest tree seedlings, probably because they offered a refuge from competition with pasture grasses. Remnant pasture trees also showed promise as a means for promoting forest regeneration. More seeds were deposited, more seedlings were established, and planted trees grew more quickly near remnant trees. These results share similarities with Handel's landfill studies, in that plantings locally affected both seed dispersal and seedling establishment.

An ambitious broad-scale experiment was recently established by Holl and colleagues to ask whether *applied nucleation* facilitates forest recovery in different landscape settings. The experiments involve three native seedling planting treatments (fig. 10.5): *plantation*, which consists of rows of trees; *islands* of three different sizes; and *control sites* with no plantings (K. Holl, pers. comm.), for a total of over 8000 tree seedlings! Each block was planted at one of 16 (1 ha) sites, in settings that differ in the amount of forest cover, in order to assess the local and landscape effects on forest ecosystem recovery. Early results on visits by avian seed dispersers suggest that both island size and tree species affect the number and duration of the visits (Fink et al., forthcoming).

Understory Vegetation in Riparian Woodlands

There may be instances where restoration site size, isolation, or context have relatively small influences on restoration success, and these cases may be able to teach us just as much about the conditions under which spatial features are likely to affect the colonization of restored sites as the ones previously discussed. For example, Holl and Crone (2004) sampled riparian forest understory species that had naturally colonized 15 riparian restoration sites along the Sacramento River in central California. This large river has been heavily modified, and by the late

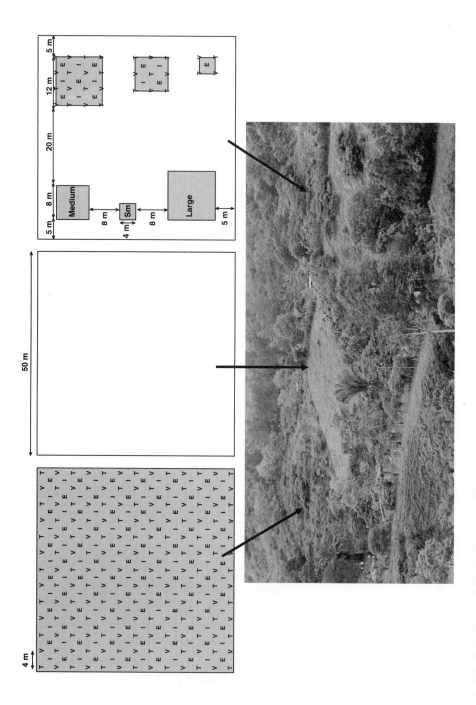

1970s, only about 5% of the native riparian vegetation remained as small isolated patches (Greco 1999). Restoration efforts along the river involve planting woody species, in the hope that understory plant species and riparian woodland animals from surrounding remnant sites will colonize these areas.

Holl and Crone (2004) observed much heterogeneity in native understory species occurrence among the restored sites, but the native understory species were generally slow to colonize restored riparian sites—remnant forests differed significantly in their understory species composition from older restored sites. The richness and abundance of native species in the restored sites was significantly negatively affected both by the presence of exotic species and by isolation from the remnant forest. However, the majority of the variance in native richness was attributed to exotic cover. More native species occurred in restored sites that were closer to the remnant forest, but the amount of variance in native richness explained by this variable was quite low. Landscape context did affect seed dispersal to restored sites. The abundance of wind-dispersed species was highest in sites that were surrounded by less than 20% of the remnant forest, water-dispersed species were most abundant at sites closer to the river, but animal-dispersed species were not affected significantly by the composition of the surrounding landscape. In general, Holl and Crone concluded that patch size and, to a lesser extent, isolation were relatively weakly associated with understory plant species richness and cover in this system. At these restoration sites, native plant species richness was strongly governed by the presence of exotic species, rather than the spatial attributes of the sites. Exotic species distribution, in turn, was unaffected by landscape variables, but was strongly influenced by local variables (such as the amount of overstory cover), with lower exotic richness and cover in sites with high overstory cover. Efforts aimed at restoring native understory plant communities in this highly fragmented landscape may be most successful if they focus on methods for improving local conditions, such as increasing overstory cover and reducing exotic cover, rather than emphasizing the spatial location of restored sites.

Figure 10.5. (opposite) Schematic of an experimental design (*top*) and photograph of one experimental block (*bottom*) of forest restoration in Costa Rica, illustrating plantation treatment (*left*), control treatment (*middle*), and island treatment (*right*). The gray areas are planted with seedlings of four trees—*Erythrina poeppigiana* (E), *Inga edulis* (I), *Terminalia amazonia* (T), and *Vochysia guatemalensis* (V)—and the white areas are unplanted.

Wetlands

Limited colonization may also explain the differences in vegetation between restored and natural wetlands. For example, Galatowitsch and van der Valk (1996) and Seabloom and van der Valk (2003) surveyed plant assemblages in a collection of naturally-occurring and restored prairie pothole wetlands in northern Iowa. The restored wetlands had not been planted with native species, so their species composition reflected natural colonization over the less-than-10-year period since their hydrology was restored. The former study assessed the colonization of the restored wetlands by native wetland plants in its early stages (within 3 years), and the latter study evaluated plant species composition after 5 to 7 years of colonization. In both studies, despite similar hydrology in the two types of wetlands, vegetation in the restored wetlands did not reach the diversity found in the natural wetlands, but it was a subset of that found in natural wetlands. The restored wetlands had lower species richness and more annual species (indicative of disturbed areas) than the natural wetlands. Interestingly, exotic species richness was similar between the natural and restored wetlands.

Both studies concluded that the restored wetlands differed from the natural wetlands in native species composition because of limited seed dispersal to the restored wetlands. Exotic species are often superior dispersers, so their similarity in natural versus restored wetlands suggested that they were perhaps not dispersal limited and thus were able to colonize all the sites. Neither study explicitly analyzed the species composition of wetlands in relation to wetland spatial position, but a direct comparison of natural and restored wetlands within the same landscape would likely have controlled for any biases in spatial position between the natural and the restored wetlands. Spatial position may be important at several spatial scales in this landscape. As Galatowitsch and van der Valk (1996) pointed out, the overall coverage and spatial distribution of prairie pothole wetlands in this region has changed dramatically over the past 100 years. Most wetlands have been drained and converted to agricultural uses, so the few remaining natural wetlands are much more isolated than they were previously. The spatial location of natural wetlands relative to restored wetlands is therefore likely to strongly influence the colonization of restored wetlands by native plants.

Isolation may limit plant dispersal to restored sites, and it may also affect the interactions of plants with other organisms. For example, Watts and Didham (2006a, b) studied the interaction of the invertebrate community in general, and an undescribed species of herbivorous caterpillar (Lepidoptera: *Batrachedra* sp.)

in particular, on a wetland plant, *Sporadanthus ferrugineus* (Restionaceae), that inhabits New Zealand peat bogs. Experimental, potted *S. ferrugineus* plants were placed at six distances (from 30 to 800 m) from an intact bog and allowed to be colonized by invertebrates for 6, 12, and 18 weeks. The researchers observed a steady decline in the abundance and richness of invertebrate species with increasing distance from the intact bog, showing that invertebrate colonization of isolated plants was limited (Watts and Didham 2006a). For the stem-boring caterpillar, the decline in abundance with distance was sharper; at 400 to 800 m, there were few to no caterpillars present on the host plant (Watts and Didham 2006b). The authors assessed the rate of recovery of this plant-herbivore interaction by surveying the herbivore on restored habitat islands that were three or six years old and were located from 30 to 800 m from an intact bog. In the three-year-old restored islands, oviposition by the herbivore was similar at all distances, and for six-year-old islands, the levels of herbivory were similar to those in the intact bog. So, despite the strong effect of isolation on this plant-insect interaction, recovery was fairly rapid following restoration.

The disruption of plant-insect interactions in restored habitats could potentially have negative effects on plant persistence. Ramp (2005) studied pollination of the endangered herb *Lasthenia conjugens* (Asteraceae) in restored and naturally-occurring vernal pools in central California. Like prairie pothole wetlands in the midwestern United States, vernal pools are spatially discrete wetlands that have become increasingly isolated as a result of land conversion (Barbour et al. 1993). Vernal pools in California are ephemeral wetlands that host a distinct flora during the dry phase, and many of the endemic vernal plants require specialized pollinators for successful reproduction. In a complex of restored vernal pools adjacent to a collection of naturally-occurring pools, Ramp (2005) observed little or no visitation of specialist solitary bees to *L. conjugens* flowers in the restored pools, but she did observe bees in the natural pools. In restored pools, flowers were visited by gnats in the genus *Eugnoriste* (Sciaridae) instead of bees, but there were no differences in *L. conjugens* seed set between the natural and the restored pools. The restored pools may have lacked bee visits due either to their relatively small floral displays or to limited dispersal by the bees from natural pools—in other mark-recapture studies, bees were observed to forage only up to 20 m from the site of capture (Thorp 1990)—since the restored pools in this study were located 10–110 m from natural pools. Although the specialized plant-pollinator interaction was disrupted in the restored pools, gnats appeared to compensate for this by providing pollinator services. Further exploration of these sorts of functional redundancies in species' roles within restored ecosystems may shed light on how habitat modification affects relationships between biological diversity and ecosystem services.

Habitat for Butterflies and Moths

Because plant-insect interactions may be substantially altered in restored sites that are relatively isolated from native habitats, efforts explicitly aimed at restoring animal populations may also fall short if dispersal is hampered by a lack of landscape connectivity. Animal reintroductions have primarily involved species that have declined due to overexploitation by humans, such as the gray wolf in the Yellowstone National Park ecosystem in the western United States, or exotic predators, such as the brush-tailed bettong in South Australia. In these cases the habitat is largely intact, but new populations must be established from the release of translocated individuals or those bred in captivity. But probably the most frequent restoration projects geared explicitly toward restoring habitat to encourage the recovery of animal populations are those for butterflies and moths (Lepidoptera) that inhabit fragmented landscapes, such as the Fender's blue butterfly studied by Schultz and Crone (2005). The successful establishment of populations may be critically dependent on habitat connectivity; for example, Hanski (1999) observed that "reintroductions of British butterflies to isolated sites have generally produced only temporary success at best . . . [while] butterfly introductions in patch networks have produced better results" (pp. 190–191).

To investigate the effects of landscape spatial structure on the success of recovery efforts, Keith Summerville and colleagues studied moths in restored oak savanna (Summerville, Steichen, and Lewis 2005) and tallgrass prairie (Summerville, Bonte, and Fox 2007) habitats in central Iowa. Both habitat types have declined to less than 1% of their historic range in Iowa, so habitat restoration is a crucial component of recovery for the species that occupy these habitats. Researchers found that for all moth species that occurred in oak savannas, richness was generally higher in large patches that had minimal edge and were dominated by oaks. Surprisingly, the richness of oak-specialist moths as a group was higher in isolated patches, but this was probably because patches with high connectivity tended to be dominated by tree species that were not suitable hosts for these highly specialized species. In this case, the quality of patches and their connections influenced species occurrence patterns significantly (Summerville, Steichen, and Lewis 2005). For moths sampled from tallgrass prairie, richness increased with the age of the restored site, suggesting that colonization exceeds extinction in these sites, allowing species to accumulate over time (Summerville, Bonte, and Fox 2007). Moreover, the species composition of established restored sites was more akin to remnant sites than to newly restored sites. Differences in moth communities among the sites were related to particular combinations of ecological traits that were shared among species. For example, species that had long flight periods, produced several generations per year, and were regionally abundant tended to occur at most sites, including newly restored prairies, but those with more restricted dispersal occurred primarily on remnant sites (Summerville, Conoan, and Steichen 2006). Taken together, these results support the notion that habitat spatial characteristics, as well as habitat quality, are likely to influence the success of restoration efforts.

Do butterflies colonize restored prairies in the same way that moths do? Shepherd and Debinski (2005) studied butterflies that occur in the same region of central Iowa that was described in Summerville's moth studies. They examined butterfly and plant richness and abundance in three different settings: in remnant prairies surrounded by non-prairie habitat, in restored prairies that were integrated within a large complex of restored prairie, and in isolated restored prairies that were surrounded by non-prairie habitat. All restored sites were between 4 and 11 years old. Butterfly species richness was highest in the remnant prairies and lowest in the isolated restored prairies. Correspondingly, plant diversity was highest in the remnants and lowest in the isolated restored sites. But a quite interesting result from this study was that butterfly species richness was similar in the integrated restoration sites and the remnant prairies. Additionally, the richness, but not the abundance, of seven habitat-sensitive species (i.e., those that are typically found in grasslands with relatively little human disturbance) was similar between the integrated restoration sites and the remnant prairies. These results suggest that isolation affects the success of tallgrass prairie restoration for butterflies, since isolated restored prairies had been colonized by fewer species than the integrated restored sites. The remnant prairies in this study may be large enough to support viable populations of habitat specialists, even though they are surrounded by non-prairie habitat, but restored sites clearly benefited from being surrounded by other prairie habitat.

SYNTHESIS

Ecological theory related to fragmented habitats has been usefully applied in considering problems in ecological restoration. This review concentrated primarily on research that emphasized a spatial component to restoration, so it cannot provide a fair evaluation of whether local or landscape factors are more likely to influence restoration success. Several studies have clearly demonstrated strong effects of landscape spatial position on the colonization of restored sites; other studies, however, have shown weaker effects of landscape spatial features. Further modeling efforts would be useful in applying theory to restored situations, such as evaluating which habitat patches should be selected for species introductions and how many patches should be used for introductions (e.g., expanding on the study by Schultz and Crone 2005). For example, C. Thomas and Hanski (1997) suggested that as a broad generalization, introduction at multiple sites would be likely to be more successful for relatively sedentary species with high rates of population increase, but which are still vulnerable to random variation in environmental conditions (such as many arthropod species). But single-site introductions may be more appropriate for species with low population growth rates, for species more strongly affected by random variation in demographic rather than environmental parameters, and for instances when only a small number of individuals are available for introduction, as is the case for many vertebrate species (C. Thomas and Hanski 1997).

Despite some progress, we are still in the early stages of this research agenda, and much work remains to be done. For example, it would be worthwhile to compare naturally occurring and restored sites within a single metapopulation or metacommunity analysis, both to see how dynamics differ and to determine the relative contribution of each to overall population dynamics. Further, it would be useful to investigate whether species interactions vary systematically in restored versus natural fragments, depending on their spatial context. There is a dearth of long-term evaluations of restoration sites in order to determine whether they are successful and are contributing to regional population dynamics, so more effort should be devoted to such studies, in particular with regard to spatial features. There are some examples of the long-term monitoring of restoration sites (e.g., Petranka et al. 2007 for salamanders and frogs in North Carolina), but these analyses could be taken further by relating local population dynamics with habitat spatial characteristics. Another critical research question is whether habitat restoration is likely to enhance population recovery. This is generally assumed to be true, since most species' declines are due to habitat loss. But further simulation and empirical studies are needed to confirm or refute the generality of the findings of Schrott, With, and King (2005) regarding this assumption.

Finally, restoration that enhances habitat connectivity may not always be desirable, especially if it increases the spread of pathogens, parasites, exotic species, or natural disturbances, such as fire (see also chapter 6). For example, Holdenrieder et al. (2004) cautioned against anthropogenic activities that may promote the spread of plant diseases. One such instance was the novel juxtaposition of native forests and reforested areas, which may enhance functional connectivity among host trees in a way that allows pathogens to reach new hosts (see Perkins and Matlack 2002, discussed in chapter 9). This situation involved the creation of a new landscape pattern, rather than the restoration of a historic pattern, so unexpected consequences may occur. However, in cases where restoration efforts are aimed at replacing lost habitats according to historic spatial configurations, it seems safe to assume that these efforts would be beneficial, or at least neutral (e.g., Schrott, With, and King 2005), but not harmful.

CHAPTER ELEVEN

Ecological Planning

plan, *n*.: a scheme, program, or method worked out
beforehand for the accomplishment of an objective.
　　　　　　　　　　　—American Heritage Dictionary (2006)

Landscapes are constantly changing. As is often the case with our own lives, some of those changes are carefully planned and some are unplanned. The designation of a new national park or a major suburban housing development is usually the result of thoughtful planning and decision-making over several years, with input from many stakeholders. But arsonous wildfires that burn out of control and haphazard development around tourist towns are clearly unplanned occurrences. Ecological planning, as it is considered in this chapter, covers a wide array of activities, ranging from the development of algorithms to optimize the design of conservation reserve networks that will maximize native biological diversity, to designs for housing developments that reduce urban sprawl by creating compact neighborhoods with protected open space, to regional plans that project alternative future scenarios of land use change. The common denominator of all these efforts is the integration of ecological knowledge with intentional human actions to direct spatial patterns of environmental change. All of these efforts rely, to a large extent, on scientific research relating patterns of biological diversity and the consequences of habitat loss and fragmentation.

Ecological planning can be thought of as a continuum of activities that integrate ecology with land use change. At one end of this spectrum are landscape architects and planners, who are primarily engaged in designing landscapes for human activities, but often give high priority to native habitats and species. Ecological thinking entered the fields of landscape architecture and planning in the late 19th and early 20th centuries; at the time, these ideas were considered revolutionary. But many landscape architecture and planning programs that exist today rely heavily on ecological information in formulating innovative designs (Ahern, Leduc, and York 2006). At the other end are conservation biologists, who focus primarily on choosing appropriate sites and management strategies for protected areas that are designed to have relatively little human intrusion. Ecological scientists formally developed spatial principles for the design of conservation reserves in the 1970s, but they were active in nature preservation much earlier— especially in the late 19th and early 20th centuries, under the influence of people like John Muir and Teddy Roosevelt. These two ends of the ecological planning spectrum diverge substantially in their primary goals, but they may ultimately have closely convergent outcomes. Although the ends of this spectrum are still quite recognizable, the middle area of overlap is rapidly growing. Landscape ecology lies toward this middle ground, which originated in Europe in the 1930s and was embraced in the United States and elsewhere by the early 1980s (Forman and Godron 1981, 1986). With its explicit emphasis on the causes and consequences of ecological patterns and processes, including human influences on landscapes, landscape ecology offers a formal means of integrating many disparate activities related to ecological planning, including those in ecology, geography, resource management, conservation biology, and landscape architecture and planning. Other ecological planning activities, such as community-based conservation, may be even broader, incorporating ecology and planning with social science disciplines such as anthropology, political science, social psychology, and economics (Gibson, McKean, and Ostrom 2000; Russell and Harshbarger 2003).

The goal of this chapter is to illustrate some of the ways in which ecological research in fragmented landscapes has been integrated with ecological planning. Efforts all along the spectrum have an immense potential to preserve, protect, restore, and halt declines in biological diversity. There are many different ways in which this integration has been accomplished, stemming from both the planning and design professions and the biological sciences. Because of this plethora of approaches, the intent here is to provide a summary of the diversity of efforts to translate research on habitat loss and fragmentation into actions that are beneficial for biodiversity and for human well-being. The chapter begins with a few examples from the planning end of the spectrum, and then continues with a few examples from the conservation biology end. It concludes with examples that

are somewhat difficult to assign to either end, fitting more easily into the growing center of the continuum. Although these planning approaches have been
placed in three arbitrary bins, the boundaries between them are often quite
blurry, and there may be significant overlap among them.

MAPS

Before delving into these different planning approaches, it is worth taking a moment to talk about maps. We all know that a good map is essential, especially if
we are traveling down an unfamiliar dark road. But good maps are also vital to
any planning exercise. Maps are records of the spatial locations of things, but because they are abstractions of reality, they must always omit some details. The
maps used in planning often include information from a wide variety of sources.
For example, a planning exercise may necessitate the use of historical paper
maps of property boundaries or land uses, or hand-drawn maps of traditional
hunting and fishing areas. These can be digitized and integrated with other
spatially-referenced information. Most other geographic data are readily available
in digital form from government agencies or non-governmental organizations,
and include digital aerial photos, satellite images, maps of water bodies, cities,
towns, buildings, roads, zoning regulations, vegetation types, and species occurrences. Maps may also depict threats to biodiversity and the values of ecosystem
services in reference to particular spatial locations (e.g., W. Turner et al. 2007).

   All of these different pieces of information about a place, also called *data layers*, can be incorporated into a *geographic information system* (GIS), which is a
computerized map, or what I like to call a *smart map*, because the computer re-
members the spatial locations of all of the different features. The user can perform GIS analyses that relate the spatial locations of these features to one another, something that is impossible to do with a paper map. For example, a
relatively straightforward exercise would be for the user to assemble data layers
of the woodland vegetation cover and roads for a particular area and then calculate the average size of the woodland patches, as well as the range of distances of
different-sized woodland patches to the roads. This sounds easy, but in reality the
preparation and analysis of spatial information are complex tasks. In the above
example, the user must first define what a woodland patch is, before the GIS can
calculate average patch size. And the designation of a patch may depend on the
goals of the project and the organisms being studied, which brings us to the issue of spatial scale.

   An important consideration in mapping in general, and in ecological planning in particular, is the issue of scale. I presented hierarchy theory and the relevance of spatial and temporal scales in the first two chapters of this book, but in the context of this chapter, scale is meaningful when compiling data that will be used in a planning process. To continue the example above, a woodland patch, through a butterfly's eyes, may be a quite different thing than a woodland patch through a wolf's eyes. So defining the scale of the map and the analysis is critical. Additionally, some key features, such as vernal pools or prairie pothole wetlands that are relatively small, may not show up on maps that represent vegetation types at a coarse scale. They may need to be mapped individually in order to address specific requirements of the analysis. Ultimately, questions of scale and representation arise in any computerized mapping study, so the careful consideration of these issues will be essential to ensure effective outcomes.

THE PLANNING PERSPECTIVE

Landscape architecture and planning are disciplines focused on landscape change. Jack Ahern (2005), professor of landscape architecture and planning at the University of Massachusetts, had this reply when asked to offer his perspective on landscape ecology: "Landscape architects change landscapes. We build things and places—in the process altering landscape patterns, sometimes profoundly" (US-IALE web site). Ecological concepts and principles have long been considered in landscape architectural design and planning (summarized by Spirn 1985; Zube 1986; Ndubisi 1997, 2002). Probably the most well-known historic example of ecological design is Frederick Law Olmsted's plan for an urban open-space system in Boston, consisting of a series of connected parks, wetlands, ponds, and tree-lined boulevards, and referred to as the "emerald necklace." Not only was this open space system a true gem for the city by providing spaces for public enjoyment, but it also restored a natural wetland ecosystem that was badly polluted (Spirn 1985; Zube 1986), which helped to improve the health and well-being of Boston's urban population. Olmsted's design involved transforming the foul-smelling mud flats within the city to constructed salt marshes that would enhance water quality and store flood waters. In essence, this was a 19th century urban ecological restoration project.

Lesser-known landscape architects of this era were similarly forward thinking. For example, in 1883, H. W. S. Cleveland designed a regional park system for Minneapolis–St. Paul, Minnesota, that integrated this urban structure into and protected the natural systems of the Mississippi River, including many lakes in the region. Charles Eliot, an apprentice of Olmsted, designed a comprehensive open-space system for Boston in which he systematically classified different types of forest landscapes and incorporated both hydrological and natural features into a green infrastructure for the region (Zube 1986; Ndubisi 1997). Many of the elements that Eliot designed are highly valued today as both public open spaces and refuges for native biological diversity—including the 2600 acre Middlesex Fells and the 7000 acre Blue Hills reservation—despite some invasion by exotic species due to human introduction and trampling (Drayton and Primack 1996). In the mid-20th century, many landscape architects and planners were influenced by the writings of scientist and conservationist Aldo Leopold, who noted the critical interdependence of humans with nature and advocated a land ethic to ensure the health of the land (Leopold 1949). For instance, landscape architect Phil Lewis combined aesthetics and ecology in his delineation of what he referred to as "environmental corridors." for the entire state of Wisconsin. These corridors linked natural features such as water, wetlands, and significant topography with human perceptions of visually-pleasing scenery, and Lewis suggested that these areas be prioritized for environmental protection (cited in Ndubisi 1997, 2002).

McHarg's Design with Nature

In the environmental turmoil of the 1960s, Ian McHarg published his book *Design with Nature* (1969), which became a classic in landscape architecture and planning. This work was so popular and influential because it introduced a new method for the systematic analysis of spatial relationships of human activities and natural systems, essentially serving as one of the precursors to modern geographic information systems. McHarg's method consisted of "identifying both social and natural processes as social values" (p. 33), and he used this method in planning parkways, suburbs, and entire river basins. Using the technology available in the 1960s, McHarg took clear acetate sheets and laid them over maps that included information on natural resources and cultural features. He then used separate acetate sheets, one for each feature, and traced road networks, housing developments, lakes, rivers, streams, woodlands, and so on. He chose dark gray to color those areas that were unsuitable for building (e.g., steep slopes or wetlands), lighter gray for areas that were marginally suitable, and left suitable areas uncolored. McHarg then superimposed these sheets on one another on a light table, and the areas that were the darkest were the ones to be avoided. McHarg's planning efforts resulting from these *suitability analyses* were typically broad in scope, encompassing cities or regions, and offered a rational, systematic way to formulate landscape plans for these large, complex areas.

One substantial ecological planning project that demonstrated the utility of McHarg's suitability analysis method was a design for a new town on an 18,000 acre site north of Houston called the Woodlands (McHarg and Sutton 1975; McHarg, Johnson, and Berger 1979; Spirn 1985; D. Smith 1993). McHarg and Sutton (1975) called it the "first city plan produced by ecological planning" (p. 78). The site was a flat, oak-pine forested coastal plain with poorly drained soils, so a contentious issue was how to preserve the woodland while simultaneously drain-ing land for development. McHarg's team arrived at a natural water management system that would provide the least disruption to the natural hydrological regime, supporting essential ground-water recharge with the use of designed swales, ponds, wooded floodplains, golf courses, and even lawns. The design also advo-cated the preservation of both forest (about 1900 acres were specified in the com-pleted plan) and wooded corridors for wildlife movement that were between 100 and 600 feet wide (McHarg, Johnson, and Berger 1979; D. Smith 1993). Spirn (1985) reported the results of stormwater monitoring at the Woodlands and noted that increased runoff from the site after construction was about two-thirds less than the amount generated by a typical suburban development in Houston. In our current parlance of the value of ecosystem services, clearly the natural ar-eas within this development provided water management services for the resi-dents of this community. The benefits to native plants and animals are not so clear, since they have not been routinely measured at this site since its develop-ment (D. Smith 1993).

Planning Alternative Futures

McHarg's method of identifying opportunities and constraints for landscape in-tervention became widely used by his planning team and others in the decades following the publication of Design with Nature. Landscape planner Carl Steinitz and colleagues at the Harvard Graduate School of Design transformed the acetate-sheet technology of suitability analyses into computer-generated maps that eco-logical planners could use to perform sophisticated spatial analyses without colored markers and light tables (Tomlin 1990; Ndubisi 2002). In addition to these technological advances, Steinitz pioneered and popularized the method of alternative futures planning, which is effectively a method of modeling landscape change (see chapter 9) and is now broadly used in landscape planning (Steinitz, Arias, and Shearer 2003) and conservation biology (Peterson, Cumming, and Carpenter 2003). The premise of this method is to map the significant cultural and natural resources within a region, and then project changes in those resources according to alternative scenarios that make different assumptions about the patterns and intensities of land use change. These scenarios are then presented to community stakeholders and become useful tools for implementing decisions regarding landscape change that originate with the collective desires of the constituents. Such alternative futures models have been developed by Steinitz and colleagues for many locations, including Monroe County in eastern Pennsylvania, which includes the Pocono Mountains (D. White et al. 1997); the region of Camp Pendleton in Southern California (Steinitz et al. 1997); the Upper San Pedro River Basin, which encompasses parts of Arizona and the Mexican state of Sonora (Steinitz, Arias, and Shearer 2003); and for conservation and development in Panama's Coiba National Park (Steinitz et al. 2005).

Each of the scenarios proposed in alternative futures modeling can be evaluated quantitatively for changes in key cultural or biological variables, including land cover types, large patches, or habitat for particular species (D. White et al. 1997; Steinitz, Arias, and Shearer 2003; Steinitz et al. 2005). For example, in the Monroe County, Pennsylvania, study, six alternative scenarios were proposed, ranging from the continuation of existing development practices (*plan-trend*), to rampant, uncontrolled development (*build-out*), to a conservation-oriented scenario that advocated conserving all of the currently undeveloped land in the county (*park*) (D. White et al. 1997). These six scenarios were evaluated for their projected impacts on species richness and the abundance of habitat for terrestrial vertebrates. The proportion of habitat lost was greatest for the scenarios that projected the most development and least for the conservation-oriented scenario, as expected. Changes in terrestrial vertebrate species richness were projected to be quite low and similar among the various scenarios, but the analysis admittedly did not include much ecological detail about the sensitivities of species to landscape change or the projected population dynamics of different species. However, the proportion of habitat at risk for each species group was quite high in several scenarios (fig. 11.1), and was over 50% for herpetofauna in the two scenarios involving the greatest amount of development. Because habitat abundance was projected to decline by up to 50% in the most developed scenarios, it is likely that the species would also decline, based on what we know about species' responses to habitat loss discussed throughout this book.

The method of alternative futures modeling has been employed in many diverse landscapes to identify both the social and biological aspects of landscape change scenarios and to guide future development. For example, this method was used to project human population growth and land use change in the Mojave Desert of California (Hunter et al. 2003), changes in agricultural landscapes in

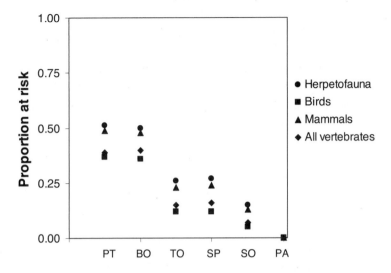

Figure 11.1. The proportion of habitat at risk, caused by proposed alternative development scenarios, for three classes of vertebrate fauna in Monroe County, Pennsylvania. The scenarios are PT = Plan-trend, BO = Build-out, TO = Township, SP = Spine, SO = Southern, and PA = Park. See the text for descriptions of PT, BO, and PA. The remaining three scenarios were intermediate in the amount of land to be developed and are described fully in D. White et al. (1997). Redrawn from D. White et al. (1997).

Iowa (Santelmann et al. 2004), residential development in the Greater Yellowstone Ecosystem (Gude, Hansen, and Jones 2007), and for assessing projected changes in biodiversity according to different development scenarios for 25 European Union countries (Verboom et al. 2007).

Greenways

Within the large regional landscapes planned by McHarg and Steinitz, greenways may provide green infrastructure in which ecological patterns and processes are incorporated with human activities. Little (1990) broadly described *greenways* as "linear parks, open spaces and protected areas in cities, suburbs or the countryside" (p. 4). They are typically linear, vegetated landscape features that may provide opportunities for outdoor recreation, have the potential to function as wildlife habitat or movement corridors, and, if located along streams or rivers, may help to maintain water quality (D. Smith and Hellmund 1993; Fábos and Ahern 1995; Mortberg and Wallentinus 2000; Fábos and Ryan 2006; Mason et al. 2007).

The design and planning of greenways originated within the design professions, with workers such as Olmsted and Cleveland in the late 19th century (Zube 1995). The concept and practice of greenway design resurged in the 1970s with increased public awareness of environmental issues and demands for outdoor recreation. As a design tool, greenways may be a "formative device for stitching together fragmenting cities and their urbanizing hinterlands" (Walmsley 1995, p. 81). The design of greenways may be particularly challenging, but it may also have quite meaningful outcomes in highly urbanized areas. For example, Tan (2006) discussed design and planning for greenways in Singapore, an island city-state of four million residents with one of the highest population densities in the world. The key design strategy for this greenway network was to identify and maximize the use of under-utilized land, that is, land that was not considered economically valuable, such as existing drainage ditches and river systems. Linear features were designed to connect parks throughout the island. These park connectors include simple asphalt paths lined with dense plantings of native and ornamental trees and shrubs, which provide recreational pathways for people and also promote social interaction. In many instances the greenways follow the 11 water courses and tributaries that drain from the forested center of the island in all directions to the sea. These greenways may also provide functional connectivity for native plants and animals, although that has not been evaluated (Tan 2006).

New Urbanism

Greenways are often featured design elements in proposals for new urban and suburban developments that subscribe to the principles of new urbanism (Krieger 1991; Walmsley 1995). *New urbanism,* sometimes called neotraditional town planning, emerged in response to the vast, sprawling, post-war suburban developments of the mid-20th century in the United States. In their response to the traditional suburb, new urbanists have advocated (1) the design of neighborhoods in which residents could easily walk to shops, schools, parks, and work, thereby limiting the use of automobiles, (2) preservation of open spaces, and (3) the judicious use of water in landscaping, particularly in arid ecosystems.

The scale of new urbanist developments ranges widely, so the opportunity for integrating them with ecological knowledge regarding habitat loss and fragmentation also varies broadly. For example, the design proposal for Avalon Park, a large development for 25,000 inhabitants adjacent to the Econlockahatchee River and associated wetlands east of Orlando, Florida, was large enough to consider issues of habitat fragment size and connectivity (Krieger 1991). Within the 9500 acre development, 5000 acres were to be reserved for green space, including the river corridor, wetlands, greenways, playing fields, and parks. In this design the river, wetlands, and adjacent upland areas provided the backbone for the town plan. The protection of the river and its associated wetlands would potentially reduce the negative impacts of the proposed development on water quality and biological diversity. Although the Avalon Park proposal addressed the importance of maintaining large, connected fragments, there is potential for negative edge effects along the interface between the preserved and developed areas. This project is in progress, so it is impossible to currently evaluate whether the plan will succeed in protecting native plants and animals associated with the river and its wetlands. But the designation of nearly half the area as green space highlights the philosophy of integrating nature with design.

Loreto Bay, on the Baja peninsula of Mexico, is another large new urbanist development that is in progress. The development portion of the 8800 acre site sits along the shores of the Sea of Cortez, while the western 75% of the acreage is designed to remain undeveloped as a nature reserve (Loreto Bay Company 2007). The plans call for the maintenance and enhancement of biodiversity at the site, including restoration of both upland habitat, to maintain downstream water quality, and mangrove estuaries, to provide nursery habitat for marine species and sequester carbon. The entire project promotes sustainability in terms of the use of efficient, renewable energy, water conservation, an organic farm, carbon offsets for air travel, and economic benefits to the local economy. There is a substantial potential for these developments, especially new urbanist designs that cover thousands of acres, to contribute positively to ecological sustainability.

Aesthetics and Ecology

Most human beings love order—we tend to plant vegetables and crops in straight rows and prefer manicured lawns to overgrown weed patches. Landscape architect Joan Nassauer has studied the relevance of human perceptions of landscape attractiveness in the context of landscape designs that mimic or restore natural systems (Nassauer 1992, 1995, 1997). She noted that "landscapes that evoke the sustained attention of people—that compel aesthetic experience—are more likely to be ecologically maintained in a world dominated by humans" (1997, p. 81). However, she also realizes that to many people, "ecological quality tends to look messy" (1995, p. 161), which means that landscape designs with a greater benefit for biodiversity, for example, may not be viewed positively by people who perceive the designs to be unattractive or unkempt. These negative perceptions may pose a significant challenge to designers involved in ecologically oriented projects, who must convince their constituents to accept these designs despite their rather untidy appearance (Nassauer 1992, 1995, 1997). For example, at the residential scale, Nassauer (1995) suggested that an unmown lawn might be considered more acceptable by suburban neighbors if there were "cues to care" present on the property. The presence of a narrow mown strip along streets or walkways, a well-maintained fence on the property boundary, or bird houses and feeders may all indicate that the proprietor is taking good care of the property. So, Nassauer reasoned, if these cues are present, neighbors would be more willing to accept a wildflower meadow filled with native prairie plants in place of a closely-cropped bluegrass lawn.

Beyond the neighborhood, public perception of the attractiveness of natural areas within urban contexts may similarly thwart ecological design efforts. For example, wetlands are less likely to be appreciated by the public as scenic natural areas than, say, a forest or a flowing stream. So wetland projects in urban areas may face neglect and even degradation by neighbors and visitors if they are deemed to be unattractive, unkempt, or unsafe. In a study of wetland restoration projects within the Minneapolis–St. Paul metropolitan area in Minnesota, Nassauer (2004) compared cultural and ecological measures of wetlands by surveying visitors, neighbors, planners, and managers of the wetlands. The overall goal was to see whether wetlands of greater ecological quality, in this case defined as high plant and bird species richness, were also perceived to be more attractive. For neighbors, the vast majority (76.7%) of the differences in their perception of wetland attractiveness were explained by cultural variables, including a highly visible mown area on the edge of the wetland, signage, walkways, and seating. Bird species richness explained a significant, but very small (about 2%) proportion of the overall variance in the perception of attractiveness. Plant species richness was not associated significantly with attractiveness. In all cases, those sites that appeared to be attended to were perceived as being more attractive.

The results of this urban wetland study were consistent with Nassauer's (2004) hypothesis that perceived attractiveness is *not* related to the biological integrity of wetlands. The importance of cultural sensitivities to landscape beauty is clearly meaningful for conservation and restoration activities, especially for vegetation types that are not immediately perceived to be beautiful, such as prairies or wetlands. But perhaps in this context, "beauty is truth, truth beauty" (Keats 1884). If visitors to a restored wetland learn about its significance as a habitat for rare plant and animal species, perhaps it will be perceived as more beautiful than a park sprayed with herbicides and fertilizers to keep the grass green. Then the ecological service, beauty, could be valued and enjoyed by visitors.

THE BIOLOGICAL PERSPECTIVE

The first part of this chapter has emphasized ways in which the professions of landscape architecture and planning have approached ecological planning. Many features, however, are shared with the efforts advocated by biologists at the other end of the ecological planning spectrum. From the biologists' viewpoint, the primary goals of ecological planning underscore the protection of biological diversity and ecosystem services. Based on island biogeography theory and other observations of natural systems, early conservation biologists advocated particular rules of design for natural preserves. For example, Diamond (1975) and E. Wilson and Willis (1975) suggested that preserves should be continuous rather than fragmented, large rather than small, connected by habitat corridors to other reserves, circular rather than linear in shape, and clumped rather than arranged linearly. These simple reserve design rules made sense, given what was known about species' responses to habitat loss and fragmentation, and they have broadly guided conservation planning.

The circular-versus-linear rule pointed out a potential for negative edge effects that may denigrate the reserve interior. Janzen (1986) highlighted the dangers associated with this phenomenon, which he called the *eternal external threat*. That is, reserves are surrounded by other land uses, which harbor influences—such as exotic species, water pollution, or hunters—that are likely to permeate the reserves and may reduce the viability of populations within the reserves. As an example of this problem, Woodroffe and Ginsberg (1998) analyzed probabilities of extinction for carnivores in protected areas in relation to home-range size. They found that wide-ranging carnivores were more vulnerable to extinction than those with smaller ranges. This was not because they had smaller populations, but because wide-ranging carnivores were more likely to encounter reserve boundaries, where human-induced mortality was higher than in protected areas. Data such as these essentially provide tests of the reliability of the simple reserve design rules proposed three decades ago. But in addition to these rules, several sophisticated methods have been developed since that time to choose sites and strategies for protecting biodiversity and ecosystem services.

Biological approaches to conservation planning vary substantially, but generally fall into the categories of *where* to do conservation and *how* to do conservation (Redford et al. 2003); this section will discuss both types of approaches. The *where* category includes efforts to set geographical priorities for conservation, and the *how* category emphasizes developing and implementing strategies to accomplish conservation goals. Several books summarize or provide case studies of different approaches. Groves (2003, discussed below) gives an excellent overview of the field of biologically based conservation planning. Three recent books review the scientific basis for implementing habitat corridors and connectivity and outline case studies from sites throughout the world: A. Anderson and Jenkins 2006; Crooks and Sanjayan 2006; and Hilty, Lidicker, and Merenlender 2006. And Margules and Sarkar (2007) summarized the systematic conservation planning approach, discussed below.

Systematic Conservation Planning

One of the key challenges in ecological planning for conservation is to determine where conservation action should occur. The method of formulating systematic rules and iterative algorithms to follow in identifying the best places for conservation was largely developed by Australian and South African conservation biologists, starting in the 1980s. By the late 1980s, these methods were sufficiently developed to dedicate a special issue of the journal *Biological Conservation* to the subject of using systematic methods to determine where reserves should be located (Margules 1989). Contributions to this issue focused on methods for establishing databases of spatial locations of species and communities and for using those databases to identify priority areas for conservation in Australia. Such databases as have been established for systematic conservation planning generally include information on (1) the species richness or the endemism of all the taxa for which data are available, (2) vegetation types, and (3) aquatic ecosystems such as lakes, rivers, streams, and wetlands. Conservation planners conceive of the second step in using these databases to determine specific areas for conservation as going "beyond opportunism" and using defensible principles for selecting priority regions. These principles include complementarity, flexibility, and irreplaceability (Pressey et al. 1993), and they can be applied in situations where the planners are given a particular number of sites to be chosen (Pykälä and Heikkinen 2005) or a constrained budget (Polasky, Camm, and Garber-Yonts 2001). Together, these two steps in conservation planning have provided a systematic means to choose portfolios of conservation sites that would represent the full suite of species and communities in a region.

Briefly, the principle of *complementarity* dictates that reserve selection proceed in a stepwise manner, such that the features of each reserve that is added to the network are complementary to those in the reserves that were already chosen. This is meant to ensure that a reserve network is as efficient as possible in protecting biological diversity, yet is not redundant. *Flexibility* refers to the generation of several alternative networks that can be evaluated by planners, in order to determine which one would be most easily implemented in a particular region. And *irreplaceability* captures unique sites for which certain conservation options would be lost if they were not included in the reserve network (Pressey et al. 1993). According to the developers of systematic conservation planning, reserve networks have two primary functions: "They should sample or represent the biodiversity of each region and they should separate this biodiversity from processes that threaten its persistence" (Margules and Pressey 2000, p. 243).

The systematic conservation planning approach has been used extensively by many conservation scientists all over the world to identify priority areas for conservation (see Margules and Sarkar 2007 for a summary of methods and examples). Recent elaborations on this method include explicit maps of various threats to biodiversity, including urban expansion, agricultural conversion, and invasive species (Rouget et al. 2003, discussed in chapter 9), as well as maps of ecosystem services (Chan et al. 2006; Naidoo and Ricketts 2006; W. Turner et al. 2007) and those that emphasize cost efficiency (K. Wilson et al. 2006). For example, for the Mbaracayu Forest Biosphere Reserve in eastern Paraguay, part of the Atlantic forest ecoregion, Naidoo and Ricketts (2006) mapped the spatial distribution of five ecosystem services to evaluate the costs and benefits of conservation in this region. The benefits to conserving this region included values gained through sustainable bushmeat consumption, sustainable timber harvest, the discovery of pharmaceutical compounds (also called *bioprospecting*), existence value, and carbon storage. The major cost was the lost opportunity for local communities to convert forest to agricultural production. Interestingly, when only the three services that had direct benefits to local communities were included in the analysis (bushmeat, timber, and bioprospecting), the benefits of conservation exceeded the costs only in the core area of the biosphere reserve where agricultural activities were restricted. With the addition of existence value and carbon storage, benefits exceeded opportunity costs for almost the entire area (Naidoo and Ricketts 2006).

Gap Analysis

Other efforts to map biological features in the context of the protection of biological diversity were designed explicitly to reveal critical areas that were not currently protected. The Gap Analysis program was initiated by the U.S. Fish and Wildlife Service in the 1980s to provide a method for assessing the representation of vegetation types and species in areas designated for protection (Scott et al. 1987, 1993). The first gap analysis was conducted for the state of Hawaii, where researchers suspected that endangered forest birds were not adequately protected by the existing network of parks and reserves. The concept was simple: map the spatial distribution of the bird species, then map the areas under protection, then combine the maps to see if they overlap or if there are significant gaps between where the birds occur and where the protected areas are. The power of this simple exercise became quite clear—the analysis revealed that only 5% of the area covered by these endangered birds was within the boundaries of protected areas. The stark results of this analysis prompted the planning of additional forest reserves and protected areas in Hawaii for these birds, as well as for other species.

Given the success of the Hawaii analysis for identifying critical gaps in species protection, a gap analysis program has now been initiated in all 50 states, as well as in a few other countries besides the United States. Each state is involved in mapping actual vegetation types and actual or predicted distributions of all vertebrate species. Some state efforts also include maps of invertebrates, including butterflies or ants, as well as aquatic habitats and species. The program is administered by the U.S. Geological Survey, and each state has its own office and staff for compiling and analyzing information. The gap analysis program also collaborates extensively with partners to obtain and distribute information, as well as to integrate the conservation efforts of government agencies and nongovernmental conservation organizations. The maps produced by this program are incredibly useful in answering two basic questions. Where are areas of high diversity? And are those areas currently protected? If there are clearly conservation gaps, then the analysis can guide planning efforts to prioritize areas that have high diversity but are currently unprotected.

HCPs and NCCPs

Most of the ecological planning efforts discussed so far are voluntary, meaning they are not legally mandated, but other conservation planning actions are at least partially required by law. The Endangered Species Act, or ESA—created in 1973 and amended in 1978 and 1982—is the most powerful piece of legislation in the United States to protect biodiversity. The ESA prohibits the *take* of listed species, which means that it is illegal to harm, harass, injure, or kill an endangered species. Moreover, a 1995 U.S. Supreme Court ruling confirmed that indirect harm to listed species via habitat destruction also constitutes take and is prohibited by the ESA. There are two critical exceptions to the take prohibition, however. First, Section 7 of the ESA stipulates that on federal lands, take is allowed if the actions involved are considered not to jeopardize the species as a whole. Second, Section 10(a) of the ESA includes a provision that allows for the loss of habitat or of a species if that loss is incidental to development or other routine land use or land management activities by a private landowner. The landowner must submit a habitat conservation plan (HCP) that specifies the actions that will be taken to minimize and mitigate for the negative consequences of the proposed project, and in return they are issued an *incidental take permit*. Noss, O'Connell, and Murphy (1997) provide an excellent summary of the scientific and legal aspects of the HCP process.

During the 1980s and early 1990s, most HCPs were submitted by individual landowners to compensate for actions that would affect individual listed species on relatively small parcels of land. By the mid-1990s, the U.S. Fish and Wildlife Service realized that multiple listed species often occurred on land parcels simultaneously, prompting the need for multiple-species HCPs. One of the first such MSHCPs was initiated for the Coachella Valley in Riverside County, California, which built upon the HCP previously established for the Coachella Valley fringe-toed lizard, *Uma inornata* (Barrows 1996). There was also a growing recognition of several limitations of HCPs. For example, mitigation actions required monitoring, but it was not always clear that monitoring was completed or that remedial actions were taken if mitigation was unsuccessful (Harding et al. 2001). Critics of HCPs also argued for greater involvement by independent scientists in the preparation and review of conservation plans. And many realized that conservation would be likely to be more effective in broad-scale networks of conserved lands rather than in small, piece-by-piece efforts.

Several of these limitations of HCPs were addressed by an innovative conservation planning effort developed in the state of California. Like Florida and Texas, California is a state with large numbers of endemic and endangered species, so it was particularly well positioned to implement broad-scale, multi-species conservation planning efforts. In 1991, the California Department of Fish and Game lobbied the state legislature to pass the Natural Community Conservation Planning Act (NCCP), which imposed more stringent standards than HCPs for mitigation in response to the incidental take of species protected by the ESA. Specifically, the NCCP Act stipulated that the planning process would be voluntary, would include many species over large areas, would streamline the permitting process for private landowners that needed incidental take permits, and would involve independent scientific review.

The first NCCP was initiated for the Southern California coastal sage scrub region, which included parts of five counties, encompassed 6000 square miles, and covered over 100 species of plants and animals (California Department of Fish and Game 1999). Within this large region, plans for sub-regions were prepared and adopted by cities and counties. The NCCP process is designed to proactively protect large areas of conserved lands while allowing development in this rapidly growing state, rather than taking a more reactive, opportunistic approach to individual development projects one by one. The NCCP process is now in progress in many regions of California, and all of these conservation plans involve the protection of multiple species in large regional landscapes. These planning efforts integrate scientific information on species distributions, abundances, and movement patterns with planning schemes that combine basic ideas from conservation biology about maximizing fragment area, minimizing edge effects, and promoting functional connectivity among patches. It is too early to tell whether these efforts will be successful—most of the plans developed so far are designed to be in place for 30 to 50 years, and most of the conservation and restoration actions are still in progress. But the broad, integrative approach of the NCCP process versus individual HCPs seems destined to be an improvement, and it could serve as a model for other states to follow in their conservation planning efforts.

Conservation by Design

The Nature Conservancy (TNC) is a large non-governmental conservation organization based in the United States, but with conservation programs in several regions outside the country, including Asia, Central and South America, the Caribbean, and Africa. TNC uses and has further refined many of the tools developed by systematic conservation planning to identify and prioritize conservation areas. In the context of the planning approaches discussed in this chapter, TNC focuses on choosing areas in which to protect biological diversity and the processes that generate and maintain that diversity (the *where* of conservation), and on developing strategies to implement conservation actions (the *how* of conservation). TNC developed a scheme called Conservation by Design that involves setting conservation priorities, creating strategies, implementing those strategies, and then assessing whether the strategies have effectively achieved the conservation goals (Poiani et al. 1998; Groves 2003; Nature Conservancy 2006). The scheme involves multiple scales of analysis and action, including appraisals of global conservation priorities, ecoregional assessments, and conservation action plans that are conducted within ecoregions. Craig Groves spent five years directing conservation planning efforts for TNC and has summarized many of the strategies and methods used in this process (Groves 2003). Groves described the result of prioritization efforts as a *conservation blueprint*, defined as "a map that identifies conservation areas of the planning region and associated information on the conservation targets contained in these areas" (p. 216). With blueprints in hand, TNC conservation biologists can move forward to implement strategies for building conservation networks.

Much of the conservation planning activity with TNC occurs as the result of ecoregional assessments, which identify key areas for conservation within biophysically defined ecoregions. One such ecoregion is the Osage Plains/Flint Hills prairie ecoregion of eastern Kansas and northeastern Oklahoma, which is featured in the latest Conservation by Design publication (Nature Conservancy 2006) and which, you may recall, is where chapter 1 of this book began. Within this ecoregion, conservation projects focus on the protection of prairie species in reserves, as well as on community-based conservation projects. These latter programs engage local landowners and stakeholder organizations from the farming and ranching communities to promote compatible ranching practices that maintain and enhance rare species native to the tallgrass prairie, including greater prairie chickens (fig. 11.2). Strategies include collaborations with private landowners and communities to implement conservation easements, the management of invasive exotic species, and assessments of suitable levels of prescribed fire and grazing to achieve long-term economic and biological viability of the tallgrass prairie ecosystem.

Living Landscapes Program

Several of the approaches discussed so far prioritize areas for protection that have the highest number of species, with the rationale that these "hotspots" of diversity are the most valuable for conservation. A slightly different method of prioritizing sites is to focus on particular species whose presence indicates that the ecosystem is viable, such as those based on umbrella, keystone, flagship, or focal species (e.g., Mills, Soulé, and Doak 1993; Simberloff 1998). The Wildlife Conservation Society's (WCS) Living Landscapes program developed an approach based on the requirements of landscape species, but one that differs slightly from previously used single-species designations, such as umbrella or flagship species (Sanderson et al. 2002; Coppolillo et al. 2004). This approach blends biologically

Figure 11.2. Top: Scene from the Osage Plains/Flint Hills prairie ecoregion near Cassoday, Kansas. *Bottom:* Prairie chicken pride. Photos by S. K. Collinge.

based spatial data with information on the locations, types, and intensities of human land use, in order to design protected-area networks that will hopefully prevent antagonistic interactions between people and wildlife while sustaining biological diversity and human well-being.

Landscape species are defined as "biological species that use large, ecologically diverse areas and often have significant impacts on the structure and function of ecosystems" (Redford et al. 2000, quoted in Sanderson et al. 2002, p. 43). To be designated as such, landscape species must have populations that (1) occupy large, ecologically diverse areas, (2) usually have direct impacts on ecological structure and function, (3) are susceptible to human activities, and (4) may be socioeconomically meaningful. The general methodology for moving from the choice of a landscape species to a geographically based conservation plan proceeds in several steps. First, the requirements of the landscape species are mapped on the landscape. This area includes a population of the landscape species that is large enough to be "ecologically functional" (Redford 1992; Sanderson et al. 2002; Soulé et al. 2003, 2005; see also chapter 7). So the minimum population size in this context is not what is required for viability in demographic or genetic terms, but what is required for that species to perform its ecological role within that ecosystem (see Sanderson 2006 for a summary of different approaches to setting population target levels). Setting the target size of an ecologically functional population may be particularly difficult in practice, since we rarely know enough about the strength and redundancy of species interactions (Sanderson 2006). Second, human activities within the proposed conservation area are defined and mapped. Third, the first two maps are combined to identify areas of spatial overlap. The final steps use these maps to identify priority areas for conservation, based on the requirements of both landscape species and human inhabitants.

Before this general framework is applied, landscape species must be selected according to a systematic procedure (Sanderson et al. 2002). This process was developed and tested for two WCS conservation sites—an area in the northwestern Bolivian Andes known as the Madidi landscape, and an area in the northern Congo Republic known as the Ndoki-Likouala landscape. Specific details can be found in Coppolillo et al. (2004), but the general procedure can be summarized as follows: (1) score each species according to each of five criteria (area used, heterogeneity of habitat use, vulnerability, functionality, and socioeconomic status), (2) add subsequent high-scoring species that provide complementarity to the species already chosen, and then (3) complete a sensitivity analysis to determine whether any of the five criteria had a disproportionate effect on the choice of a particular species. For the Madidi site, this process resulted in the selection of six landscape species: white-lipped peccary, spectacled bear, condor, catfish, jaguar, and vicuna. For the central African site, five landscape species were selected: elephant, chimp, bongo, forest buffalo, and dwarf crocodile.

The landscape species framework is now being applied by WCS at 12 sites throughout the world. Staff are working to identify landscape species, set population targets, and map the biological and human landscapes for site-conservation planning. For example, in the Adirondack Park landscape of northern New York State, the designated landscape species are black bear, marten, common loon, moose, three-toed woodpecker, and wood turtle. WCS research efforts in this region are focused on identifying the biological needs of these species, assessing threats to their continued persistence, and identifying potential areas of conflict between wildlife and humans.

The Wildlands Project

The Wildlands Project emerged as a collective vision of conservation biologists Michael Soulé and Reed Noss and conservation activist Dave Foreman (Foreman et al. 1992; Noss 1992) as a means of providing wall-to-wall coverage of the North American continent to protect native biodiversity. Their slogan—"reconnect, restore, rewild"—emphasizes their goals of re-establishing keystone species, such as carnivores, protecting large areas of native habitat, and providing functional landscape connectivity from Canada to Mexico and from the Atlantic to the Pacific. The rewilding aspect of the Wildlands project is predicated on the knowledge that carnivores play critical functional roles in regulating populations, communities, and ecosystems (Soulé and Terborgh 1999; Soulé et al. 2003, 2005). Because most large carnivore populations in North America have been overexploited over the past two centuries, prey populations in many areas have increased to the point where they pose threats to ecosystems through their heavy grazing, such as elk in Colorado's Rocky Mountain National Park or white-tailed deer throughout the northeastern United States. Several regionally based conservation groups are working together and joining with local and regional conservation partners and government agencies to develop broad-scale, long-term conservation plans that converge in their emphasis on big, connected areas that support viable carnivore populations (Noss 2003).

One of the first case studies proposed as part of the Wildlands project was for the Oregon Coast Range (Noss 1993; Noss and Cooperrider 1994). This reserve design embodied the principles of the Wildlands project to conserve large, connected wild places—and included substantial core areas of coastal temperate rainforests that supported spotted owls, marbled murrelets, and other rare species; important watersheds; multiple-use buffer zones; and regional connections among core areas. In the past, this area supported carnivores such as grizzly bears, gray wolves, Pacific fishers, and wolverines, but all are regionally extinct (Noss and Cooperrider 1994). This ambitious effort resulted in a plan that designated 50% of the region as protected at some level, which was estimated by Noss (1992) as the amount needed in any region to restore and maintain biological diversity.

Ongoing regional efforts associated with the Wildlands project include the Southern Rockies Ecosystem Project (SREP), which calls for the restoration and connection of a conservation network that extends from southern Wyoming to northern New Mexico, a distance of approximately 500 miles (Southern Rockies Ecosystem Project 2004). Within this ecoregion, SREP scientists assessed ecosystem health and integrity by mapping historic and current human settlements and land uses; the diversity of plants, animals and natural communities; terrestrial, riparian, and aquatic ecosystems; and protected areas. Combining these maps showed the areas of greatest threat as well as those having the potential for restoration. This assessment provided a baseline for exploring two strategies for conservation within this region: the establishment of a biologically comprehensive reserve system for the region, including existing protected areas, and the use of ecosystem management principles, such as prescribed fire, that can help restore these ecosystems to a range of conditions consistent with their evolutionary history.

Interdisciplinary Approaches

To call the following examples interdisciplinary implies that the approaches already discussed were not, but that is not quite accurate. The beginning of this chapter noted that the disciplinary boundaries of many ecological planning activities were blurry, and that the approaches fell along a continuum. All of the projects discussed so far contain some elements of ecology and planning. But the following examples illustrate endeavors to explicitly join efforts across disciplines to create ecologically meaningful plans. The first example relates to the design and construction of transportation infrastructure, primarily roads. The second explores conservation development as a strategy for reducing the loss of biodiversity and ecosystem services in rural landscapes that are undergoing residential development. The third integrates projections of shifts in the distribution of human households in southwestern China with the availability of forests for giant pandas.

It is estimated that one million vertebrates are killed on roads in the United States every day (Forman and Alexander 1998). This is an astounding number, even though many of these animals may be small, abundant species whose populations are not severely affected by mortality due to road kill. But the sheer magnitude and extent of this phenomenon suggests that *road ecology* is a fundamental aspect of ecological planning (Forman et al. 2003). Roads are not only routes for fast-moving cars that can collide with animals, but they may also disrupt the breeding behavior of nearby animals, serve as barriers to animal movement, reduce habitat area, fragment the landscape, act as corridors for movements of exotic species, and change water flows (Forman and Alexander 1998, Forman et al. 2003). And all of these issues relate to the ecological consequences of habitat loss and fragmentation that have been discussed in this book.

The disruption of animal movement is one aspect of roads that has been the subject of creative collaboration between ecologists and transportation planners. Because roads may effectively cut off animals' access to particular areas, many road corridors now have built-in wildlife passages, including culverts, underpasses, and overpasses, to secure animal movement across roads (Forman 2004). These wildlife structures have been constructed in Europe, North America, and Australia; recently, in China, railway bridges and wildlife passages were constructed to allow the passage of the highly endangered Tibetan antelope across the Qinghai–Tibet railway (Xia and Yang 2007). The location of these features, as well as their design, may influence their use by animals. Clevenger and Waltho (2005) evaluated whether underpasses and overpasses constructed along highways in Banff National Park, Alberta, Canada, facilitated the movement of wide-ranging large carnivores and studied which features of the structures influenced their use. They used both track stations and infrared cameras to record the use of crossing structures by wildlife over a three-year period. Their analyses showed that the structural features of underpasses and overpasses strongly influenced their use; crossing structures that were relatively short, wide, and high tended to be used by grizzly bears, wolves, elk, and deer, whereas longer, narrow passages were used by black bear and cougars (Clevenger and Waltho 2005). The authors concluded that, given the differences in species' responses to crossing structures, future road construction projects should include a diversity of wildlife passages to facilitate movement. Transportation planning with wildlife in mind is an exciting area of research and implementation, combining a knowledge of engineering with ecological studies of animal movement behavior and the animals' responses to habitat loss and fragmentation.

Ideas about improving the conservation value of residential landscapes integrate landscape architecture and planning with ecology and conservation (Arendt 1996; Dramstad, Olson, and Forman 1996), and these ideas partly contribute to the philosophical principles of what Krieger (1991) and others have termed new urbanism. These principles generally apply to urban or suburban landscapes with relatively high-density residential development and a relatively low potential for contributing to broad-scale conservation. Recent trends in residential development in the United States and elsewhere, however, involve the conversion of rural landscapes into low-density residential development, sometimes called exurban development or rural sprawl (Theobald, Miller, and Hobbs 1997; Theobald 2004). Because this type of development often involves habitat loss and the fragmentation of otherwise relatively undisturbed lands, it has a substantial potential to severely reduce biodiversity and ecosystem services (Pejchar et al. 2007).

Given that many current patterns of exurban development are ecologically harmful, collaborative efforts to motivate landowners or municipalities toward less-damaging spatial patterns of development could contribute substantially to conservation goals. An interdisciplinary team of researchers at Stanford University examined biophysical, economic, and institutional opportunities and constraints, seeking to implement exurban development that is more compatible with conservation (Pejchar et al. 2007). As they envision it, "conservation development is a potential but rarely realized development strategy that integrates conservation of biodiversity and ecosystem services with development" (p. 71). Conventional patterns of exurban development typically do not consider a property's conservation values when designing development plans, but conservation development does prioritize the ecological values of a site, in order to promote development that is compatible with the protection of biodiversity and ecosystem services.

Biophysical considerations in development include manipulating site selection, housing density, and land management to determine which conditions produce larger or smaller amounts of ecological harm. Land management, for example, affected grasshopper abundance in an exurban landscape in southern Arizona significantly; abundance was higher where homeowners kept a few livestock than for homes without livestock (Bock, Jones, and Bock 2006). Economic analyses showed that conservation development can be, but not always is, more beneficial economically to developers than conventional development. For example, in conservation developments, residents typically have broader views of open space, so individual houses have a higher property value than those in conventional developments, a factor that could compensate for lost revenue from the sale of fewer houses (Pejchar et al. 2007). But there can be institutional impediments, since conservation development in some instances may require the developer to obtain variances to zoning regulations, which may greatly impede the development process and effectively provide disincentives for the developer. The authors noted that because large amounts of land are undergoing exurban development, formal strategies should be pursued to encourage developments that support biodiversity and ecosystem services while providing benefits to human communities.

The footprint of residential development continues to expand via exurban growth, and in some cases directly threatens the persistence of wildlife within adjacent nature reserves. For example, in southwestern China, the rapid increase in the number of human households adjacent to the Wolong Nature Reserve (the largest remaining protected area for giant pandas) jeopardizes the quality of suitable forests for this critically endangered species (Liu et al. 2001; Linderman et al. 2005; Viña et al. 2007). Liu and colleagues have constructed simulation models that integrate human demography, resource use, and the distribution and abundance of forests used by giant pandas to project the ecological consequences of alternative land-use-change scenarios (Linderman et al. 2005). The results provide useful guidelines for decision-making in this region. For example, one of the clearest outcomes of their model projections was that increased fuel-wood consumption by local communities inside the reserve degrades forest areas for the pandas, so the implementation of alternative fuel sources (other than wood) is likely to be especially beneficial for maintaining intact forests for these animals (fig. 11.3).

SYNTHESIS

Ecological planning encompasses a wide variety of activities that emanate from the fields of landscape planning and conservation biology. Traditionally, biologists' expertise was useful for identifying areas of particularly high value for biodiversity, and planners were able to integrate biological information with other types of data on social needs and desires—such as areas suitable for housing and transportation infrastructure—to determine the spatial locations and features of human interventions. Disparate methods have given way to more interdisciplinary, comprehensive approaches in recent years, ones that fully integrate human

Figure 11.3. Top: A giant panda (*Ailuropoda melanoleuca*) at the captive breeding center in Wolong Nature Reserve, Yunnan Province, China. *Bottom:* A pile of fuel wood at a household in Wolong Nature Reserve. Photos by S. K. Collinge.

well-being with the protection of biodiversity. Planning generally occurs at relatively broad spatial scales, which makes sense given the magnitude and time frame of proposed actions. But because landscape change occurs one parcel at a time, each of our individual actions are also worthy of attention, since they collectively influence the patterns and outcomes of such changes.

Because human beings are ultimately dependent on nature for survival, ecological planning schemes that simultaneously preserve biodiversity, provide ecosystem services, and promote human well-being are the ones that will be the most successful. The many diverse approaches to ecological planning considered here attend to significant aspects of biological and human needs. All of these approaches involve long-term plans and actions, so it is too early to tell whether their intended benefits will be realized. But our scientific understanding of the consequences of landscape change for ecological systems is vast, and we can creatively use this storehouse of knowledge to guide our efforts. Ewers and Didham (2006) pointed out that in many landscapes, reserve networks are not yet established, so we still have the opportunity to act on what we know about species' responses to habitat loss and fragmentation to do effective conservation planning. And Nassauer (2006) asserted that conservation biology and landscape planning would benefit by building closer relationships, as both fields need to think more broadly about the full array of landscapes in which humans and other species reside. She advocated that we "creatively examin[e] other land uses, near and far from reserves, as well as other plausible landscape matrix futures" (p. 678) to accomplish the goals of maintaining biodiversity and human well-being.

Some Final Thoughts

The care of the Earth is our most ancient and most worthy
and, after all, our most pleasing responsibility. To cherish
what remains of it, and to foster its renewal, is our only
hope.

—*Wendell Berry (1977)*

Modern humans have extensively altered the surface of the Earth. Human actions—whether taken to cultivate food, or build roads, dwellings, and dams, or gather fuel and fiber—have left a visible legacy in the scattered remnants of native ecosystems. As the research summarized in this book illustrates, this kind of landscape alteration generally disrupts ecological systems to the point where biological diversity and ecosystem functions are greatly diminished. Although there will always be more that we can learn about the ecological consequences of habitat loss and fragmentation, the main motivation for writing this synthesis was to summarize our current knowledge so that students and practitioners could work toward successful efforts to protect and re-assemble natural ecosystems. The present brief chapter highlights key concepts learned from this body of research, suggests some promising future research directions, and describes a few examples where research results have been applied toward conservation efforts.

KEY CONCEPTS

The conclusion of each of the chapters in this book presented a brief synthesis of the key concepts and findings relevant to the particular topic covered in that chapter. Rather than reiterating each of those conclusions, this précis instead notes five major findings that are worthy of special recognition. First, it is clear from observations and models of landscape change that there are a variety of spatial patterns and processes of landscape transformation, and each of these may pose particular consequences for individuals, species, communities, and ecosystems. For example, the discussion in the first few chapters regarding definitions of habitat loss versus fragmentation (or subdivision) highlights the necessity of accurately describing the processes of landscape change. This discussion is crucial, not only because scientists should strive to be precise in their inferences from research findings, but also because the interventions that society chooses to implement—such as providing corridors between habitat patches or increasing the area of such patches—are based on those research conclusions, and the successes or failures of those actions are likely to depend to a significant extent on whether the ecological consequences of landscape change *are* primarily due to the loss or fragmentation of habitat.

A second major finding from this field of study is that landscape change can and does affect species interactions significantly. Because species are intimately linked to one another in complex interaction webs via competition, predation, parasitism, and mutualisms, shifts in the abundance of particular players in these roles may result in cascading effects on other species. Some of these interactions may become stronger and others weaker, depending on changes in resource distribution and the abundance of interacting species. In particular, recent work on parasites and pathogens in relation to human-induced landscape changes clearly shows the ecological relevance of such shifts, even to the point of increasing the human risk of exposure to harmful parasites and pathogens.

Third, the construction of models of ecological reality has been especially useful in our understanding of fragmented landscapes, since, in many cases, field experiments are impossible, unethical, or harmful to species and ecosystems. Models can help clarify what the outcomes of particular landscape changes are likely to be, given specific assumptions about species' responses to those changes. Moreover, simulation models are quite helpful when they project alternative scenarios of land use change and estimate the ecological consequences of each scenario. These kinds of analyses assist decision-makers in arriving at more informed choices regarding the likely outcomes of particular policies pertaining to land or resource use, such as zoning, harvest quotas, or habitat mitigation requirements.

Fourth, ecological restoration and conservation planning provide a wide variety of promising approaches to repair or mitigate the damage to ecosystems caused by habitat loss and fragmentation. However, in designing restoration projects for declining populations or species, it may be necessary to first examine the assumption that restoration is likely to increase population sizes (e.g., see Schrott, With, and King 2005). The spatial arrangement of restored habitat does appear to matter in some contexts. This suggests that, at least in some cases, there is symmetry between expectations of species' responses to habitat loss and fragmentation versus predictions of their responses to habitat restoration.

Lastly, a fifth major finding that is worthy of a more detailed discussion is that landscape change alters the delivery of ecosystem services, sometimes in unexpected ways. Most research to date on the ecology of fragmented landscapes has focused primarily on the consequences for biodiversity—on genes, populations, communities, and ecosystems. But there are vital links between changes in diversity at each of these levels and the delivery of ecosystem services. A meaningful next step would be to develop a fuller understanding of the relationship between landscape pattern and ecological services. For example, the discussion of landscape context in chapter 5 gives rise to some pertinent questions in this regard. How does habitat spatial configuration contribute to the maintenance of water quality and quantity? And, in turn, can spatial patterns of ecological restoration positively influence water quality and quantity?

One of the most famous examples of the water filtration properties of natural landscapes comes from the New York City watershed, which provides clean drinking water to that city's millions of residents, largely through the acquisition and protection of forested land in the upper reaches of the watershed (reviewed in Daily and Ellison 2002). Instead of building a new and vastly expensive water filtration plant, city officials elected to purchase and protect land in the watershed from degradation, thereby ensuring the high quality of New York City's water supply. From this project the question arises, Would a thoughtful design of the spatial configuration of these protected lands enhance the provisioning of this vital ecosystem service? Based on what we know about land use and water quality, the answer is probably yes, given that research on landscape context has emphasized how the spatial arrangement of particular land uses may influence ecological processes, including the flow of nutrients and pollutants into surface water bodies (e.g., Soranno et al. 1996; Canham et al. 2004).

Such attention to these sorts of questions will allow us to expand our knowledge of the effects of habitat spatial configuration on biodiversity to include its influences on the delivery of ecosystem services. Some innovative examples of this type of research effort are described below.

PROMISING RESEARCH DIRECTIONS

In the past few years, several international conservation organizations have explicitly expanded their conservation campaigns to include the notion that protecting nature is essential to human well-being, because nature provides vital ecosystem services on which all life depends. Intact ecosystems harbor high biological diversity, and they are also better able to provide services such as carbon sequestration, flood control, water filtration, and crop pollination. Recent research efforts to support this mission have included (1) identifying and mapping the spatial locations of particular ecosystem services at scales ranging from local (Naidoo and Ricketts 2006) to global (W. Turner et al. 2007), (2) superimposing maps of ecosystem services with maps of biodiversity to examine their regions of overlap (W. Turner et al. 2007), and (3) investigating landscape spatial configurations that enhance the provisioning of ecosystem services (Ricketts, Williams, and Mayfield 2006; Kremen et al. 2007).

Most relevant to the topic of this book are efforts to clarify the influence of landscape spatial configuration on the delivery of ecosystem services. As an example, Ricketts, Williams, and Mayfield (2006) reviewed three case studies of agroecosystems where the pollination of a particular crop plant—in these instances, coffee, watermelon, and kiwi—was enhanced by the presence of native habitat adjacent to agricultural fields. The native habitat provided nesting resources for pollinators (native bees), and the agricultural fields provided foraging habitat. The authors argued that structural connectivity (in this case, the distance of forest remnants to foraging resources) had a significant effect on the delivery of this critical ecosystem service. They also emphasized that connectivity among *different* habitat types may affect ecosystem services such as pollination significantly. But this review also noted two examples where proximity to native habitat did not influence pollination, so clearly more research needs to be done in order to evaluate the conditions under which this phenomenon is likely to occur. For example, it is likely that landscape connectivity positively affects other services that involve species interactions, such as seed dispersal, although there are not many studies as of yet with data to provide supporting evidence.

In a comprehensive study of the effects of land use change on coffee production, Priess et al. (2007) highlighted "the limited availability of empirical data that quantify the relationship between land use patterns and ecosystem services" (p. 407). Their study used spatially explicit models to evaluate four simulated land use scenarios for an agricultural landscape in Sulawesi, Indonesia, where the main crop was coffee and the ecosystem service was the pollination of this crop. The scenarios projected changes in both total forest cover and the spatial distribution of forest, which provides critical habitat for the insects that pollinate coffee. They calculated the economic value of retaining forests as pollinator habitat and showed that the retention of forest patches clearly positively influenced economic returns from coffee production.

A recent comprehensive review of research projects focused on pollination and landscape connectivity found results from sixteen different crop species on five continents (Ricketts et al. 2008). Taken together, these studies showed that pollinator species richness and visitation to crop plants generally tend to decline with the distance from native habitat, especially in tropical ecosystems. But the impacts of reduced pollinator visitation on seed and fruit set were less clear. Further study of the links between landscape spatial configuration, crop pollination, and economic returns should prove quite beneficial in informing the discussion of the best way(s) to maintain this critical ecosystem service in highly modified landscapes.

A second promising research direction for students and scholars of fragmented landscapes is to quantify the effects of *habitat degradation*, in addition to habitat loss and fragmentation, on biodiversity and ecosystem services. With notable exceptions, most investigators have treated the landscape as binary—either a location is suitable habitat for a species, or it is not. The emphasis on fragment size and isolation has meant that most researchers have considered habitat fragments to be more or less equal, except for their area and connectivity. But we know that this is a simplification, and that much of the variation in species' responses to landscape alteration may come from subtle changes in the quality of the fragments. The effects of habitat quality on population dynamics were formalized in the concept of source-sink dynamics (Pulliam 1988; With and King 2001), but demographic rates are not often measured across different fragments or landscapes. In a powerful illustration of this concept, Doak (1995) modeled the population dynamics of grizzly bears in Yellowstone National Park and concluded that subtle changes in habitat quality could have dramatic consequences for these dynamics. Additionally, land use or management practices in areas adjacent to habitat remnants may strongly influence individuals, populations, communities, and ecosystem processes (reviewed in chapter 5). For example, the simulation model results of Fahrig (2001) showed that management of the matrix was the most promising avenue for enhancing population survival in fragmented landscapes.

Third, there are still numerous opportunities for creative interdisciplinary research that integrates scientific knowledge with policies and economic incentives that will motivate desired actions regarding land use change at many spatial scales. Many of the conservation planning activities reviewed here inform decisions taken by large government agencies or conservation organizations, but they do not necessarily stimulate responses by private industry or individual property holders. So in addition to these broad-scale efforts, there is room for creative solutions that will provide incentives and information for stakeholders ranging from private homeowners to large landowners, and from small businesses to large multinational corporations. For example, research by Liu and colleagues in the forests of southwestern China suggests a need for alternative fuels to maintain human livelihoods and yet protect habitat for pandas. This information, combined with incentives for households to switch fuel sources, will be meaningful at the local scale and will also help to protect an animal species that is adored around the world. There is an increasing need to raise awareness at each level—from local to global—of the consequences of particular land use actions that restore rather than degrade biodiversity and ecosystem services.

INCORPORATING RESULTS WITH ACTION

Given all of the studies on the ecology of fragmented landscapes, we should ask whether this knowledge has been successfully implemented into conservation action. There are clearly examples where scientific findings from this field of study have been usefully incorporated into meaningful activities. For example, the general concepts about maintaining large, well-connected areas of native habitat have made their way into conservation plans at local, regional, national, and international levels. But, as noted in the discussions of animal and plant movement (see chapter 7), the concept of landscape connectivity goes beyond structural links among habitat fragments. This broader view of connectivity has not been fully appreciated and implemented—probably because it is much more difficult, often requiring specific, detailed studies of animal movement, with models of that movement then being added into complex spatial mapping exercises (e.g., Theobald 2006).

In the context of ecological restoration, few projects have included perspectives from studies of landscape spatial configuration. There are noteworthy exceptions, however. For example, Schultz and Crone's (2005) studies of the Fender's blue butterfly in Oregon have integrated field observations with mathematical models to make explicit recommendations regarding patches of prairie habitat to be acquired and restored.

Clearly there are many opportunities to meaningfully incorporate what has been a few decades of scientific research on fragmented landscapes into actions that will stem the tide of biodiversity losses and the degradation of ecosystem services. My hope is that those who read this book will find knowledge and motivation that will then inspire action. Pick your favorite region or locale, or your favorite part of this field of study, and take a step toward meaningful change, even though it may seem small or insignificant. When she was in her mid-80s, my grandmother planted fruit and nut trees on her Kansas farm, knowing she would never taste their products. It is that kind of hopeful, forward-thinking act—to ensure Earth's viability—that must propel us forward.

Literature Cited

Aars, J., H.P. Andreassen, and R.A. Ims. 1995. Root voles: Litter sex ratio variation in fragmented habitat. *Journal of Animal Ecology* 64:459–472.

Aars, J., E. Johannesen, and R.A. Ims. 1999. Demographic consequences of movements in subdivided root vole populations. *Oikos* 85:204–216.

Åberg, J., G. Jansson, J.E. Swenson, and P. Angelstam. 1995. The effect of matrix on the occurrence of hazel grouse (*Bonasa bonasia*) in isolated habitat fragments. *Oecologia* 103:265–269.

Åberg, J., J.E. Swenson, and H. Andrén. 2000. The dynamics of hazel grouse (*Bonasa bonasia* L.) occurrence in habitat fragments. *Canadian Journal of Zoology* 78:352–358.

Aguilar, R., L. Ashworth, L. Galetto, and M.A. Aizen. 2006. Plant reproductive susceptibility to habitat fragmentation: Review and synthesis through a meta-analysis. *Ecology Letters* 9:968–980.

Ahern, J. 2005. What is landscape ecology? Essay as part of the US-IALE (United States Regional Association of the International Association for Landscape Ecology) web site. www.usiale.org/whatisle/ahern.htm.

Ahern, J., E. Leduc, and M.L. York. 2006. *Biodiversity planning and design: Sustainable practices.* Island Press, Washington, DC.

Aizen, M.A., L. Ashworth, and L. Galetto, 2002. Reproductive success in fragmented habitats: Do compatibility systems and pollination specialization matter? *Journal of Vegetation Science* 13:885–892.

Aizen, M.A., and P. Feinsinger. 1994. Forest fragmentation, pollination, and plant reproduction in a Chaco dry forest, Argentina. *Ecology* 75:330–351.

Akçakaya, H.R., M.A. McCarthy, and J.L. Pearce. 1995. Linking landscape data with population viability analysis: Management options for the helmeted honeyeater *Lichenostomus melanops cassidix*. *Biological Conservation* 73:169–176.

Akçakaya, H.R., V.C. Radeloff, D.J. Mlandenoff, and H.S. He. 2004. Integrating landscape and metapopulation modeling approaches: Viability of the sharp-tailed grouse in a dynamic landscape. *Conservation Biology* 18:526–537.

Alexander, G. and J.R. Hilliard Jr. 1969. Altitudinal and seasonal distribution of Orthoptera in the Rocky Mountains of northern Colorado. *Ecological Monographs* 39:385–431.

Allan, B.F., F. Keesing, and R.S. Ostfeld. 2003. Effect of forest fragmentation on Lyme disease risk. *Conservation Biology* 17:267–272.

Allan, J.D. 2004. Landscapes and riverscapes: The influence of land use on stream ecosystems. *Annual Review of Ecology and Systematics* 35:257–284.

Allee, W.C. 1951. *The social life of animals*. Beacon Press, Boston, MA.

Allen, T.F.H., and T.W. Hoekstra. 1992. *Toward a unified ecology*. Columbia University Press, New York, NY.

Allen, T.F.H., and T.B. Starr. 1982. *Hierarchy: Perspectives for ecological complexity*. University of Chicago Press, Chicago, IL.

Allendorf, F.W., and G. Luikart. 2007. *Conservation and the genetics of populations*. Blackwell, Malden, MA.

Amarasekare, P. 2004. Spatial dynamics of mutualistic interactions. *Journal of Animal Ecology* 73:128–142.

American Heritage dictionary of the English language. 2006. 4th ed. Houghton Mifflin, Boston, MA.

Andel, J. van, and J. Aronson, eds. 2005. *Restoration ecology: The new frontier*. Blackwell, Malden, MA.

Anderson, A.B., and C.N. Jenkins. 2006. *Applying nature's design: Corridors as a strategy for biodiversity conservation*. Columbia University Press, New York, NY.

Anderson, P.K., A.A. Cunningham, N.G. Patel, F.J. Morales, P.R. Epstein, and P. Daszak. 2004. Emerging infectious diseases of plants: Pathogen pollution, climate change, and agrotechnology drivers. *Trends in Ecology and Evolution* 19:535–544.

Andreassen, H.P., S. Halle, and R.A. Ims. 1996. Optimal width of movement corridors for root voles: Not too narrow and not too wide. *Journal of Applied Ecology* 33:63–70.

Andreassen, H.P., K. Hertzberg, and R.A. Ims. 1998. Space-use responses to habitat fragmentation and connectivity in the root vole *Microtus oeconomus*. *Ecology* 79:1223–1235.

Andreassen, H.P., and R.A. Ims. 2001. Dispersal in patchy vole populations: Role of patch configuration, density dependence, and demography. *Ecology* 82:2911–2926.

Andrén, H. 1992. Corvid density and nest predation in relation to forest fragmentation: A landscape perspective. *Ecology* 73:794–804.

———. 1994. Effects of habitat fragmentation on birds and mammals in landscapes with different proportions of suitable habitat: A review. *Oikos* 71:355–366.

Arendt, R.G. 1996. *Conservation design for subdivisions*. Island Press, Washington, DC.

Armitage, K.B. 1991. Social and population dynamics of yellow-bellied marmots: Results from long-term research. *Annual Review of Ecology and Systematics* 22:379–407.

Arnold, G.W. 1995. Incorporating landscape pattern into conservation programs. In Hansson, Fahrig, and Merriam, 309–337.

Arrhenius, O. 1921. Species and area. *Journal of Ecology* 9:95–99.

Ås, S. 1993. Are habitat islands islands? Woodliving beetles (Coleoptera) in deciduous forest fragments in boreal forest. *Ecography* 16:219–228.

Asquith, N.M., S.J. Wright, and M.J. Clauss. 1997. Does mammal community composition control recruitment in neotropical forests? Evidence from Panama. *Ecology* 78:941–946.

Bach, C.E., and D. Kelly. 2004. Effects of forest edges on herbivory in a New Zealand mistletoe, *Alepis flavida*. *New Zealand Journal of Ecology* 28:195–205.

Baker, W.L. 1989. A review of models of landscape change. *Landscape Ecology* 2:111–133.

Baker, W.L., and D.J. Mladenoff. 1999. Progress and future directions in spatial modeling of forest landscapes. In *Spatial modeling of forest landscape change: Approaches and applications*, ed. D.J. Mladenoff and W.L. Baker, 333–349. Cambridge University Press, Cambridge, UK.

Banks, S.C., D.B. Lindenmayer, J. Ward, and A.C. Taylor. 2005. The effects of habitat fragmentation via forestry plantation establishment on spatial genotypic structure in the small marsupial carnivore, *Antechinus agilis*. *Molecular Ecology* 14:1667–1680.

Barbour, M., B. Pavlik, F. Drysdale, and S. Lindstrom. 1993. *California's changing landscapes*. California Native Plant Society, Sacramento, CA.

Barrows, C. 1996. An ecological model for the protection of a dune ecosystem. *Conservation Biology* 10:888–891.

Beier P., and R.F. Noss. 1998. Do habitat corridors provide connectivity? *Conservation Biology* 12:1241–1252.

Beier, P., M. Van Drielen, and B.O. Kankam. 2002. Avifaunal collapse in West African forest fragments. *Conservation Biology* 16:1097–1111.

Beissinger, S.R. and D.R. McCullough, eds. 2002. *Population viability analysis*. University of Chicago Press, Chicago, IL.

Beissinger, S.R., J.R. Walters, D.G. Catanzaro, K.G. Smith, J.B. Dunning, S.M. Haig, B.R. Noon, and B.M. Stith. 2006. Modeling approaches in avian conservation and the role of field biologists. *Auk* 123(Suppl.):1–56.

Beissinger, S.R., and M.I. Westphal. 1998. On the use of demographic models of population viability in endangered species management. *Journal of Wildlife Management* 62:821–841.

Bellamy, P.E., S.A. Hinsley, and I. Newton. 1996. Factors influencing bird species numbers in small woods in south-east England. *Journal of Applied Ecology* 33:249–262.

Bender, D.J., T.A. Contreras, and L. Fahrig. 1998. Habitat loss and population decline: A meta-analysis of the patch size effect. *Ecology* 79:517–533.

Bender, D.J., and L. Fahrig. 2005. Matrix structure obscures the relationship between interpatch movement and patch size and isolation. *Ecology* 86:1023–1033.

Bender, D.J., L. Tischendorf, and L. Fahrig. 2003. Using patch isolation metrics to predict animal movement in binary landscapes. *Landscape Ecology* 18:17–39.

Benítez-Malvido, J., and A. Lemus-Albor. 2005. The seedling community of tropical rain forest edges and its interaction with herbivores and pathogens. *Biotropica* 37:301–313.

Benítez-Malvido, J., and M. Martínez-Ramos. 2003. Impact of forest fragmentation on understory plant species richness in Amazonia. *Conservation Biology* 17:389–400.

Bennett, A.F. 1990. *Habitat corridors: Their role in wildlife management and conservation*. Department of Conservation and Environment, Melbourne, Australia.

Bennett, A.F., K. Henein, and G. Merriam. 1994. Corridor use and the elements of corridor quality: Chipmunks and fencerows in a farmland mosaic. *Biological Conservation* 68:155–165.

Bennett, B.C. 1997. Vegetation on the city of Boulder Open Space grasslands. Ph.D. diss., University of Colorado.

Berglund, H., and B.G. Jonsson. 2003. Nested plant and fungal communities: The importance of area and habitat quality in maximizing species capture in boreal old-growth forests. *Biological Conservation* 112:319–328.

Berry, M.E., and C.E. Bock. 1998. Effects of habitat and landscape characteristics on avian breeding distributions in Colorado foothills shrub. *Southwestern Naturalist* 43:453–461.

Berry, M.E., C.E. Bock, and S.L. Haire. 1998. Abundance of diurnal raptors on open space grasslands in an urbanized landscape. *The Condor* 100:601–608.

Berry, W. 1977. The Unsettling of America: Culture and agriculture. Sierra Club Books, San Francisco, CA.

Bertness, M.D., and R.M. Callaway. 1994. Positive interactions in communities. *Trends in Ecology and Evolution* 9:191–193.

Bierregaard, R.O., Jr., T.E. Lovejoy, V. Kapos, A.A. dos Santos, and R.W. Hutchings. 1992. The biological dynamics of tropical rainforest fragments. *Bioscience* 42:859–866.

Biesmeijer, J.C., S.P.M. Roberts, M. Reemer, R. Ohlemüller, M. Edwards, T. Peeters, A.P. Schaffers, S.G. Potts, R. Kleukers, C.D. Thomas, J. Settele, and W.E. Kunin. 2006. Parallel declines in pollinators and insect-pollinated plants in Britain and the Netherlands. *Science* 313:351–354.

Bilton, D.T, J.R. Freeland, and B. Okamura. 2001. Dispersal in freshwater invertebrates. *Annual Review of Ecology and Systematics* 32:159–181.

Blake, J.G., and J.R. Karr. 1987. Breeding birds of isolated woodlots: Area and habitat relationships. *Ecology* 68:1724–1734.

Blaustein, A.R., and P.T.J. Johnson. 2003. The complexity of deformed amphibians. *Frontiers in Ecology and the Environment* 1:87–94.

Bock, C.E., J.H. Bock, and B.C. Bennett. 1999. Songbird abundance in grasslands at a suburban interface on the Colorado high plains. In *Ecology and conservation of grassland birds of the Western Hemisphere: Proceedings of a conference, Tulsa, Oklahoma, October 1995*, ed. P.D. Vickery and J.R. Herkert, 131–136. Studies in Avian Biology 19. Cooper Ornithological Society, Camarillo, CA.

Bock, C.E., Z.F. Jones, and J.H. Bock. 2006. Grasshopper abundance in an Arizona rangeland undergoing exurban development. *Rangeland Ecology and Management* 59:640–647.

Bock C.E., K.T. Vierling, S.L. Haire, J.D. Boone, and W.W. Merkle. 2002. Patterns of rodent abundance on open-space grasslands in relation to suburban edges. *Conservation Biology* 16:1653–1658.

Boecklen, W.J. 1986. Effects of habitat heterogeneity on species-area relationships of forest birds. *Journal of Biogeography* 13:59–68.

———. 1997. Nestedness, biogeographic theory, and the design of nature reserves. *Oecologia* 112:123–142.

Bogaert, J., D. Salvador-Van Eysenrode, I. Impens, and P. Van Hecke. 2001. The interior-to-edge breakpoint distance as a guideline for nature conservation policy. *Environmental Management* 27:493–500.

Bolger, D.T., T.A. Scott, and J.T. Rotenberry. 2001. Use of corridor-like landscape structures by bird and small mammal species. *Biological Conservation* 102:213–224.

Bolker, B., and B. Grenfell. 1995. Space, persistence and dynamics of measles epidemics. *Philosophical Transactions of the Royal Society of London, B* 348:309–320.

Boone, R.B., and M.L. Hunter Jr. 1996. Using diffusion models to simulate the effects of land use on grizzly bear dispersal in the Rocky Mountains. *Landscape Ecology* 11:51–64.

Borgella, R. 1995. Population size, survivorship, and movement rates of resident birds in Costa Rican forest fragments. Master's thesis, Cornell University.

Borgella, R., A.A. Snow, and T.A. Gavin. 2001. Species richness and pollen loads of hummingbirds using forest fragments in southern Costa Rica. *Biotropica* 33:90–109.

Boström, C., E.L. Jackson, and C.A. Simenstad. 2006. Seagrass landscapes and their effects on associated fauna: A review. *Estuarine, Costal and Shelf Science* 68:383–403.

Bowler, D.E., and T.G. Benton. 2005. Causes and consequences of animal dispersal strategies: Relating individual behaviour to spatial dynamics. *Biological Reviews* 80:205–225.

Bowne, D.R., and M.A. Bowers. 2004. Interpatch movements in spatially structured populations: A literature review. *Landscape Ecology* 19:1–20.

Bowne, D.R., M.A. Bowers, and J.E. Hines. 2006. Connectivity in an agricultural landscape as reflected by interpond movements of a freshwater turtle. *Conservation Biology* 20:780–791.

Box, G.E.P. and N.R. Draper. 1987. *Empirical model-building and response surfaces.* John Wiley and Sons, New York, NY.

Bradley, C.A., and S. Altizer. 2007. Urbanization and the ecology of wildlife diseases. *Trends in Ecology and Evolution* 22:95–102.

Braschler, B., and B. Baur. 2005. Experimental small-scale grassland fragmentation alters competitive interactions among ant species. *Oecologia* 143:291–300.

Brinkerhoff, R.J., N.M. Haddad, and J.L. Orrock. 2005. Corridors and olfactory predator cues affect small mammal behavior. *Journal of Mammalogy* 86:662–669.

Brittingham, M.C., and S.A. Temple. 1983. Have cowbirds caused forest songbirds to decline? *BioScience* 33:31–35.

Brosset, A., P. Charles-Dominique, A. Cockle, J.F. Cosson, and D. Masson. 1996. Bat communities and deforestation in French Guiana. *Canadian Journal of Zoology* 74:1974–1982.

Brothers, T.S. 1993. Fragmentation and edge effects in central Indiana old-growth forests. *Natural Areas Journal* 13:268–275.

Brothers, T.S., and A. Spingarn. 1992. Forest fragmentation and alien plant invasion of central Indiana old-growth forests. *Conservation Biology* 6:91–100.

Brotons, L., and S. Herrando. 2001. Factors affecting bird communities in fragments of pine forests in the north-western Mediterranean basin. *Acta Oecologica* 22:21–31.

Brown, D.G., K.M. Johnson, T.R. Loveland, and D.M. Theobald. 2005. Rural land-use trends in the conterminous United States, 1950–2000. *Ecological Applications* 15:1851–1863.

Brown, J.H., and A. Kodric-Brown. 1977. Turnover rates in insular biogeography: Effect of immigration on extinction. *Ecology* 58:445–449.

Brown, J.K.M., and M.S. Hovmøller. 2002. Aerial dispersal of pathogens on the global and continental scales and its impact on plant disease. *Science* 297:537–541.

Brown, K.S., Jr., and R.W. Hutchings. 1997. Disturbance, fragmentation, and the dynamics of diversity in Amazonian forest butterflies. In Laurance and Bierregaard, 91–110.

Brownstein, J.S., D.K. Skelly, T.R. Holford, and D. Fish. 2005. Forest fragmentation predicts local scale heterogeneity of Lyme disease risk. *Oecologia* 146:469–475.

Bruhl, C.A., T. Eltz, and K.E. Linsenmair. 2003. Size does matter: Effects of tropical rainforest fragmentation on the leaf litter ant community in Sabah, Malaysia. *Biodiversity and Conservation* 12:1371–1389.

Bruna, E.M. 1999. Seed germination in rain forest fragments. *Nature* 402:139.

———. 2003. Are plant populations in fragmented habitats recruitment limited? Tests with an Amazonian herb. *Ecology* 84:932–947.

Bruna, E.M., and W.J. Kress. 2002. Habitat fragmentation and the demographic structure of an Amazonian understory herb (*Heliconia acuminata*). *Conservation Biology* 16:1256–1266.

Bruna, E.M., O. Nardy, S.Y. Strauss, and S.P. Harrison. 2002. Experimental assessment of *Heliconia acuminata* growth in a fragmented Amazonian landscape. *Journal of Ecology* 90:639–649.

Bruna, E.M., and M.K. Oli. 2005. Demographic effects of habitat fragmentation on a tropical herb: Life table response experiments. *Ecology* 86:1816–1824.

Bruna, E.M., H.L. Vasconcelos, and S. Heredia. 2005. The effect of habitat fragmentation on communities of mutualists: Amazonian ants and their host plants. *Biological Conservation* 124:209–216.

Bruno, J.F., L.E. Petes, C.D. Harvell, and A. Hettinger. 2003. Nutrient enrichment can increase the severity of coral diseases. *Ecology Letters* 6:1056–1061.

Bruun, H.H. 2000. Patterns of species richness in dry grassland patches in an agricultural landscape. *Ecography* 23:641–650.

Buchmann, S.L., and G.B. Nabhan. 1997. *The forgotten pollinators*. Island Press, Washington, DC.

Bulger, J.B., N.J. Scott, and R.B. Seymour. 2003. Terrestrial activity and conservation of adult California red-legged frogs *Rana aurora draytonii* in coastal forests and grasslands. *Biological Conservation* 110:85–95.

Burel, F. 1989. Landscape structure effects on carabid beetles spatial patterns in western France. *Landscape Ecology* 2:215–226.

Burgess, L., and D.M. Sharpe, eds. 1981. *Forest island dynamics in man-dominated landscapes*. Springer-Verlag, New York, NY.

Burkey, T.V. 1997. Metapopulation extinction in fragmented landscapes: Using bacteria and protozoa communities as model ecosystems. *American Naturalist* 150:568–591.

Butler, M.J., IV. 2003. Incorporating ecological process and environmental change into spiny lobster population models using a spatially-explicit, individual-based approach. *Fisheries Research* 65:63–79.

Cadenasso, M.L., S.T.A. Pickett, K.C. Weathers, and C.G. Jones. 2003. A framework for a theory of ecological boundaries. *BioScience* 53:750–758.

Calabrese, J.M., and W.F. Fagan. 2004. A comparison-shopper's guide to connectivity metrics. *Frontiers in Ecology and the Environment* 2:529–536.

Caldwell, I.R., and V.O. Nams. 2006. A compass without a map: Tortuosity and orientation of eastern painted turtles (*Chrysemys picta picta*) released in unfamiliar territory. *Canadian Journal of Zoology* 84:1129–1137.

Cale, P.G. 2003. The influence of social behaviour, dispersal and landscape fragmentation on population structure in a sedentary bird. *Biological Conservation* 109:237–248.

Caley, M.J., K.A. Buckley, and G.P. Jones. 2001. Separating ecological effects of habitat fragmentation, degradation, and loss on coral commensals. *Ecology* 82:3435–3448.

California Department of Fish and Game. 1999. *Natural Community Conservation Planning, 1991–1998: A partnership for conservation.* Natural Community Conservation Planning Program, Habitat Conservation Branch, California Department of Fish and Game, Sacramento, CA.

Cam, E., J.D. Nichols, J.E. Hines, J.R. Sauer, R. Alpizar-Jara, and C.H. Flather. 2002. Disentangling sampling and ecological explanations underlying species-area relationships. *Ecology* 83:1118–1130.

Cane, J.H., R.L. Minckley, L.J. Kervin, T.H. Roulston, and N.M. Williams. 2006. Complex responses within a desert bee guild (Hymenoptera: Apiformes) to urban habitat fragmentation. *Ecological Applications* 16:632–644.

Canham, C.D., M.L. Pace, M.J. Papaik, A.G.B. Primack, K.M. Roy, R.J. Maranger, R.P. Curran, and D.M. Spada. 2004. A spatially explicit watershed-scale analysis of dissolved organic carbon in Adirondack lakes. *Ecological Applications* 14:839–854.

Cappuccino, N., D. Lavertu, Y. Gergeron, and J. Régnière. 1998. Spruce budworm impact, abundance, and parasitism rate in a patchy landscape. *Oecologia* 114:236–242.

Carlo, T.A. 2005. Interspecific neighbors change seed dispersal pattern of an avian-dispersed plant. *Ecology* 86:2440–2449.

Carroll, C., D.G. Miquelle, and G. Dale. 2006. Spatial viability analysis of Amur tiger *Panthera tigris altaica* in the Russian Far East: The role of protected areas and landscape matrix in population persistence. *Journal of Applied Ecology* 43:1056–1068.

Castellón, T.D., and K.E. Sieving. 2006. An experimental test of matrix permeability and corridor use by an endemic understory bird. *Conservation Biology* 20:135–145.

Chalfoun, A.D., F.R. Thompson III, and M.J. Ratnaswamy. 2002. Nest predators and fragmentation: A review and meta-analysis. *Conservation Biology* 16:306–318.

Chambers, J.C., and J.A. MacMahon. 1994. A day in the life of a seed: Movements and fates of seeds and their implications for natural and managed systems. *Annual Review of Ecology and Systematics* 25:263–292.

Chan, K.M.A., M.R. Shaw, D.R. Cameron, E.C. Underwood, and G.C. Daily. 2006. Conservation planning for ecosystem services. *PLoS Biology* 4:2138–2152.

Chapman, C.A. M.L. Speirs, T.R. Gillespie, T. Holland, and K.M. Austad. 2006. Life on the edge: Gastrointestinal parasites from the forest edge and interior primate groups. *American Journal of Primatology* 68:397–409.

Chapman, C.A., M.D. Wasserman, T.R. Gillespie, M.L. Speirs, M.J. Lawes, T.L. Saj, and T.E. Ziegler. 2006. Do food availability, parasitism, and stress have synergistic effects on red colobus populations living in forest fragments? *American Journal of Physical Anthropology* 131:525–534.

Chase, J.M. 2005. Towards a really unified theory for metacommunities. *Functional Ecology* 19:182–186.

Chen, J., J.F. Franklin, and T.A. Spies. 1992. Vegetation responses to edge environments in old-growth Douglas-fir forests. *Ecological Applications* 2:387–396.

Chetkiewicz, C.-L.B., C. Cassady St. Clair, and M.S. Boyce. 2006. Corridors for conservation: Integrating pattern and process. *Annual Review of Ecology, Evolution, and Systematics* 37:317–342.

Chisholm, R.A., and B.A. Wintle. 2007. Incorporating landscape stochasticity into population viability analysis. *Ecological Applications* 17:317–322.

Chittaro, P.M. 2002. Species-area relationships for coral reef fish assemblages of St. Croix, US Virgin Islands. *Marine Ecology Progress Series* 233:253–261.

Cleaveland, S., M.G.J. Appel, W.S.K. Chalmers, C. Chillingworth, M. Kaare, and C. Dye. 2000. Serological and demographic evidence for domestic dogs as a source of canine distemper virus infection for Serengeti wildlife. *Veterinary Microbiology* 72:217–227.

Cleaveland, S., T. Mlengeya, M. Kaare, D. Haydon, T. Lembo, M.K. Laurenson, and C. Packer. 2007. The conservation relevance of epidemiological research into carnivore viral diseases in the Serengeti. *Conservation Biology* 21:612–622.

Clevenger, A.P., B. Chruszcz, and K. Gunson. 2001. Drainage culverts as habitat linkages and factors affecting passage by mammals. *Journal of Applied Ecology* 38:1340–1349.

Clevenger, A.P., and N. Waltho. 2005. Performance indices to identify attributes of highway crossing structures facilitating movement of large mammals. *Biological Conservation* 121:453–464.

Clewell, A. 2000. Restoring natural capital. *Restoration Ecology* 8:1.

Clewell, A., and J. Aronson. 2006. Motivations for the restoration of ecosystems. *Conservation Biology* 20:420–428.

Collinge, S.K. 1996. Ecological consequences of habitat fragmentation: Implications for landscape architecture and planning. *Landscape and Urban Planning* 36:59–77.

———. 1998. Spatial arrangement of habitat patches and corridors: Clues from ecological field experiments. *Landscape and Urban Planning* 42:157–168.

———. 2000. Effects of grassland fragmentation on insect species loss, recolonization, and movement patterns. *Ecology* 81:2211–2226.

Collinge, S.K., and R.T.T. Forman. 1998. A conceptual model of land conversion processes: Predictions and evidence from a microlandscape experiment with grassland insects. *Oikos* 82:66–84.

Collinge, S.K., M. Holyoak, J.T. Marty, and C. Barr. 2001. Riparian habitat fragmentation and population persistence of the Valley Elderberry Longhorn Beetle (Coleoptera: Cerambycidae) in northern California. *Biological Conservation* 100:103–113.

Collinge, S.K., W.C. Johnson, C. Ray, R. Matchett, J. Grensten, J.F. Cully Jr., K.L. Gage, M.Y. Kosoy, J.E. Loye, and A.P. Martin. 2005. Landscape structure and plague occurrence in black-tailed prairie dogs on grasslands of the western USA. *Landscape Ecology* 20:941–955.

Collinge, S.K., and T.M. Palmer. 2002. The influences of patch shape and boundary contrast on insect response to fragmentation in California grasslands. *Landscape Ecology* 17:647–656.

Collinge, S.K., K.L. Prudic, and J.C. Oliver. 2003. Effects of local habitat characteristics and landscape context on grassland butterfly diversity. *Conservation Biology* 17:178–187.

Collinge, S.K., and C. Ray, eds. 2006. *Disease ecology: Community structure and pathogen dynamics.* Oxford University Press, Oxford, UK.

Collinge, S.K., C. Ray, and J.F. Cully Jr. 2008. Effects of disease on keystone species, dominant species, and their communities. In *Infectious disease ecology: Effects of ecosystems on disease and of disease on ecosystems*, ed. R.S. Ostfeld, F. Keesing, and V. Eviner, 189–213. Princeton University Press, Princeton, NJ.

Connor, E.F., and E.D. McCoy. 1979. Statistics and biology of the species-area relationship. *American Naturalist* 113:791–833.

Connor, E.F., and D. Simberloff. 1979. The assembly of species communities: Chance or competition. *Ecology* 60:1132–1140.

Conrad, P.A., M.A. Miller, C. Kreuder, E.R. James, J. Mazet, H. Dabritz, D.A. Jessup, F. Gulland, and M.E. Grigg. 2005. Transmission of *Toxoplasma*: Clues from the study of sea otters as sentinels of *Toxoplasma gondii* flow into the marine environment. *International Journal for Parasitology* 35:1125–1168.

Cook, R.R. 1995. The relationship between nested subsets, habitat subdivision, and species diversity. *Oecologia* 101:204–210.

Cook, W.M., J. Yao, B.L. Foster, R.D. Holt, and L.B. Patrick. 2005. Secondary succession in an experimentally fragmented landscape: Community patterns across space and time. *Ecology* 86:1267–1279.

Coppolillo, P.B., H. Gomez, F. Maisels, and R. Wallace. 2004. Selection criteria for suites of landscape species as a basis for site-based conservation. *Biological Conservation* 115:419–430.

Cox, J., and R.T. Engstrom. 2001. Influence of the spatial pattern of conserved lands on the persistence of a large population of red-cockaded woodpeckers. *Biological Conservation* 100:137–150.

Craig, D.P., C.E. Bock, B.C. Bennett, and J.H. Bock. 1999. Habitat relationships among grasshoppers (Orthoptera: Acrididae) at the western limit of the Great Plains in Colorado. *American Midland Naturalist* 142:314–327.

Crooks, K.R., and M. Sanjayan, eds. 2006. *Connectivity conservation*. Cambridge University Press, Cambridge, UK.

Crooks, K.R., and M.E. Soulé. 1999. Mesopredator release and avifaunal extinctions in a fragmented system. *Nature* 400:563–566.

Crooks, K.R., A.V. Suarez, and D.T. Bolger. 2004. Avian assemblages along a gradient of urbanization in a highly fragmented landscape. *Biological Conservation* 115:451–462.

Crooks, K.R., A.V. Suarez, D.T. Bolger, and M.E. Soulé. 2001. Extinction and colonization of birds on habitat islands. *Conservation Biology* 15:159–172.

Curran, L.M., I. Caniago, G.D. Paoli, D. Astianti, M. Kusneti, M. Leighton, C.E. Nirarita, and H. Haeruman. 1999. Impact of El Niño and logging on canopy tree recruitment in Borneo. *Science* 286:2184–2188.

Cushman, S.A. 2006. Effects of habitat loss and fragmentation on amphibians: A review and prospectus. *Biological Conservation* 128:231–240.

Cutler, A. 1991. Nested faunas and extinction in fragmented habitats. *Conservation Biology* 5:496–505.

Daily, G.C., P.R. Ehrlich, and N.M. Haddad. 1993. Double keystone bird in a keystone species complex. *Proceedings of the National Academy of Sciences* 90:592–594.

Daily, G.C., and K. Ellison. 2002. *The new economy of nature*. Island Press, Washington, DC.

Dale, V.H., R.V. O'Neill, F. Southworth, and M. Pedlowski. 1994. Modeling effects of land management in the Brazilian Amazonian settlement of Rondônia. *Conservation Biology* 8:196–206.

Damschen, E.I., N.M. Haddad, J.L. Orrock, J.J. Tewksbury, and D.J. Levey. 2006. Corridors increase plant species richness at large scales. *Science* 313:1284–1286.

Danielson, B.J., and M.W. Hubbard. 2000. The influence of corridors on the movement behavior of individual *Peromyscus polionotus* in experimental landscapes. *Landscape Ecology* 15:323–331.

Daszak, P., A.A. Cunningham, and A.D. Hyatt. 2000. Emerging infectious diseases of wildlife: Threats to biodiversity and human health. *Science* 287:443–449.

Daszak, P., R.K. Plowright, J.H. Epstein, J. Pulliam, S. Abdul Rahman, H.E. Field, A. Jamaluddin, S.H. Sharifah, C.S. Smith, K.J. Olival, S. Luby, K. Halpin, A.D. Hyatt, A.A. Cunningham, and the Henipavirus Ecology Research Group (HERG). 2006. The emergence of Nipah and Hendra virus: Pathogen dynamics across a wildlife-livestock-human continuum. In Collinge and Ray 2006, 186–201.

Davies, K.F., and C.R. Margules. 1998. Effects of habitat fragmentation on carabid beetles: Experimental evidence. *Journal of Animal Ecology* 67:460–471.

Davies, K.F., C.R. Margules, and J.F. Lawrence. 2000. Which traits of species predict population declines in experimental forest fragments? *Ecology* 81:1450–1461.

———. 2004. A synergistic effect puts rare, specialized species at greater risk of extinction. *Ecology* 85:265–271.

Davies, K.F., B.A. Melbourne, and C.R. Margules. 2001. Effects of within- and between-patch processes on community dynamics in a fragmentation experiment. *Ecology* 82:1830–1846.

Debinski, D.M., and R.D. Holt. 2000. A survey and overview of habitat fragmentation experiments. *Conservation Biology* 14:342–355.

de Blois, S., G. Domon, and A. Bouchard. 2002. Factors affecting plant species distribution in hedgerows of southern Quebec. *Biological Conservation* 105:355–367.

Delcourt, H.R., P.A. Delcourt, and T. Webb III. 1983. Dynamic plant ecology: The spectrum of vegetational change in space and time. *Quaternary Science Reviews* 1:153–175.

den Boer, P.J. 1981. On the survival of populations in a heterogeneous and variable environment. *Oecologia* 50:39–53.

Deng, W., and G. Zheng. 2004. Landscape and habitat factors affecting Cabot's tragopan *Tragopan caboti* occurrence in habitat fragments. *Biological Conservation* 117:25–32.

Denslow, J.S. 1987. Tropical rain-forest gaps and tree species diversity. *Annual Review of Ecology and Systematics* 18:431–451.

Derraik, J.G.B., and D. Slaney. 2007. Anthropogenic environmental change, mosquito-borne diseases and human health in New Zealand. *EcoHealth* 4:72–81.

Diamond, J. 1975. The island dilemma: Lessons of modern biogeographic studies for the design of nature reserves. *Biological Conservation* 7:129–146.

Dickie, I.A., and P.B. Reich. 2005. Ectomycorrhizal fungal communities at forest edges. *Journal of Ecology* 93:244–255.

Dickie, I.A., S.A. Schnitzer, P.B. Reich, and S.E. Hobbie. 2007. Is oak establishment in old-fields and savanna openings context dependent? *Journal of Ecology* 95:309–320.

Didham, R.K. 1998. Altered leaf-litter decomposition rates in tropical forest fragments. *Oecologia* 116:397–406.

Didham, R.K., P.M. Hammond, J.H. Lawton, P. Eggleton, and N.E. Stork. 1998. Beetle species responses to tropical forest fragmentation. *Ecological Monographs* 68:295–323.

Diffendorfer, J.E., M.S. Gaines, and R.D. Holt. 1995. Habitat fragmentation and movements of three small mammals (*Sigmodon, Microtus,* and *Peromyscus*). *Ecology* 76:827–839.

Dingle, H. 1996. *Migration: The biology of life on the move.* Oxford University Press, New York, NY.

Dixon J.D., M.K. Oli, M.C. Wooten, T.H. Eason, J.W. McCown, and D. Paetkau. 2006. Effectiveness of a regional corridor in connecting two Florida black bear populations. *Conservation Biology* 20:155–162.

Doak, D.F. 1995. Source-sink models and the problem of habitat degradation: General models and applications to the Yellowstone grizzly. *Conservation Biology* 9:1370–1379.

Doak, D.F., and L.S. Mills. 1994. A useful role for theory in conservation. *Ecology* 75:615–626.

Dobson, A.P., A.D. Bradshaw, and A.J.M. Baker. 1997. Hopes for the future: Restoration ecology and conservation biology. *Science* 277:515–522.

Dobson, A.P., I. Cattadori, R.D. Holt, R.S. Ostfeld, F. Keesing, K. Krichbaum, J.R. Rohr, S.E. Perkins, and P.J. Hudson. 2006. Sacred cows and sympathetic squirrels: The importance of biological diversity to human health. *PLoS Medicine* 3:714–718.

Dobson, A.P., and M. Meagher. 1996. The population dynamics of brucellosis in the Yellowstone National Park. *Ecology* 77:1026–1036.

Donlan, C.J., J. Knowlton, D.F. Doak, and N. Biavaschi. 2005. Nested communities, invasive species and Holocene extinctions: Evaluating the power of a potential conservation tool. *Oecologia* 145:475–485.

Donoso, D.S., A.A. Grez, and J.A. Simonetti. 2003. Effects of forest fragmentation on the granivory of differently sized seeds. *Biological Conservation* 115:63–70.

Donovan, T.M., P.W. Jones, E.M. Annand, and F.R. Thompson III. 1997. Variation in local-scale edge effects: Mechanisms and landscape context. *Ecology* 78:2064–2075.

Dooley, J.L., Jr., and M.A. Bowers. 1998. Demographic responses to habitat fragmentation: Experimental tests at the landscape and patch scale. *Ecology* 79:969–980.

Dover, J., and T. Sparks. 2000. A review of the ecology of butterflies in British hedgerows. *Environmental Management* 60:51–63.

Drakare, S., J.J. Lennon, and H. Hillebrand. 2006. The imprint of the geographical, evolutionary and ecological context on species-area relationships. *Ecology Letters* 9:215–227.

Dramstad, W.E., J.D. Olson, and R.T.T. Forman. 1996. *Landscape ecology principles in landscape architecture and land-use planning.* Island Press, Washington, DC.

Drayton, B., and R.B. Primack. 1996. Plant species lost in an isolated conservation area in metropolitan Boston from 1894 to 1993. *Conservation Biology* 10:30–39.

Drechsler, M., K. Frank, I. Hanski, R.B. O'Hara, and C. Wissel. 2003. Ranking metapopulation extinction risk: From patterns in data to conservation management decisions. *Ecological Applications* 13:990–998.

Driscoll, M.J.L., and T.M. Donovan. 2004. Landscape context moderates edge effects: Nesting success of wood thrushes in central New York. *Conservation Biology* 18:1330–1338.

Duncan, B.W., V.L. Larson, and P.A. Schmalzer. 2004. Historic landcover and recent landscape change in the north Indian River Lagoon Watershed, Florida, USA. *Natural Areas Journal* 24:198–215.

Dunford, W., and K. Freemark. 2005. Matrix matters: Effects of surrounding land uses on forest birds near Ottawa, Canada. *Landscape Ecology* 20:497–511.

Dunning, J.B., Jr., D.J. Stewart, B.J. Danielson, B.R. Noon, T.L. Root, R.H. Lamberson, and E.E. Stevens. 1995. Spatially explicit population models: Current forms and future uses. *Ecological Applications* 5:3–11.

Edenius, L., and K. Sjöberg. 1997. Distribution of birds in natural landscape mosaics of old-growth forests in northern Sweden: Relations to habitat area and landscape context. *Ecography* 20:425–431.

Edwards, T., C.R. Schwalbe, D.E. Swann, and C.S. Goldberg. 2004. Implications of anthropogenic landscape change on inter-population movements of the desert tortoise (*Gopherus agassizii*). *Conservation Genetics* 5:485–499.

Enoksson, B., P. Angelstam, and K. Larsson. 1995. Deciduous forest and resident birds: The problem of fragmentation within a coniferous forest landscape. *Landscape Ecology* 10:267–275.

Estrada, A., R. Coates-Estrada, and D. Meritt Jr. 1993. Bat species richness and abundance in tropical rain forest fragments and agricultural habitats at Los Tuxtlas, Mexico. *Ecography* 16:309–318.

Ewers, R.M., and R.K. Didham. 2006. Confounding factors in the detection of species responses to habitat fragmentation. *Biological Reviews* 81:117–142.

Ezenwa, V. 2004. Parasite infection rates of impala (*Aepyceros melampus*) in fenced game reserves in relation to reserve characteristics. *Biological Conservation* 118:397–401.

Ezenwa, V., M.S. Godsey, R.J. King, and S.C. Guptill. 2006. Avian diversity and West Nile virus: Testing associations between biodiversity and infectious disease risk. *Proceedings of the Royal Society of London, B* 273:109–117.

Fábos, J.G., and J. Ahern, eds. 1995. *Greenways: The beginning of an international movement.* Elsevier, Amsterdam, The Netherlands.

Fábos, J.G., and R.L. Ryan. 2006. An introduction to greenway planning around the world. *Landscape and Urban Planning* 76:1–6.

Fahrig, L. 1997. Relative effects of habitat loss and fragmentation on population survival. *Journal of Wildlife Management* 61:603–610.

———. 1998. When does fragmentation of breeding habitat affect population survival? *Ecological Modeling* 105:273–292.

———. 2001. How much habitat is enough? *Biological Conservation* 100:65–74.

———. 2002. Effect of habitat fragmentation on the extinction threshold: A synthesis. *Ecological Applications* 12:346–353.

———. 2003. Effects of habitat fragmentation on biodiversity. *Annual Review of Ecology and Systematics* 34:487–515.

Fahrig, L., and G. Merriam. 1985. Habitat patch connectivity and population survival. *Ecology* 66:1762–1768.

————. 1994. Conservation of fragmented populations. *Conservation Biology* 8:50–59.

Falk, D.A., M.A. Palmer, and J.B. Zedler, eds. 2006. *Foundations of restoration ecology: The science and practice of ecological restoration.* Island Press, Washington, DC.

Farnsworth, M.L., L.L. Wolfe, N.T. Hobbs, K.P. Burnham, E.S. Williams, D.M. Theobald, M.M. Conner, and M.W. Miller. 2005. Human land use influences chronic wasting disease prevalence in mule deer. *Ecological Applications* 15:119–126.

Farwig, N., K. Böhning-Gaese, and B. Bleher. 2006. Enhanced seed dispersal of *Prunus africana* in fragmented and disturbed forests? *Oecologia* 147:238–252.

Feer, F., and Y. Hingrat. 2005. Effects of forest fragmentation on a dung beetle community in French Guiana. *Conservation Biology* 19:1103–1112.

Fenton, A., and A.B. Pedersen. 2005. Community epidemiology framework for classifying disease threats. *Emerging Infectious Diseases* 11:1815–1821.

Fernández-Juricic, E. 2004. Spatial and temporal analysis of the distribution of forest specialists in an urban-fragmented landscape (Madrid, Spain): Implications for local and regional bird conservation. *Landscape and Urban Planning* 69:17–32.

Ferraz, G., J.D. Nichols, J.E. Hines, P.C. Stouffer, R.O. Bierregaard Jr., and T.E. Lovejoy. 2007. A large-scale deforestation experiment: Effects of patch area and isolation on Amazon birds. *Science* 315:238–241.

Ferreras, P. 2001. Landscape structure and asymmetrical inter-patch connectivity in a metapopulation of the endangered Iberian lynx. *Biological Conservation* 100:125–136.

Ferreras, P., J.J. Aldama, J.F., Beltrán, and M. Delibes. 1992. Rates and causes of mortality in a fragmented population of Iberian lynx *Felis pardina* Temminck, 1824. *Biological Conservation* 61:197–202.

Fèvre, E.M., B.M.D.C. Bronsvoort, K.A. Hamilton, and S. Cleaveland. 2006. Animal movements and the spread of infectious diseases. *Trends in Microbiology* 14:125–131.

Ficetola, G.F., and F. De Bernardi. 2004. Amphibians in a human-dominated landscape: The community structure is related to habitat features and isolation. *Biological Conservation* 119:219–230.

Field, S.G., M. Lange, H. Schulenburg, T.P. Velavan, and N.K. Michiels. 2007. Genetic diversity and parasite defense in a fragmented urban metapopulation of earthworms. *Animal Conservation* 10:162–175.

Fink, R.D., C.A. Lindell, E.B. Morrison, R.A. Zahawi, and K.D. Holl. Forthcoming. Patch size and tree species influence the number and duration of bird visits in forest restoration plots in southern Costa Rica. *Restoration Ecology.* doi:10.1111/j.1526-100X.2008.00383.

Fischer, J., and D.B. Lindenmayer. 2005. Perfectly nested or significantly nested: An important difference for conservation management. *Oikos* 109:485–494.

Fleishman, E., and R. MacNally. 2002. Topographic determinants of faunal nestedness in Great Basin butterfly assemblages: Applications to conservation planning. *Conservation Biology* 16:422–429.

Fleishman, E., C. Ray, P. Sjögren-Gulve, C.L. Boggs, and D.D. Murphy. 2002. Assessing the roles of patch quality, area, and isolation in predicting metapopulation dynamics. *Conservation Biology* 16:706–716.

Fletcher, R.J. 2005. Multiple edge effects and their implications in fragmented landscapes. *Journal of Animal Ecology* 74:342–352.

Fonseca, G.A.B. da, and J.G. Robinson. 1990. Forest size and structure: Competitive and predatory effects on small mammal communities. *Biological Conservation* 53:265–294.

Ford, H.A., G.W. Barrett, D.A. Saunders, and H.F. Recher. 2001. Why have birds in the woodlands of Southern Australia declined? *Biological Conservation* 97:71–88.

Foreman, D., J. Davis, D. Johns, R.F. Noss, and M. Soulé. 1992. The Wildlands Project mission statement. *Wild Earth* (Special issue):3–4.

Forman, R.T.T. 1995. *Land mosaics: The ecology of landscapes and regions.* Cambridge University Press, Cambridge, UK.

———. 2004. Road ecology's promise: What's around the bend? *Environment* 46:9–21.

Forman, R.T.T., and L.E. Alexander. 1998. Roads and their major ecological effects. *Annual Review of Ecology and Systematics* 29:207–231.

Forman, R.T.T., A.E. Galli, and C.F. Leck. 1976. Forest size and avian diversity in New Jersey woodlots with some land use implications. *Oecologia* 26:1–8.

Forman, R.T.T., and M. Godron. 1981. Patches and structural components for a landscape ecology. *BioScience* 31:733–740.

———. 1986. *Landscape ecology.* Wiley, New York, NY.

Forman, R.T.T., D. Sperling, J.A. Bissonette, A.P. Clevenger, C.D. Cutshall, V.H. Dale, L. Fahrig, R. France, C.R. Goldman, K. Heanue, J.A. Jones, F.J. Swanson, T. Turrnetine, and T.C. Winter. 2003. *Road ecology: Science and solutions.* Island Press, Washington, DC.

Forney, K.A., and M.E. Gilpin. 1989. Spatial structure and population extinction: A study with *Drosophila* flies. *Conservation Biology* 3:45–51.

Foster, D.R. 1992. Land-use history (1730–1990) and vegetation dynamics in central New England, USA. *Journal of Ecology* 80:753–772.

Foster, D.R., M. Fluet, and E.R. Boose. 1999. Human or natural disturbance: Landscape-scale dynamics of the tropical forests of Puerto Rico. *Ecological Applications* 9:555–572.

Franklin, J.F., and R.T.T. Forman. 1987. Creating landscape patterns by forest cutting: Ecological consequences and principles. *Landscape Ecology* 1:5–18.

Fritz, R., and G. Merriam. 1993. Fence row habitats for plants moving between farmland forests. *Biological Conservation* 64:141–148.

Fry, G., and A.R. Main. 1993. Restoring seemingly natural communities on agricultural land. In Saunders, Hobbs, and Ehrlich 1993, 225–241.

Fukamachi, K., S. Iida, and T. Nakashizuka. 1996. Landscape patterns and plant species diversity of forest reserves in the Kanto region, Japan. *Vegetatio* 124:107–114.

Galatowitsch, S.M., and A.G. van der Valk. 1996. The vegetation of restored and natural prairie wetlands. *Ecological Applications* 6:102–112.

Galetti, M., C.I. Donatti, A.S. Pires, P.R. Guimarães Jr., and P. Jordano. 2006. Seed survival and dispersal of an endemic Atlantic forest palm: The combined effects of defaunation and forest fragmentation. *Botanical Journal of the Linnean Society* 151:141–149.

Game, M. 1980. Best shape for nature reserves. *Nature* 287:630–632.

García, D., and N.P. Chacoff. 2007. Scale-dependent effects of habitat fragmentation on hawthorn pollination, frugivory, and seed predation. *Conservation Biology* 21:400–411.

Gardner, R.H., B.T. Milne, M.G. Turner, and R.V. O'Neill. 1987. Neutral models for the analysis of broad-scale landscape pattern. *Landscape Ecology* 1:19–28.

Gardner, R.H., and R.V. O'Neill. 1991. Pattern, process and predictability: The use of neutral models for landscape analysis. In Turner and Gardner, 289–307.

Gardner, R.H., R.V. O'Neill, and M.G. Turner. 1993. Ecological implications of landscape fragmentation. In *Humans as components of ecosystems*, ed. M.J. McDonnell and S.T.A. Pickett, 208–226. Springer-Verlag, New York, NY.

Gardner, R.H., and D.L. Urban. 2007. Neutral models for testing landscape hypotheses. *Landscape Ecology* 22:15–29.

Gascon, C., and T.E. Lovejoy. 1998. Ecological impacts of forest fragmentation in central Amazonia. *Zoology* 101:273–280.

Gascon, C., T.E. Lovejoy, R.O. Bierregaard Jr., J.R. Malcolm, P.C. Stouffer, H.L. Vasconcelos, W.F. Laurance, B. Zimmerman, M. Tocher, and S. Borges. 1999. Matrix habitat and species richness in tropical forest remnants. *Biological Conservation* 91:223–229.

Gates, J.E., and L.W. Gysel. 1978. Avian nest dispersion and fledging success in field-forest ecotones. *Ecology* 59:871–883.

Gehring, T.M., and R.K. Swihart. 2004. Home range and movements of long-tailed weasels in a landscape fragmented by agriculture. *Journal of Mammalogy* 85:79–86.

Gergel, S.E., and M.G. Turner. 2002. *Learning landscape ecology: A practical guide to concepts and techniques*. Springer-Verlag, New York, NY.

Gerhardt, F., and S.K. Collinge. 2007. Abiotic constraints eclipse biotic resistance in determining invasibility along experimental vernal pool gradients. *Ecological Applications* 17:922–933.

Ghazoul, J. 2005. Pollen and seed dispersal among dispersed plants. *Biological Reviews* 80:413–443.

Ghazoul, J., and M. McLeish. 2001. Reproductive ecology of tropical forest trees in logged and fragmented habitats in Thailand and Costa Rica. *Plant Ecology* 153:335–345.

Gibb, H., and D.F. Hochuli. 2002. Habitat fragmentation in an urban environment: Large and small fragments support different arthropod assemblages. *Biological Conservation* 106:91–100.

Gibbs, J.P., and J. Faaborg. 1990. Estimating the viability of ovenbird and Kentucky warbler populations in forest fragments. *Conservation Biology* 4:193–196.

Gibbs, S.E.J., M.C. Wimberly, M. Madden, J. Masour, M.J. Yabsley, and D.E. Stallknecht. 2006. Factors affecting the geographic distribution of West Nile virus in Georgia, USA: 2002–2004. *Vector-borne and Zoonotic Diseases* 6:73–82.

Gibson, C.C., M.A. McKean, and E. Ostrom, eds. 2000. *People and forests: Communities, institutions, and governance*. MIT Press, Cambridge, MA.

Gignac, L.D., and M.R.T. Dale. 2005. Effects of fragment size and habitat heterogeneity on cryptogam diversity in the low-boreal forest of western Canada. *Bryologist* 108:50–66.

Gilbert, F.S., A. Gonzalez, and I. Evans-Freke. 1998. Corridors maintain species richness in the fragmented landscapes of a microecosystem. *Proceedings of the Royal Society of London, B* 265:577–582.

Gilbert, K.A., and E.Z.F. Setz. 2001. Primates in a fragmented landscape: Six species in central Amazonia. In *Lessons from Amazonia: Ecology and conservation of a fragmented forest*, ed. R.O. Bierregaard Jr., C. Gason, T.E. Lovejoy, and R. Mesquita, 262–270. Yale University Press, New Haven, CT.

Gillespie, T.R., and C.A. Chapman. 2006. Prediction of parasite infection dynamics in primate metapopulations based on attributes of forest fragmentation. *Conservation Biology* 20:441–448.

Gillespie, T.R., C.A. Chapman, and E.C. Greiner. 2005. Effects of logging on gastrointestinal parasite infections and infection risk in African primates. *Journal of Applied Ecology* 42:699–707.

Gillis, E.A., and V.O. Nams. 1998. How red-backed voles find habitat patches. *Canadian Journal of Zoology* 76:791–794.

Gilpin, M.E. 1996. Metapopulations and wildlife conservation: Approaches to modeling spatial structure. In McCullough 1996, 11–27.

Godefroid, S., and N. Koedam. 2003. How important are large vs. small forest remnants for the conservation of the woodland flora in an urban context? *Global Ecology and Biogeography* 12:287–298.

Gog, J., R. Woodroffe, and J. Swinton. 2002. Disease in endangered metapopulations: The importance of alternative hosts. *Proceedings of the Royal Society of London, B* 269:671–676.

Gómez-Mendoza, L.E. Vega-Peña, M.I. Ramírez, J.L. Palacio-Prieto, and L. Galicia. 2006. Projecting land-use change processes in the Sierra Norte of Oaxaca, Mexico. *Applied Geography* 26:276–290.

Gonzalez, A. 2005. Local and regional community dynamics in fragmented landscapes. In Holyoak, Leibold, and Holt 2005, 146–169.

Goodsell, P.J., and S.D. Connell. 2002. Can habitat loss be treated independently of habitat configuration? Implications for rare and common taxa in fragmented landscapes. *Marine Ecology Progress Series* 239:37–44.

Gotelli, N.J. 1995. *A primer of ecology.* Sinauer Associates, Sunderland, MA.

Graff, P., M.R. Aguiar, and E.J. Chaneton. 2007. Shifts in positive and negative plant interactions along a grazing intensity gradient. *Ecology* 88:188–199.

Grant, T.A., and G.B. Berkey. 1999. Forest area and avian diversity in fragmented aspen woodland of North Dakota. *Wildlife Society Bulletin* 27:904–914.

Grashof-Bokdam, C. 1997. Forest species in an agricultural landscape in the Netherlands: Effects of habitat fragmentation. *Journal of Vegetation Science* 8:21–28.

Grenfell, B.T., O.N. Bjørnstad, and B.F. Finkenstadt. 2002. Dynamics of measles epidemics: Scaling noise, determinism, and predictability with the TSIR model. *Ecological Monographs* 72:185–202.

Grenfell, B.T., O.N. Bjørnstad, and J. Kappey. 2001. Travelling waves and spatial hierarchies in measles epidemics. *Nature* 414:716–723.

Grillas, P., P. Gauthier, N. Yavercovski, and C. Perennou. 2004. *Issues relating to conservation, functioning, and management.* Vol. 1 of *Mediterranean temporary pools.* Station biologique de la Tour du Valat, Arles, France.

Grimm, N.B., J.M. Grove, S.T.A. Pickett, and C.L. Redman. 2000. Integrated approaches to long-term studies of urban ecological systems. *BioScience* 50:571–584.

Groom, M.J. 1998. Allee effects limit population viability of an annual plant. *American Naturalist* 151:487–496.

———. 2001. Consequences of subpopulation isolation for pollination, herbivory, and population growth in *Clarkia concinna concinna* (Onagraceae). *Biological Conservation* 100:55–63.

Groom, M.J., and N. Schumaker. 1993. Evaluating landscape change: Patterns of worldwide deforestation and local fragmentation. In Kareiva, Kingsolver, and Huey, 24–44.

Groves, C.R. 2003. *Drafting a conservation blueprint: A practitioner's guide to planning for biodiversity.* Island Press, Washington, DC.

Gude, P.H., A.J. Hansen, and D.A. Jones. 2007. Biodiversity consequences of alternative future land use scenarios in Greater Yellowstone. *Ecological Applications* 17:1004–1018.

Guirado, M., J. Pino, and F. Roda. 2006. Understorey plant species richness and composition in metropolitan forest archipelagos: Effects of forest size, adjacent land use and distance to the edge. *Global Ecology and Biogeography* 15:50–62.

Gustafson, E.J. 1998. Quantifying landscape spatial pattern: What is the state of the art? *Ecosystems* 1:143–156.

Gustafson, E.J., and R.H. Gardner. 1996. The effect of landscape heterogeneity on the probability of patch colonization. *Ecology* 77:94–107.

Haas, C.A. 1995. Dispersal and use of corridors by birds in wooded patches on an agricultural landscape. *Conservation Biology* 9:845–854.

Haddad, N.M. 1999. Corridor and distance effects on interpatch movements: A landscape experiment with butterflies. *Ecological Applications* 9:612–622.

Haddad, N.M., and K.A. Baum. 1999. An experimental test of corridor effects on butterfly densities. *Ecological Applications* 9:623–633.

Haddad, N.M., D.R. Bowne, A. Cunningham, B.J. Danielson, D.J. Levey, S. Sargent, and T. Spira. 2003. Corridor use by diverse taxa. *Ecology* 84:609–615.

Haila, Y. 1990. Toward an ecological definition of an island: A northwest European perspective. *Journal of Biogeography* 17:561–568.

———. 2002. A conceptual genealogy of fragmentation research: From island biogeography to landscape ecology. *Ecological Applications* 12:321–334.

Haines, A.M., M.E. Tewes, L.L. Laack, J.S. Horne, and J.H. Young. 2006. A habitat-based population viability analysis for ocelots (*Leopardus pardalis*) in the United States. *Biological Conservation* 132:424–436.

Haire, S.L., C.E. Bock, B.S. Cade, and B.C. Bennett. 2000. The role of landscape and habitat characteristics in limiting abundance of grassland nesting songbirds in an urban open space. *Landscape and Urban Planning* 48:65–82.

Hannon, S.J., and F.K.A. Schmiegelow. 2002. Corridors may not improve the conservation value of small reserves for most boreal birds. *Ecological Applications* 12:1457–1468.

Hansen, M.J., and A.P. Clevenger. 2005. The influence of disturbance and habitat on the presence of non-native plant species along transport corridors. *Biological Conservation* 125:249–259.

Hanski, I. 1994. A practical model of metapopulation dynamics. *Journal of Animal Ecology* 63:151–162.

————. 1999. *Metapopulation ecology*. Oxford University Press, Oxford, UK.

Hanski, I., and O.E. Gaggiotti, eds. 2004. *Ecology, genetics, and evolution of metapopulations*. Elsevier Academic Press, Burlington, MA.

Hanski, I., and M. Gilpin. 1991. Metapopulation dynamics: Brief history and conceptual domain. *Biological Journal of the Linnean Society* 42:3–16.

————, eds. 1997. *Metapopulation biology*. Academic Press, San Diego, CA.

Hanski, I., A. Moilanen, T. Pakkala, and M. Kuussaari. 1996. The quantitative incidence function model and persistence of an endangered butterfly metapopulation. *Conservation Biology* 10:578–590.

Hanski, I., and D. Simberloff. 1997. The metapopulation approach, its history, conceptual domain, and application to conservation. In Hanski and Gilpin 1997, 5–26.

Hansson, L. 1991. Dispersal and connectivity in metapopulations. *Biological Journal of the Linnean Society* 42:89–103.

Hansson, L., L. Fahrig, and G. Merriam, eds. 1995. *Mosaic landscapes and ecological processes*. Chapman and Hall, London, UK.

Harcourt, A.H., and D.A. Doherty. 2005. Species-area relationships of primates in tropical forest fragments: A global analysis. *Journal of Applied Ecology* 42:630–637.

Harding, E.K., E.E. Crone, B.D. Elderd, J.M. Hoekstra, A.J. McKerrow, J.D. Perrine, J. Regetz, L.J. Rissler, A.G. Stanley, E.L. Walters, and the NCEAS Habitat Conservation Plan Working Group. 2001. The scientific foundations of habitat conservation plans: A quantitative assessment. *Conservation Biology* 15:488–500.

Harris, L.D. 1984. *The fragmented forest*. University of Chicago Press, Chicago, IL.

Harris, L.D., and J. Scheck. 1991. From implications to applications: The dispersal corridor principle applied to the conservation of biological diversity. In Saunders and Hobbs, 189–220.

Harris, L.D., and G. Silva-López. 1992. Forest fragmentation and the conservation of biological diversity. In *Conservation biology: The theory and practice of nature conservation, preservation and management*, ed. P.L. Fiedler and S.K. Jain, 198–237. Chapman and Hall, New York, NY.

Harrison, R.L. 1992. Toward a theory of inter-refuge corridor design. *Conservation Biology* 6:293–295.

Harrison, S. 1989. Long distance dispersal and colonization in the bay checkerspot butterfly, *Euphydryas editha bayensis*. *Ecology* 70:1236–1243.

————. 1991. Local extinction in a metapopulation context: An empirical evaluation. *Biological Journal of the Linnean Society* 42:73–88.

————. 1997. How natural habitat patchiness affects the distribution of diversity in Californian serpentine chaparral. *Ecology* 78:1898–1906.

————. 1999. Local and regional diversity in a patchy landscape: Native, alien and endemic herbs on serpentine soils. *Ecology* 80:70–80.

Harrison, S., and E. Bruna. 1999. Habitat fragmentation and large-scale conservation: What do we know for sure? *Ecography* 22:225–232.

Harrison, S., and L. Fahrig. 1995. Landscape pattern and population conservation. In Hansson, Fahrig, and Merriam, 293–308.

Harrison, S., D.D. Murphy, and P.R. Ehrlich. 1988. Distribution of the bay checkerspot butterfly, *Euphydryas editha bayensis*: Evidence for a metapopulation model. *American Naturalist* 132:360–382.

Harrison, S., and C. Ray, 2002. Plant population viability and metapopulation-level processes. In *Population viability analysis*, ed. S. Beissinger and D. McCullough, 109–122. University of Chicago Press, Chicago, IL.

Harrison, S., K. Rice, and J. Maron. 2001. Habitat patchiness promotes invasion by alien grasses on serpentine soil. *Biological Conservation* 100:45–53.

Harrison, S., H.D. Safford, J.B. Grace, J.H. Viers, and K.F. Davies. 2006. Regional and local species richness in an insular environment: Serpentine plants in California. *Ecological Monographs* 76:41–56.

Harrison, S., and A.D. Taylor. 1997. Empirical evidence for metapopulation dynamics. In Hanski and Gilpin 1997, 27–42.

Harte, J., T. Blackburn, and A. Ostling. 2001. Self-similarity and the relationship between abundance and range size. *American Naturalist* 157:374–386.

Hartley, M.J., and M.L. Hunter. 1998. A meta-analysis of forest cover, edge effects, and artificial nest predation rates. *Conservation Biology* 12:465–469.

Harvell, C.D., K. Kim, J.M. Burkholder, R.R. Colwell, P.R. Epstein, D.J. Grimes, E.E. Hofmann, E.K. Lipp, A.D.M.E. Osterhaus, R.M. Overstreet, J.W. Porter, G.W. Smith, and G.R. Vasta. 1999. Emerging marine diseases: Climate links and anthropogenic factors. *Science* 285:1505–1510.

Harvell, C.D., C.E. Mitchell, J.R. Ward, S. Altizer, A.P. Dobson, R.S. Ostfeld, and M.D. Samuel. 2002. Climate warming and disease risks for terrestrial and marine biota. *Science* 296:2158–2162.

Harvey, C.A. 2000. Colonization of agricultural windbreaks by forest trees: Effects of connectivity and remnant trees. *Ecological Applications* 10:1762–1773.

Haydon D.T., S. Cleaveland, L.H. Taylor, and M.K. Laurenson. 2002. Identifying reservoirs of infection: A conceptual and practical challenge. *Emerging Infectious Diseases* 8:1468–1473.

Henderson, M.T., G. Merriam, and J. Wegner. 1985. Patchy environments and species survival: Chipmunks in an agricultural mosaic. *Biological Conservation* 31:95–105.

Henein, K., and G. Merriam. 1990. The elements of connectivity where corridor quality is variable. *Landscape Ecology* 4:157–170.

Herkert, J.R. 1994. The effects of habitat fragmentation on midwestern grassland bird communities. *Ecological Applications* 4:461–471.

Hess, G.R. 1994. Conservation corridors and contagious disease: A cautionary note. *Conservation Biology* 8:256–262.

———. 1996a. Disease in metapopulation models: Implications for conservation. *Ecology* 77:1617–1632.

———. 1996b. Linking extinction to connectivity and habitat destruction in metapopulation models. *American Naturalist* 148:226–236.

Hilborn, R. 2006. Salmon-farming impacts on wild salmon. *Proceedings of the National Academy of Sciences* 103:15277.

Hilderbrand, R.H., A.C. Watts, and A.M. Randle. 2005. The myths of restoration ecology. *Ecology and Society* 10:19. www.ecologyandsociety.org/vol10/iss1/art19/.

Hill, C.J. 1995. Linear strips of rain forest vegetation as potential dispersal corridors for rain forest insects. *Conservation Biology* 9:1559–1566.

Hill, J.K., C.L. Hughes, C. Dytham, and J.B. Searle. 2006. Genetic diversity in butterflies: Interactive effects of habitat fragmentation and climate-driven range expansion. *Biology Letters* 2:152–154.

Hilty, J.A., W.Z. Lidicker Jr., and A.M. Merenlender. 2006. *Corridor ecology: The science and practice of linking landscapes for biodiversity conservation.* Island Press, Washington, DC.

Hilty, J.A., and A.M. Merenlender. 2004. Use of riparian corridors and vineyards by mammalian predators in northern California. *Conservation Biology* 18:126–135.

Hinsley, S.A., P.E. Bellamy, B. Enoksson, G. Fry, L. Gabrielsen, D. McCollin, and A. Schotman. 1998. Geographical and land-use influences on bird species richness in small woods in agricultural landscapes. *Global Ecology and Biogeography Letters* 7:125–135.

Hobbs, R.J. 1992. The role of corridors in conservation: Solution or bandwagon? *Trends in Ecology and Evolution* 7:389–392.

Hobson, K.A., E.M. Bayne, and S.L. Van Wilgenburg. 2002. Large-scale conversion of forest to agriculture in the boreal plains of Saskatchewan. *Conservation Biology* 16:1530–1541.

Hoekstra, J.M., T.M. Boucher, T.H. Ricketts, and C. Roberts. 2005. Confronting a biome crisis: Global disparities of habitat loss and protection. *Ecology Letters* 8:23–29.

Hokit, D.G., B.M. Stith, and L.C. Branch. 1999. Effects of landscape structure in Florida scrub: A population perspective. *Ecological Applications* 9:124–134.

Holdenrieder, O., M. Pautasso, P.J. Weisberg, and D. Lonsdale. 2004. Tree diseases and landscape processes: The challenge of landscape pathology. *Trends in Ecology and Evolution* 19:446–452.

Holderegger, R., and H.W. Wagner. 2006. A brief guide to landscape genetics. *Landscape Ecology* 21:793–796.

Holl, K.D., and E.E. Crone. 2004. Local vs. landscape factors affecting restoration of riparian understorey plants. *Journal of Applied Ecology* 41:922–933.

Holl, K.D., M.E. Loik, E.H.V. Lin, and I.A. Samuels. 2000. Restoration of tropical rain forest in abandoned pastures in Costa Rica: Overcoming barriers to dispersal and establishment. *Restoration Ecology* 8:339–349.

Holland, R.F., and S. Jain. 1981. Insular biogeography of vernal pools in the Central Valley of California. *American Naturalist* 117:24–37.

Holmquist, J.G. 1998. Permeability of patch boundaries to benthic invertebrates: Influences of boundary contrast, light level, and faunal density and mobility. *Oikos* 81:558–566.

Holt, R.D., M. Holyoak, and M.A. Leibold. 2005. Future directions in metacommunity ecology. In Holyoak, Leibold, and Holt 2005, 465–489.

Holt, R.D., J.H. Lawton, G.A. Polis, and N.D. Martinez. 1999. Trophic rank and the species-area relationship. *Ecology* 80:1495–1504.

Holt, R.D., G.R. Robinson, and M.S. Gaines. 1995. Vegetation dynamics in an experimentally fragmented landscape. *Ecology* 76:1610–1624.

Holway, D.A. 1999. Competitive mechanisms underlying the displacement of native ants by the invasive Argentine ant. *Ecology* 80:238–251.

Holway D.A., A.V. Suarez, and T.J. Case. 1998. Loss of intraspecific aggression in the success of a widespread invasive social insect. *Science* 282:949–952.

Holyoak, M., and S.P. Lawler. 1996. Persistence of an extinction-prone predator-prey interaction through metapopulation dynamics. *Ecology* 77:1867–1879.

Holyoak, M., M.A. Leibold, and R.D. Holt, eds. 2005. *Metacommunities: Spatial dynamics and ecological communities.* University of Chicago Press, Chicago, IL.

Honnay, O., M. Hermy, and P. Coppin. 1999. Nested plant communities in deciduous forest fragments: Species relaxation or nested habitats? *Oikos* 84:119–129.

Hoopes, M.F., R.D. Holt, and M. Holyoak. 2005. The effects of spatial processes on two species interactions. In Holyoak, Leibold, and Holt 2005, 35–67.

Hopkins, P.J., and N.R. Webb. 1984. The composition of the beetle and spider faunas on fragmented heathlands. *Journal of Applied Ecology* 21:935–946.

Hrudey, S.E., and E.J. Hrudey. 2007. Published case studies of waterborne disease outbreaks: Evidence of a recurrent threat. *Water Environment Research* 79:233–245.

Hubbell, S.P. 2001. *The unified neutral theory of biodiversity and biogeography.* Princeton University Press, Princeton, NJ.

Huffaker, C.B. 1958. Experimental studies on predation: Dispersion factors and predator-prey oscillations. *Hilgardia* 27(14):343–383.

Hunter, L.M., M.D. Gonzalez, M. Stevenson, K.S. Karish, R. Toth, T.C. Edwards, R.J. Lilieholm, and M. Cablk. 2003. Population and land use change in the California Mojave: Natural habitat implications of alternative futures. *Population Research and Policy Review* 22:373–397.

Huxel, G.R. 2000. The effect of the Argentine ant on the threatened valley elderberry longhorn beetle. *Biological Invasions* 2:81–85.

Huxel, G.R., and A. Hastings. 1999. Habitat loss, fragmentation, and restoration. *Restoration Ecology* 7:309–315.

Huxel, G.R., M. Holyoak, T.S. Talley, and S.K. Collinge. 2003. Perspectives on the recovery of the threatened valley elderberry longhorn beetle. In *California riparian systems: Processes and floodplains management, ecology, and restoration,* ed. P.M. Faber, 457–462. Riparian Habitat Joint Venture, Sacramento, CA.

Ims, R.A. 1995. Movement patterns related to spatial structures. In Hansson, Fahrig, and Merriam, 85–109.

Ims, R.A., and N.G. Yoccoz. 1997. Studying transfer processes in metapopulations: Emigration, migration, and colonization. In Hanski and Gilpin 1997, 247–264.

Irlandi, E.A., W.G. Ambrose Jr., and B.A. Orlando. 1995. Landscape ecology and the marine environment: How spatial configuration of seagrass habitat influences growth and survival of the bay scallop. *Oikos* 72:307–313.

Irwin, R.E., A.K. Brody, and N.M. Waser. 2001. The impact of floral larceny on individuals, populations, and communities. *Oecologia* 129:161–168.

Isaak, D.J., R.F. Thurow, B.E. Rieman, and J.B. Dunham. 2007. Chinook salmon use of spawning patches: Relative roles of habitat quality, size, and connectivity. *Ecological Applications* 17:352–364.

Janzen, D.H. 1966. Coevolution of mutualism between ants and acacias in Central America. *Evolution* 20:249–275.

———. 1983. No park is an island: Increase in interference from outside as park size increases. *Oikos* 41:402–410.

———. 1986. The eternal external threat. In Soulé, 286–303.

———. 1988. Management of habitat fragments in a dry tropical forest: Growth. *Annals of the Missouri Botanical Garden* 75:105–116.

Jennersten, O. 1988. Pollination in *Dianthus deltoides* (Caryophyllaceae): Effects of habitat fragmentation on visitation and seed set. *Conservation Biology* 2:359–366.

Jessup, D.A., M. Miller, J. Ames, M. Harris, C. Kreuder, P.A. Conrad, and J.A.K. Mazet. 2004. Southern sea otter as a sentinel of marine ecosystem health. *EcoHealth* 1:239–245.

Johnson, P.T.J., and J.M. Chase. 2004. Parasites in the food web: Linking amphibian malformations and aquatic eutrophication. *Ecology Letters* 7:521–526.

Johnson, P.T.J, J.M. Chase, K.L. Dosch, R.B. Hartson, J.A. Gross, D.J. Larson, D.R. Sutherland, and S.R. Carpenter. 2007. Aquatic eutrophication promotes pathogenic infection in amphibians. *Proceedings of the National Academy of Sciences* 104:15781–15786.

Johnson, P.T.J., K.B. Lunde, E.M. Thurman, E.G. Ritchie, S.M. Wray, D.R. Sutherland, J.M. Kapfer, T.J. Frest, J. Bowerman, and A.R. Blaustein. 2002. Parasite infection linked to amphibian malformation in the western USA. *Ecological Monographs* 72:151–168.

Johnson, W.C., and S.K. Collinge. 2004. Landscape effects on black-tailed prairie dog colonies. *Biological Conservation* 115:487–497.

Joly, P., C. Miaud, A. Lehmann, and O. Grolet. 2001. Habitat matrix effects on pond occupancy in newts. *Conservation Biology* 15:239–248.

Jules, E.S. 1998. Habitat fragmentation and demographic change for a common plant: Trillium in old-growth forest. *Ecology* 79:1645–1656.

Jules, E.S., M.J. Kauffman, W.D. Ritts, and A.L. Carroll. 2002. Spread of an invasive pathogen over a variable landscape: A nonnative root rot on Port Orford cedar. *Ecology* 83:3167–3181.

Jules, E.S., and B.J. Rathcke. 1999. Mechanisms of reduced Trillium recruitment along edges of old-growth forest fragments. *Conservation Biology* 13:784–793.

Jules, E.S., and P. Shahani. 2003. A broader ecological context to habitat fragmentation: Why matrix habitat is more important than we thought. *Journal of Vegetation Science* 14:459–464.

Jump, A.S., and J. Peñuelas. 2006. Genetic effects of chronic habitat fragmentation in a wind-pollinated tree. *Proceedings of the National Academy of Sciences* 103:8096–8100.

Kahmen, A., and E.S. Jules. 2005. Assessing the recovery of a long-lived herb following logging: *Trillium ovatum* across a 424-year chronosequence. *Forest Ecology and Management* 210:107–116.

Kao, R.R. 2002. The role of mathematical modeling in the control of the 2001 FMD outbreak in the UK. *Trends in Microbiology* 10:279–286.

Kapos, V. 1989. Effects of isolation on the water status of forest patches in the Brazilian Amazon. *Journal of Tropical Ecology* 5:173–185.

Kapos, V., E. Wandelli, J.L. Camargo, and G. Ganade. 1997. Edge-related changes in environment and plant responses due to forest fragmentation in central Amazonia. In Laurance and Bierregaard, 33–44.

Kareiva, P.M. 1987. Habitat fragmentation and the stability of predator-prey interactions. *Nature* 326:388–390.

Kareiva, P.M., J.G. Kingsolver, and R.B. Huey. 1993. *Biotic interactions and global change.* Sinauer Associates, Sunderland, MA.

Kareiva, P.M., and U. Wennergren. 1995. Connecting landscape patterns to ecosystem and population processes. *Nature* 373:299–302.

Karr, J.R. 1990. Avian survival rates and the extinction process on Barro Colorado Island, Panama. *Conservation Biology* 4:391–397.

Keats, J. *The poetical works of John Keats.* 1884. Macmillan, London, UK. Reprinted from the original editions, with notes by Francis T. Palgrave. Bartleby.com. Great Books Online. www.bartleby.com/126/ [accessed August 17, 2007].

Keesing, F., R.D. Holt, and R.S. Ostfeld. 2006. Effects of species diversity on disease risk. *Ecology Letters* 9:485–498.

Kehler, D., and S. Bondrup-Nielsen. 1999. Effects of isolation on the occurrence of a fungivorous forest beetle, *Bolitotherus cornutus*, at different spatial scales in fragmented and continuous forests. *Oikos* 84:35–43.

Kim, K., and C.D. Harvell. 2004. The rise and fall of a six-year coral-fungal epizootic. *American Naturalist* 164 (Suppl.):S52-S63.

King, A.W., and K.A. With. 2002. Dispersal success on spatially structured landscapes: When do spatial pattern and dispersal behavior really matter? *Ecological Modelling* 147:23–39.

King, J.L., M.A. Simovich, and R.C. Brusca. 1996. Species richness, endemism and ecology of crustacean assemblages in northern California vernal pools. *Hydrobiologia* 328: 85–116.

King, R.S., M.E. Baker, D.F. Whigham, D.E. Weller, T.E. Jordan, P.F. Kazyak, and M.K. Hurd. 2005. Spatial considerations for linking watershed land cover to ecological indicators in streams. *Ecological Applications* 15:137–153.

Kinnaird, M.F. 1998. Evidence for effective seed dispersal by the Sulawesi red-knobbed hornbill, *Aceros cassidix*. *Biotropica* 30:50–55.

Kinzig, A.P., D. Tilman, and S. Pacala, eds. 2002. *The functional consequences of biodiversity: Empirical progress and theoretical extensions.* Princeton University Press, Princeton, NJ.

Kirchner, F., J.-B. Ferdy, C. Andalo, B. Colas, and J. Moret. 2003. Role of corridors in plant dispersal: An example with the endangered *Ranunculus nodiflorus*. *Conservation Biology* 17:401–410.

Kitching, R.L. 2001. Food webs in phytotelmata: "Bottom-up" and "top-down" explanations for community structure. *Annual Review of Entomology* 46:729–760.

Kitron, U. 1998. Landscape ecology and epidemiology of vector-borne diseases: Tools for spatial analysis. *Journal of Medical Entomology* 35:435–445.

Klein, B.C. 1989. Effects of forest fragmentation on dung and carrion beetle communities in central Amazonia. *Ecology* 70:1715–1725.

Kneitel, J.M., and T.E. Miller. 2003. Dispersal rates affect species composition in meta-communities of *Sarracenia purpurea* inquilines. *American Naturalist* 162:165–171.

Koelle, K., and J. Vandermeer. 2005. Dispersal-induced desynchronization: From meta-populations to metacommunities. *Ecology Letters* 8:167–175.

Kolozsvary, M.B., and R.K. Swihart. 1999. Habitat fragmentation and the distribution of amphibians: Patch and landscape correlates in farmland. *Canadian Journal of Zoology* 77:1288–1299.

Kong, F.H., and N. Nakagoshi. 2006. Spatial-temporal gradient analysis of urban green spaces in Jinan, China. *Landscape and Urban Planning* 78:147–164.

Krauss, J., A.M. Klein, I. Steffan-Dewenter, and T. Tscharntke. 2004. Effects of habitat area, isolation, and landscape diversity on plant species richness of calcareous grasslands. *Biodiversity and Conservation* 13:1427–1439.

Kremen, C., R.L. Bugg, N. Nicola, S.A. Smith, R.W. Thorp, and N.M Williams. 2002. Native bees, native plants and crop pollination in California. *Fremontia* 30:41–49.

Kremen, C., N.M. Williams, M.A. Aizen, B. Gemmill-Harren, G. LeBuhn, R. Minckley, L. Packer, S.G. Potts, T. Roulston, I. Steffan-Dewenter, D.P. Vázquez, R. Winfree, L. Adams, E.E. Crone, S.S. Greenleaf, T.H. Keitt, A.M. Klein, J. Regetz, and T.H. Ricketts. 2007. Pollination and other ecosystem services produced by mobile organisms: A conceptual framework for the effects of land-use change. *Ecology Letters* 10:299–314.

Kremen, C., N.M. Williams, R.L. Bugg, J.P. Fay, and R.W. Thorp. 2004. The area requirements of an ecosystem service: Crop pollination by native bee communities in California. *Ecology Letters* 7:1109–1119.

Kremen, C., N.M. Williams, and R.W. Thorp. 2002. Crop pollination from native bees at risk from agricultural intensification. *Proceedings of the National Academy of Sciences* 99:16812–16816.

Kreuss, A., and T. Tscharntke. 1994. Habitat fragmentation, species loss, and biological control. *Science* 264:1581–1584.

Krieger, A. 1991. *Andres Duany and Elizabeth Plater-Zyberk: Towns and town-making principles.* Rizzoli, New York, NY.

Krkošek, M., M.A. Lewis, A. Morton, L.N. Frazer, and J.P. Volpe. 2006. Epizootics of wild fish induced by farm fish. *Proceedings of the National Academy of Sciences* 103:15507–15510.

Krkošek, M., M.A. Lewis, and J.P. Volpe. 2005. Transmission dynamics of parasitic sea lice from farm to wild salmon. *Proceedings of the Royal Society of London, B* 272:689–696.

Kruckeberg, A.R. 2006. *Introduction to California soils and plants: Serpentine, vernal pools, and other geobotanical wonders.* University of California Press, Berkeley, CA.

Kruys, N., and B.G. Jonsson. 1997. Insular patterns of calicioid lichens in a boreal old-growth forest-wetland mosaic. *Ecography* 20:605–613.

Kupfer, J.A., G.P. Malanson, and S.B. Franklin. 2006. Not seeing the ocean for the islands: The mediating influence of matrix-based processes on forest fragmentation effects. *Global Ecology and Biogeography* 15:8–20.

LaDeau, S., A. Marm Kilpatrick, and P.P. Marra. 2007. West Nile virus emergence and large-scale declines of North American bird populations. *Nature* 447:710–713.

Lafferty, K., and L.R. Gerber. 2002. Good medicine for conservation biology: The intersection of epidemiology and conservation theory. *Conservation Biology* 16:593–604.

Lahti, D.C. 2001. The "edge effect on nest predation" hypothesis after twenty years. *Biological Conservation* 99:365–374.

Laiolo, P., and J.L. Tella. 2006. Landscape bioacoustics allow detection of the effects of habitat patchiness on population structure. *Ecology* 87:1203–1214.

Lamberson, R.H., B.R. Noon, C. Voss, and K.S. McKelvey. 1994. Reserve design for territorial species: The effects of patch size and spacing on the viability of the northern spotted owl. *Conservation Biology* 8:185–195.

Lampila, P., M. Monkkonen, and A. Desrochers. 2005. Demographic responses by birds to forest fragmentation. *Conservation Biology* 19:1537–1546.

Lande, R. 1987. Extinction thresholds in demographic models of territorial populations. *American Naturalist* 130:624–635.

———. 1988. Demographic models of the northern spotted owl (*Strix occidentalis caurina*). *Oecologia* 75:601–607.

Langlois, J.P., L. Fahrig, G. Merriam, and H. Artsob. 2001. Landscape structure influences continental distribution of Hantavirus in deer mice. *Landscape Ecology* 16:255–266.

La Polla, V.N., and G.W. Barrett. 1993. Effects of corridor width and presence on the population dynamics of the meadow vole (*Microtus pennsylvanicus*). *Landscape Ecology* 8:25–37.

Laurance, S.G., and W.F. Laurance. 1999. Tropical wildlife corridors: Use of linear rainforest remnants by arboreal mammals. *Biological Conservation* 91:231–239.

Laurance, W.F. 1990. Comparative responses of five arboreal marsupials to tropical forest fragmentation. *Journal of Mammalogy* 71:641–653.

———. 1991. Ecological correlates of extinction proneness in Australian tropical rain forest mammals. *Conservation Biology* 5:79–89.

———. 2000. Do edge effects occur over large spatial scales? *Trends in Ecology and Evolution* 15:134–135.

Laurance, W.F., and R.O. Bierregaard Jr., eds. 1997. *Tropical forest remnants: Ecology, management, and conservation of fragmented communities.* University of Chicago Press Chicago, IL.

Laurance, W.F., L.V. Ferreira, J.M. Rankin-de Merona, and S.G. Laurance. 1998. Rain forest fragmentation and the dynamics of Amazonian tree communities. *Ecology* 79:2032–2040.

Laurance, W.F., S.G. Laurance, L.V. Ferreira, J. Rankin-de Merona, C. Gascon, and T.E. Lovejoy. 1997. Biomass collapse in Amazonian forest fragments. *Science* 278:1117–1118.

Laurance, W.F., T.E. Lovejoy, H.L. Vasconcelos, E.M. Bruna, R.K. Didham, P.C. Stouffer, C. Gascon, R.O. Bierregaard Jr., S.G. Laurance, and E. Sampaio. 2002. Ecosystem decay of Amazonian forest fragments: A 22-year investigation. *Conservation Biology* 16:605–618.

Laurance, W.F., H.E.M. Nascimento, S.G. Laurance, A.C. Andrade, P.M. Fearnside, J.E.L. Ribeiro, and R.L. Capretz. 2006. Rain forest fragmentation and the proliferation of successional trees. *Ecology* 87:469–482.

Leach, M.K., and T.J. Givnish. 1996. Ecological determinants of species loss in remnant prairies. *Science* 273:1555–1558.

Lei, G.C., and I. Hanski. 1997. Metapopulation structure of *Cotesia melitaearum*, a specialist parasitoid of the butterfly, *Melitaea cinxia*. *Oikos* 78:91–100.

Leibold, M.A., M. Holyoak, M. Mouquet, P. Amarasekare, J.M. Chase, M.F. Hoopes, R.D. Holt, J.B. Shurin, R. Law, D. Tilman, M. Loreau, and A. Gonzalez. 2004. The metacommunity concept: A framework for multi-scale community ecology. *Ecology Letters* 7:601–613.

Leigh, E.G., Jr., S.J. Wright, E.A. Herre, and F.E. Putz. 1993. The decline of tree diversity on newly isolated tropical islands: A test of a null hypothesis and some implications. *Evolutionary Ecology* 7:76–102.

Lejeune, K.D., and T.R. Seastedt. 2001. *Centaurea* species: The forb that won the West. *Conservation Biology* 15:1568–1574.

Lekberg, Y., R.T. Koide, J.R. Rohr, L. Aldrich-Wolfe, and J.B. Morton. 2007. Role of niche restrictions and dispersal in the composition of arbuscular mycorrhizal fungal communities. *Journal of Ecology* 95:95–105.

Lennartsson, T. 2002. Extinction thresholds and disrupted plant-pollinator interactions in fragmented plant populations. *Ecology* 83:3060–3072.

Leopold, A. 1933. *Game management.* C. Scribner's Sons, New York, NY.

———. 1949. *A Sand County almanac, and sketches here and there.* Oxford University Press, New York, NY.

Levey, D.J., B.M. Bolker, J.J. Tewksbury, S. Sargent, and N.M. Haddad. 2005. Effects of landscape corridors on seed dispersal by birds. *Science* 309:146–148.

Levin, S.A. 1992. The problem of pattern and scale in ecology. *Ecology* 73:1943–1967.

Levin, S.A., and R.T. Paine. 1974. Disturbance, patch formation and community structure. *Proceedings of the National Academy of Sciences* 71:2744–2747.

Levins, R. 1966. The strategy of model building in population biology. *American Scientist* 54:421–431.

———. 1969. Some demographic and genetic consequences of environmental heterogeneity for biological control. *Bulletin of the Entomological Society of America* 15:237–240.

Levins, R., and D. Culver. 1971. Regional coexistence of species and competition between rare species. *Proceedings of the National Academy of Sciences* 68:1246–1248.

Li, H., J.F. Franklin, F.J. Swanson, and T.A. Spies. 1993. Developing alternative forest cutting patterns: A simulation approach. *Landscape Ecology* 8:63–75.

Li, H., and J. Wu. 2004. Use and misuse of landscape indices. *Landscape Ecology* 19:389–399.

Li, W.F., Y.L. Wang, F. Peng, and G.C. Li. 2005. Landscape spatial changes associated with rapid urbanization in Shenzhen, China. *International Journal of Sustainable Development and World Ecology* 12:314–325.

Lidicker, W.Z., Jr., and W.D. Koenig. 1996. Responses of terrestrial vertebrates to habitat edges and corridors. In McCullough 1996, 85–110.

Likens, G.E., F.H. Bormann, N.M. Johnson, D.W. Fisher, and R.S. Pierce. 1970. Effects of forest cutting and herbicide treatment on nutrient budgets in the Hubbard Brook watershed-ecosystem. *Ecological Monographs* 40:23–47.

Lindenmayer, D.B. 1994. Wildlife corridors and the mitigation of logging impacts on fauna in wood-production forests in south-eastern Australia: A review. *Wildlife Research* 21:323–340.

Lindenmayer, D.B., R.B. Cunningham, and C.F. Donnelly. 1993. The conservation of arboreal marsupials in the montane ash forests of the central highlands of Victoria, south-east Australia: IV. The presence and abundance of arboreal marsupials in retained linear habitats (wildlife corridors) within logged forest. *Biological Conservation* 66:207–221.

Lindenmayer, D.B., R.B. Cunningham, and M.L. Pope. 1999. A large-scale "experiment" to examine the effects of landscape context and habitat fragmentation on mammals. *Biological Conservation* 88:387–403.

Lindenmayer, D.B., R.B. Cunningham, M.L. Pope, and C.F. Donnelly. 1999. The response of arboreal marsupials to landscape context: A large-scale fragmentation study. *Ecological Applications* 9:594–611.

Lindenmayer, D.B., and J. Fischer. 2006. *Habitat fragmentation and landscape change.* Island Press, Washington, DC.

Lindenmayer, D.B., M.A. McCarthy, K.M. Parris, and M.L. Pope. 2000. Habitat fragmentation, landscape context, and mammalian assemblages in southeastern Australia. *Journal of Mammalogy* 81:787–797.

Lindenmayer, D.B., and H.A. Nix. 1993. Ecological principles for the design of wildlife corridors. *Conservation Biology* 7:627–630.

Lindenmayer, D.B., and H.P. Possingham. 1995. Modeling the viability of metapopulations of the endangered Leadbeater's possum in southeastern Australia. *Biodiversity and Conservation* 4:984–1018.

Linderman, M.A., L. An, S. Bearer, G. He, Z. Ouyang, and J. Liu. 2005. Modeling the spatio-temporal dynamics and interactions of households, landscapes, and giant panda habitat. *Ecological Modelling* 183:47–65.

Lion, S., M. van Baalen, and W.G. Wilson. 2006. The evolution of parasite manipulation of host dispersal. *Proceedings of the Royal Society of London, B* 273:1063–1071.

Little, C.E. 1990. *Greenways for America.* Johns Hopkins University Press, Baltimore, MD.

Liu, J., M. Linderman, Z. Ouyang, L. An, J. Yang, and H. Zhang. 2001. Ecological degradation in protected areas: The case of Wolong Nature Reserve for giant pandas. *Science* 292:98–101.

Lockwood, J., M. Hoopes, and M. Marchetti. 2007. *Invasion ecology.* Blackwell, Malden, MA.

LoGiudice, K., R.S. Ostfeld, K.A. Schmidt, and F. Keesing. 2003. The ecology of infectious disease: Effects of host diversity and community composition on Lyme disease risk. *Proceedings of the National Academy of Sciences* 100:567–571.

López-Barrera, F., R.H. Manson, M. González-Espinosa, and A.C. Newton. 2007. Effects of varying forest edge permeability on seed dispersal in a neotropical montane forest. *Landscape Ecology* 22:189–203.

Loreau, M., N. Mouquet, and R.D. Holt. 2003. Meta-ecosystems: A theoretical framework for a spatial ecosystem ecology. *Ecology Letters* 6:673–679.

Loreau, M., and C.L. Nolf. 1994. Spatial structure and dynamics of a population of *Abax ater*. In *Carabid beetles: Ecology and evolution*, ed. K. Desender, M. Dufrêne, M. Loreau, M.L. Luff, and J.P. Maelfait, 165–169. Series Entomologica 51. Kluwer Academic Publishers, Boston, MA.

Loreto Bay Company. 2007. *Inagural sustainability report*. Available at www.loretobay.com/files/pdf/inaugural-sustainability-report-en.pdf.

Lovei, G.L., T. Magura, B. Tothmeresz, and V. Kodobocz. 2006. The influence of matrix and edges on species richness patterns of ground beetles (Coleoptera: Carabidae) in habitat islands. *Global Ecology and Biogeography* 15:283–289.

Lovejoy, T.E., R.O. Bierregaard Jr., A.B. Rylands, J.R. Malcolm, C.E. Quintela, L.H. Harper, K.S. Brown Jr., A.H. Powell, G.V.N. Powell, H.O. Schubart, and M.B. Hays. 1986. Edge and other effects of isolation on Amazon forest fragments. In Soulé 1986, 257–285.

Lovejoy, T.E., J.M. Rankin, R.O. Bierregaard Jr., K.S. Brown Jr., L.H. Emmons, and M.E. Van der Voort. 1984. Ecosystem decay of Amazon forest remnants. In *Extinctions*, ed. M.H. Nitecki, 295–325. University of Chicago Press, Chicago, IL.

Lynch, J.F. 1987. Responses of breeding bird communities to forest fragmentation. In Saunders, Arnold, Burbidge, and Hopkins, 123–140.

MacArthur, R.H., and E.O. Wilson. 1967. *The theory of island biogeography*. Princeton University Press, Princeton, NJ.

MacClintock, L., R.F. Whitcomb, and B.L. Whitcomb. 1977. Evidence for the value of corridors and minimization of isolation in preservation of biotic diversity. *American Birds* 31:6–16.

MacDonald, M.A., and J.B. Kirkpatrick. 2003. Explaining bird species composition and richness in eucalypt-dominated remnants in subhumid Tasmania. *Journal of Biogeography* 30:1415–1426.

MacEachern, A.K., M.L. Bowles, and N.B. Pavlovic. 1994. A metapopulation approach to Pitcher's thistle (*Cirsium pitcheri*) recovery in southern Lake Michigan dunes. In *Restoration of endangered species: Conceptual issues, planning and implementation*, ed. M.L. Bowles and C.J. Whelan, 194–218. Cambridge University Press, Cambridge, UK.

Machtans, C.S., M-A. Villard, and S.J. Hannon. 1996. Use of riparian buffer strips as movement corridors by forest birds. *Conservation Biology* 10:1366–1379.

Mackelprang, R., M.D. Dearing, and S. St. Jeor. 2001. High prevalence of Sin Nombre virus in rodent populations, central Utah: A consequence of human disturbance? *Emerging Infectious Diseases* 7:480–482.

MacKenzie, D.I., J.D. Nichols, J.A. Royle, K.H. Pollock, L.L. Bailey, and J.E. Hines. 2006. *Occupancy estimation and modeling: Inferring patterns and dynamics of species occurrence*. Elsevier Academic Press, Burlingame, MA.

MacNally, R., and A.F. Bennett. 1997. Species-specific predictions of the impact of habitat fragmentation: Local extinction of birds in the box-ironbark forests of central Victoria, Australia. *Biological Conservation* 82:147–155.

MacNally, R., A.F. Bennett, and G. Horrocks. 2000. Forecasting the impacts of habitat fragmentation: Evaluation of species-specific predictions of the impact of habitat fragmentation on birds in the box-ironbark forests of central Victoria, Australia. *Biological Conservation* 95:7–29.

Mader, H.J., C. Schell, and P. Kornaker. 1990. Linear barriers to arthropod movement in the landscape. *Biological Conservation* 54:209–222.

Madriñán, L.F., A. Etter, G.D. Boxall, and A. Ortega-Rubio. 2007. Tropical alluvial forest fragmentation in the eastern lowlands of Colombia (1939–1997). *Land Degradation and Development* 18:199–208.

Malcolm, J.R. 1997. Biomass and diversity of small mammals in Amazonian forest fragments. In Laurance and Bierregaard, 207–221.

Manel, S., M.K. Schwartz, G. Luikart, and P. Taberlet. 2003. Landscape genetics: Combining landscape ecology and population genetics. *Trends in Ecology and Evolution* 18:189–197.

Mangan, S.A., A.H. Eom, G.H. Adler, J.B. Yavitt, and E.A.D. Herre. 2004. Diversity of arbuscular mycorrhizal fungi across a fragmented forest in Panama: Insular spore communities differ from mainland communities. *Oecologia* 141:687–700.

Mann, C.C., and M.L. Plummer. 1995. Are wildlife corridors the right path? *Science* 270: 1428–1430.

Margules, C.R., ed. 1989. Australian developments in conservation evaluation. Special issue, *Biological Conservation* 50.

———. 1992. The Wog-Wog habitat fragmentation experiment. *Environmental Conservation* 19:316–325.

Margules, C.R., and R.L. Pressey. 2000. Systematic conservation planning. *Nature* 405: 243–253.

Margules, C.R., and S. Sarkar. 2007. *Systematic conservation planning.* Cambridge University Press, Cambridge, UK.

Martin, T.G., S. McIntyre, C.P. Catterall, and H.P. Possingham. 2006. Is landscape context important for riparian conservation? Birds in grassy woodland. *Biological Conservation* 127:201–214.

Martínez-Morales, M.A. 2005. Landscape patterns influencing bird assemblages in a fragmented neotropical cloud forest. *Biological Conservation* 121:117–126.

Maschinski, J. 2006. Implications of population dynamics and metapopulation theory for restoration. In Falk, Palmer, and Zedler 2006, 59–88.

Mason, J., C. Moorman, G. Hess, and K. Sinclair. 2007. Designing suburban greenways to provide habitat for forest-breeding birds. *Landscape and Urban Planning* 80:153–164.

Matlack, G.R. 1993a. Microenvironment variation within and among forest edge sites in the eastern United States. *Biological Conservation* 66:185–194.

———. 1993b. Sociological edge effects: Spatial distribution of human impact in suburban forest fragments. *Environmental Management* 17:829–835.

May, R.M., J.H. Lawton, and N.E. Stork. 1995. Assessing extinction rates. In *Extinction rates*, ed. J.H. Lawton and R.M. May, 1–24. Oxford University Press, Oxford, UK.

Mazerolle, M.J. 2005. Drainage ditches facilitate frog movements in a hostile landscape. *Landscape Ecology* 20:579–590.

Mazerolle, M.J., and M.A. Villard. 1999. Patch characteristics and landscape context as predictors of species presence and abundance: A review. *Ecoscience* 6:117–124.

McAlpine, C.A., and T.J. Eyre. 2002. Testing landscape metrics as indicators of habitat loss and fragmentation in continuous eucalypt forests (Queensland, Australia). *Landscape Ecology* 17:711–728.

McCallum, H.I., and A. Dobson. 2002. Disease, habitat fragmentation, and conservation. *Proceedings of the Royal Society of London, B* 269:2041–2049.

McCallum, H.I., and M. Jones. 2006. To lose both would look like carelessness: Tasmanian devil facial tumour disease. *PLoS Biology* 4:1671–1674.

McCallum, H.I., A. Kuris, C.D. Harvell, K.D. Lafferty, G.W. Smith, and J. Porter. 2004. Does terrestrial epidemiology apply to marine systems? *Trends in Ecology and Evolution* 19:585–591.

McCarthy, M.A., and D.B. Lindenmayer. 1999. Incorporating metapopulation dynamics of greater gliders into reserve design in disturbed landscapes. *Ecology* 80:651–667.

McCarthy, M.A., C.J. Thompson, and H.P. Possingham. 2005. Theory for designing nature reserves for single species. *American Naturalist* 165:250–257.

McClanahan, T.R., and R.W. Wolfe. 1993. Accelerating forest succession in a fragmented landscape: The role of birds and perches. *Conservation Biology* 7:279–288.

McCullough, D.R., ed. 1996. *Metapopulations and wildlife conservation.* Island Press, Washington, DC.

McGarigal, K., and S.A. Cushman. 2002. Comparative evaluation of experimental approaches to the study of habitat fragmentation effects. *Ecological Applications* 12:335–345.

McGarigal, K., S.A. Cushman, M.C. Neel, and E. Ene. 2002. FRAGSTATS: Spatial pattern analysis program for categorical maps. Computer software program produced by the authors at the University of Massachusetts–Amherst, available at www.umass.edu/landeco/research/fragstats/fragstats.html.

McGarigal, K., and B. Marks. 1995. FRAGSTATS: Spatial analysis program for quantifying landscape structure. U.S.D.A. Forest Service General Technical Report PNW-GTR-351. Pacific Northwest Research Station, U.S.D.A. Forest Service, Portland, OR.

McHarg, I.L. 1969. *Design with nature.* Doubleday/Natural History Press, Garden City, NY.

McHarg, I.L., A.H. Johnson, and J. Berger. 1979. A case study in ecological planning: The Woodlands, Texas. In *Planning the uses and management of land,* ed. M.T. Beatty, G.W. Petersen, and L.D. Swindale, 935–955. American Society of Agronomy, Crop Science Society of America, and Soil Science Society of America, Madison, WI.

McHarg, I.L., and J. Sutton. 1975. Ecological plumbing for the Texas coastal plain. *Landscape Architecture* 65 (January):78–89.

McKelvey, K., B.R. Noon, and R.H. Lamberson. 1993. Conservation planning for species occupying fragmented landscapes: The case of the northern spotted owl. In Kareiva, Kingsolver, and Huey, 424–450.

McKenzie, V.J. 2007. Human land use and patterns of parasitism in tropical amphibian hosts. *Biological Conservation* 137:102–116.

McRae, B.H. 2006. Isolation by resistance. *Evolution* 60:1551–1561.

McRae, B.H., P. Peier, L.W. Dewald, L.Y. Huynh, and P. Keim. 2005. Habitat barriers limit gene flow and illuminate historical events in a wide-ranging carnivore, the American puma. *Molecular Ecology* 14:1965–1977.

McVicar, A.H. 2004. Management actions in relation to the controversy about salmon lice infections in fish farms as a hazard to wild salmonid populations. *Aquaculture Research* 35:751–758.

Mech, S.H., and J.G. Hallett. 2001. Evaluating the effectiveness of corridors: A genetic approach. *Conservation Biology* 15:467–474.

Melbourne, B.A., K.F. Davies, C.R. Margules, D.B. Lindenmayer, D.A. Saunders, C. Wissel, and C. Henle. 2004. Species survival in fragmented landscapes: Where to from here? *Biodiversity and Conservation* 13:275–284.

Merriam, G., and A. Lanoue. 1990. Corridor use by small mammals: Field measurements of three experimental types of *Peromyscus leucopus*. *Landscape Ecology* 4:123–131.

Mesquita, R.C.G., P. Delamonica, and W.F. Laurance. 1999. Effect of surrounding vegetation on edge-related tree mortality in Amazonian forest fragments. *Biological Conservation* 91:129–134.

Mettler-Cherry, P.A., M. Smith, and T.M. Keevin. 2006. Habitat characterization and geospatial metapopulation dynamics of threatened floodplain species *Boltonia decurrens* using a GIS. *Wetlands* 26:336–348.

Michalski, F., and C.A. Peres. 2005. Anthropogenic determinants of primate and carnivore local extinctions in a fragmented forest landscape of southern Amazonia. *Biological Conservation* 124:383–396.

Micheli, F., and C.H. Peterson. 1999. Estuarine vegetated habitats as corridors for predator movements. *Conservation Biology* 13:869–881.

Millennium Ecosystem Assessment. 2005. *Ecosystems and human well-being: Biodiversity synthesis.* World Resources Institute, Washington, DC.

Miller, J.R., and N.T. Hobbs. 2000. Effects of recreational trails on nest predation rates and predator assemblages. *Landscape and Urban Planning* 50:227–236.

Miller, J.R., J.A. Wiens, N.T. Hobbs, and D.M. Theobald. 2003. Effects of human settlement on bird communities in lowland riparian areas of Colorado (USA). *Ecological Applications* 13:1041–1059.

Miller, M.A., I.A. Gardner, C. Kreuder, D.M. Paradies, K.R. Worcester, D.A. Jessup, E. Dodd, M.D. Harris, J.A. Ames, A.E. Packham, and P.A. Conrad. 2002. Coastal freshwater runoff is a risk factor for *Toxoplasma gondii* infection of southern sea otters (*Enhydra lutris nereis*). *International Journal for Parasitology* 32:997–1006.

Miller, M.A., M.E. Grigg, C. Kreuder, E.R. James, A.C. Melli, P.R. Crosbie, D.R. Jessup, J.C. Boothroyd, LD. Brownstein, and P.A. Conrad. 2004. An unusual genotype of *Toxoplasma gondii* is common in California sea otters (*Enhydra lutris nereis*) and is a cause of mortality. *International Journal for Parasitology* 34:275–284.

Mills, L.S., M.E. Soulé, and D.F. Doak. 1993. The keystone-species concept in ecology and conservation. *BioScience* 43:219–224.

Milne, B.T. 1991. Lessons from applying fractal models to landscape patterns. In Turner and Gardner 1991, 199–235.

Mitchell, C.E., and A.G. Power. 2006. Disease dynamics in plant communities. In Collinge and Ray 2006, 58–72.

Mitchell, C.E., D. Tilman, and J.V. Groth. 2002. Effects of grassland species diversity, abundance, and composition on foliar fungal disease. *Ecology* 83:1713–1726.

Moilanen, A., and I. Hanski. 1998. Metapopulation dynamics: Effects of habitat quality and landscape structure. *Ecology* 79:2503–2515.

Moilanen, A., and M. Nieminen. 2002. Simple connectivity measures in spatial ecology. *Ecology* 83:1131–1145.

Moody, M.E., and R.N. Mack. 1988. Controlling the spread of plant invasions: The importance of nascent foci. *Journal of Applied Ecology* 25:1009–1021.

Moore, A.A., and M.A. Palmer. 2005. Invertebrate biodiversity in agricultural and urban headwater streams: Implications for conservation and management. *Ecological Applications* 15:1169–1177.

Morimoto, T., K. Katoh, Y. Yamaura, and S. Watanabe. 2006. Can surrounding land cover influence the avifauna in urban/suburban woodlands in Japan? *Landscape and Urban Planning* 75:143–154.

Morris, W.F., R.A. Hufbauer, A.A. Agrawal, J.D. Bever, V.A. Borowicz, G.S. Gilbert, J.L. Maron, C.E. Mitchell, I.M. Parker, A.G. Power, M.E. Torchin, and D.P. Vázquez. 2007. Direct and interactive effects of enemies and mutualists on plant performance: A meta-analysis. *Ecology* 88:1021–1029.

Mortberg, U., and H.G. Wallentinus. 2000. Red-listed forest bird species in an urban environment: Assessment of green space corridors. *Landscape and Urban Planning* 50: 215–226.

Murakami, K., H. Maenaka, and Y. Morimoto. 2005. Factors influencing species diversity of ferns and fern allies in fragmented forest patches in the Kyoto city area. *Landscape and Urban Planning* 70:221–229.

Murcia, C. 1995. Edge effects in fragmented forests: Implications for conservation. *Trends in Ecology and Evolution* 10:58–62.

Murdoch, W., J. Chesson, and P.L. Chesson. 1985. Biological control in theory and practice. *American Naturalist* 125:344–366.

Murphy, D.D. 1989. Conservation and confusion: Wrong species, wrong scale, wrong conclusions. *Conservation Biology* 3:82–84.

Murphy, H.T., and J. Lovett-Doust. 2004. Context and connectivity in plant metapopulations and landscape mosaics: Does the matrix matter? *Oikos* 105:3–14.

Naidoo, R., and T.H. Ricketts. 2006. Mapping the economic costs and benefits of conservation. *PLoS Biology* 4:2153–2164.

Nascimento, H.E.M., and W.F. Laurance. 2004. Biomass dynamics in Amazonian forest fragments. *Ecological Applications* 14 (Suppl.):127–138.

Nassauer, J.I. 1992. The appearance of ecological systems as a matter of policy. *Landscape Ecology* 6:239–250.

———. 1995. Messy ecosystems, orderly frames. *Landscape Journal* 14:161–170.

———. 1997. Cultural sustainability: Aligning aesthetics and ecology. In *Placing nature: Culture and landscape ecology*, ed. J.I. Nassauer, 67–83. Island Press, Washington, DC.

———. 2004. Monitoring the success of metropolitan wetland restorations: Cultural sustainability and ecological function. *Wetlands* 24:756–765.

———. 2006. Landscape planning and conservation biology: Systems thinking revisited. *Conservation Biology* 20:677–678.

Nature Conservancy. 2006. *Conservation by design: A strategic framework for mission success.* Nature Conservancy, Arlington, VA.

Ndubisi, F. 1997. Landscape ecological planning. In *Ecological design and planning,* ed. G.F. Thompson and F.R. Steiner, 9–44. John Wiley and Sons, New York, NY.

———. 2002. *Ecological planning: A historical and comparative synthesis.* Johns Hopkins University Press, Baltimore, MD.

Nee, S., and R.M. May. 1992. Dynamics of metapopulations: Habitat destruction and competitive coexistence. *Journal of Animal Ecology* 61:37–40,

Nepstad, D.C., C.M. Stickler, and O.T. Almeida. 2006. Globalization of the Amazon soy and beef industries: Opportunities for conservation. *Conservation Biology* 20:1595–1603.

Ness, J.H. 2004. Forest edges and fire ants alter the seed shadow of an ant-dispersed plant. *Oecologia* 138:448–454.

Newmark, W.D. 1986. Species-area relationship and its determinants for mammals in western North American national parks. *Biological Journal of the Linnean Society* 28:83–98.

———. 1987. A land-bridge perspective on mammalian extinctions in western North American parks. *Nature* 325:430–432.

———. 1991. Tropical forest fragmentation and the local extinction of understory birds in the eastern Usambara Mountains, Tanzania. *Conservation Biology* 5:67–78.

———. 1995. Extinction of mammal populations in western North American national parks. *Conservation Biology* 9:512–526.

———. 1996. Insularization of Tanzanian parks and the local extinction of large mammals. *Conservation Biology* 10:1549–1556.

———. 2001. Tanzanian forest edge microclimatic gradients: Dynamic patterns. *Biotropica* 33:2–11.

Ney-Nifle, M., and M. Mangel. 2000. Habitat loss and changes in the species-area relationship. *Conservation Biology* 14:893–898.

Norton, D.A., R.J. Hobbs, and L. Atkins. 1995. Fragmentation, disturbance, and plant distribution: Mistletoes in woodland remnants in the Western Australia wheatbelt. *Conservation Biology* 9:426–438.

Noss, R.F. 1992. The Wildlands Project: Land conservation strategy. *Wild Earth* (Special issue):10–25.

———. 1993. A bioregional conservation plan for the Oregon Coast Range. *Natural Areas Journal* 13:276–290.

Noss, R.F., and A.Y. Cooperrider. 1994. *Saving nature's legacy: Protecting and restoring biodiversity.* Island Press, Washington, DC.

Noss, R.F., M.A. O'Connell, and D.D. Murphy. 1997. *The science of conservation planning: Habitat conservation under the Endangered Species Act.* Island Press, Washington, DC.

Nour, N., E. Matthysen, and A.A. Dhondt. 1997. Effects of habitat fragmentation on foraging behaviour of tits and related species: Does niche space vary in relation to size and degree of isolation of forest fragments? *Ecography* 20:281–286.

Nupp, T.E., and R.K. Swihart. 1998. Effects of forest fragmentation on population attributes of white-footed mice and eastern chipmunks. *Journal of Mammalogy* 79:1234–1243.

O'Brien, T.G., M.F. Kinnaird, A. Nurcahyo, M. Prasetyaningrum, and M. Igbal. 2003. Fire, demography and the persistence of siamang (*Symphalangus syndactylus*: Hylobatidae) in a Sumatran rainforest. *Animal Conservation* 6:115–121.

Ochoa-Gaona, S., M. González-Espinosa, J.A. Meave, and V.S.D. Bon. 2004. Effect of forest fragmentation on the woody flora of the highlands of Chiapas, Mexico. *Biodiversity and Conservation* 13:867–884.

Oertli, B., D.A. Joye, E. Castella, R. Juge, D. Cambin, and J.-B. Lachavanne. 2002. Does size matter? The relationship between pond area and biodiversity. *Biological Conservation* 104:59–70.

O'Hara, R.B., E. Arjas, H. Toivonen, and I. Hanski. 2002. Bayesian analysis of metapopulation data. *Ecology* 83:2408–2415.

Ohlemüller, R., S. Walker, and J.B. Wilson. 2006. Local vs. regional factors as determinants of the invasibility of indigenous forest fragments by alien plant species. *Oikos* 112:493–501.

Økland, B. 1996. Unlogged forests: Important sites for preserving the diversity of mycetophilids (Diptera: Sciaroidea). *Biological Conservation* 76:297–310.

O'Neill, R.V., D.L. DeAngelis, T.F.H. Allen, and J.B. Waide. 1986. *A hierarchical concept of ecosystems.* Monographs in Population Biology 23. Princeton University Press, Princeton, NJ.

O'Neill, R.V., J.R. Krummel, R.H. Gardner, G. Sugihara, B. Jackson, D.L. DeAngelis, B.T. Milne, M.G. Turner, B. Zygmunt, S.W. Christensen, V.H. Dale, and R.L. Graham. 1988. Indices of landscape pattern. *Landscape Ecology* 1:153–162.

Orrock, J.L. 2005. Conservation corridors affect the fixation of novel alleles. *Conservation Genetics* 6:623–630.

Orrock, J.L., and B.J. Danielson. 2005. Patch shape, connectivity, and foraging by oldfield mice (*Peromyscus polionotus*). *Journal of Mammalogy* 86:569–575.

Orrock, J.L., B.J. Danielson, M.J. Burns, and D.J. Levey. 2003. Spatial ecology of predator-prey interactions: Corridors and patch shape influence seed predation. *Ecology* 84: 2589–2599.

Ostfeld, R.S., G.E. Glass, and F. Keesing. 2005. Spatial epidemiology: An emerging (or re-emerging) discipline. *Trends in Ecology and Evolution* 20:328–336.

Ostfeld, R.S., and F. Keesing. 2000a. Biodiversity and disease risk: The case of Lyme disease. *Conservation Biology* 14:722–728.

———. 2000b. The function of biodiversity in the ecology of vector-borne zoonotic diseases. *Canadian Journal of Zoology* 78:2061–2078.

Ostfeld, R.S., F. Keesing, and K. LoGiudice. 2006. Community ecology meets epidemiology: The case of Lyme disease. In Collinge and Ray 2006, 28–40.

Ostfeld, R.S., and K. LoGiudice. 2003. Community disassembly, biodiversity loss, and the erosion of an ecosystem service. *Ecology* 84:1421–1427.

Ouin, A., J.P. Sarthou, B. Bouyjou, M. Deconchat, J.P. Lacombe, and C. Monteil. 2006. The species-area relationship in the hoverfly (Diptera: Syrphidae) communities of forest fragments in southern France. *Ecography* 29:183–190.

Packer, C., S. Altizer, M. Appel, E. Brown, J. Martenson, S.J. O'Brien, M. Roelke-Parker, R.

Hofmann-Lehmann, and H. Lutz. 1999. Viruses of the Serengeti: Patterns of infection and mortality in African lions. *Journal of Animal Ecology* 68:1161–1178.

Paine, R.T. 1966. Food web complexity and species diversity. *American Naturalist* 100:65–75.

Palomares, F., M. Delibes, P. Ferreras, J.M. Fedriani, J. Calzada, and E. Revilla. 2000. Iberian lynx in a fragmented landscape: Pre-dispersal, dispersal, and post-dispersal habitats. *Conservation Biology* 14:809–818.

Palomares, F., A. Rodríguez, R. Laffitte, and M. Delibes. 1991. The status and distribution of the Iberian lynx *Felis pardina* in Coto Doñana Area, SW Spain. *Biological Conservation* 57:159–169.

Pardini, R., S.M. de Souza, R. Braga-Neto, and J.P. Metzger. 2005. The role of forest structure, fragment size and corridors in maintaining small mammal abundance and diversity in an Atlantic forest landscape. *Biological Conservation* 124:253–266.

Parker, T.H., B.M. Stansberry, C.D. Becker, and P.S. Gipson. 2005. Edge and area effects on the occurrence of migrant forest songbirds. *Conservation Biology* 19:1157–1167.

Parris, K.M. 2006. Urban amphibian assemblages as metacommunities. *Journal of Animal Ecology* 75:757–764.

Paton, P.W. 1994. The effect of edge on avian nest success: How strong is the evidence? *Conservation Biology* 8:17–26.

Patz, J.A., P. Daszak, G.M. Tabor, A.A. Aguirre, M. Pearl, J. Epstein, N.D. Wolfe, A.M. Kilpatrick, J. Foufopoulos, D. Molyneux, D.J. Bradley, and members of the Working Group on Land Use Change and Disease Emergence. 2004. *Environmental Health Perspectives* 112:1092–1098.

Peacock, M.M., and C. Ray. 2001. Dispersal in pikas (*Ochotona princeps*): Combining genetic and demographic approaches to reveal spatial and temporal patterns. In *Dispersal: Causes, consequences and mechanisms of dispersal at the individual, population and community level*, ed. J. Clobert, É. Danchin, A.A. Dhondt, and J.D. Nichols, 43–56. Oxford University Press, Oxford, UK.

Peacock, M.M., and A.T. Smith. 1997. The effect of habitat fragmentation on dispersal patterns, mating behavior, and genetic variation in a pika (*Ochotona princeps*) metapopulation. *Oecologia* 112:524–533.

Peak, R.G., and F.R. Thompson III. 2006. Factors affecting avian species richness and density in riparian areas. *Journal of Wildlife Management* 70:173–179.

Peay, K.G., T.D. Bruns, P.G. Kennedy, S.E. Bergemann, and M. Garbelotto. 2007. A strong species-area relationship for eukaryotic soil microbes: Island size matters for ectomycorrhizal fungi. *Ecology Letters* 10:470–480.

Pe'er, G., D. Saltz, and K. Frank. 2005. Virtual corridors for conservation management. *Conservation Biology* 19:1997–2003.

Peintinger, M., A. Bergamini, and B. Schmid. 2003. Species-area relationships and nestedness of four taxonomic groups in fragmented wetlands. *Basic and Applied Ecology* 4:385–394.

Pejchar, L., P.M. Morgan, M.R. Caldwell, C. Palmer, and G.C. Daily. 2007. Evaluating the potential for conservation development: Biophysical, economic, and institutional perspectives. *Conservation Biology* 21:69–78.

Peres, C.A. 2001. Synergistic effects of subsistence hunting and habitat fragmentation on Amazonian forest vertebrates. *Conservation Biology* 15:1490–1505.

Perkins, T.E., and G.R. Matlack. 2002. Human-generated pattern in commercial forests of southern Mississippi and consequences for the spread of pests and pathogens. *Forest Ecology and Management* 157:143–154.

Peterjohn, W.T., and D.L. Correll. 1984. Nutrient dynamics in an agricultural watershed: Observations on the role of a riparian forest. *Ecology* 65:1466–1475.

Peterken, G.F. 1996. *Natural woodland: Ecology and conservation in northern temperate regions.* Cambridge University Press, Cambridge, UK.

Peters, D.P.C. 2004. Selection of models of invasive species dynamics. *Weed Technology* 18:1236–1239.

Peters, H.A. 2001. *Clidemia hirta* invasion at the Pasoh forest reserve: An unexpected plant invasion in an undisturbed tropical forest. *Biotropica* 33:60–68.

Petersen, J.E., and A. Hastings. 2001. Dimensional approaches to scaling experimental ecosystems: Designing mousetraps to catch elephants. *American Naturalist* 157:324–333.

Peterson, G.D., G.S. Cumming, and S.R. Carpenter. 2003. Scenario planning: A tool for conservation in an uncertain world. *Conservation Biology* 17:358–366.

Petit, S., L. Griffiths, S.S. Smart, G.M. Smith, R.C. Stuart, and S.M. Wright. 2004. Effects of area and isolation of woodland patches on herbaceous plant species richness across Great Britain. *Landscape Ecology* 19:463–471.

Petranka, J.W., E.M. Harp, C.T. Holbrook, and J.A. Hamel. 2007. Long-term persistence of amphibian populations in a restored wetland complex. *Biological Conservation* 138:371–380.

Pharo, E., D.B. Lindenmayer, and N. Taws. 2004. The effects of large-scale fragmentation on bryophytes in temperate forests. *Journal of Applied Ecology* 41:910–921.

Pickett, S.T.A., and P.S. White. 1985. *The ecology of natural disturbance and patch dynamics.* Academic Press, New York, NY.

Picton, J.D. 1979. The application of insular biogeographic theory to the conservation of large mammals in the northern Rocky Mountains. *Biological Conservation* 15:73–79.

Piessens, K., O. Honnay, K. Nackaerts, and M. Hermy. 2004. Plant species richness and composition of heathland relics in north-western Belgium: Evidence for a rescue-effect? *Journal of Biogeography* 31:1683–1692.

Pimm, S.L. 2002. The dodo went extinct (and other ecological myths). *Annals of the Missouri Botanical Garden* 89:190–198.

Poiani, K.A., J.V. Baumgartner, S.C. Buttrick, S.L. Green, E. Hopkins, G.D. Ivey, K.P. Seaton, and R.D. Sutter. 1998. A scale-independent, site conservation planning framework in the Nature Conservancy. *Landscape and Urban Planning* 43:143–156.

Polasky, S., J.D. Camm, and B. Garber-Yonts. 2001. Selecting biological reserves cost-effectively: An application to terrestrial vertebrate conservation in Oregon. *Land Economics* 77:68–78.

Porter, J.W., P. Dustan, W.C. Jaap, K.L. Patterson, V. Kosmynin, O.W. Meier, M.E. Patterson, and M. Parsons. 2001. Patterns of spread of coral disease in the Florida Keys. *Hydrobiologia* 46:1–24.

Porneluzi, P., J.C. Bednarz, L.J. Goodrich, N. Zawada, and J. Hoover. 1993. Reproductive performance of territorial ovenbirds occupying forest fragments and a contiguous forest in Pennsylvania. *Conservation Biology* 7:618–622.

Powell, A.H., and G.V.N. Powell. 1987. Population dynamics of male euglossine bees in Amazonian forest fragments. *Biotropica* 19:176–179.

Pressey, R.L., C.J. Humphries, C.R. Margules, R.I. Vane-Wright, and P.H. Williams. 1993. Beyond opportunism: Key principles for systematic reserve selection. *Trends in Ecology and Evolution* 8:124–128.

Priess, J.A., M. Mimler, A.-M. Klein, S. Schwarze, T. Tscharntke, and I. Steffan-Dewenter. 2007. Linking deforestation scenarios to pollination services and economic returns in coffee agroforestry systems. *Ecological Applications* 17:407–417.

Pringle, C.M. 1991. U.S.–Romanian environmental reconnaissance of the Danube delta. *Conservation Biology* 5:442–445.

Pryke, S.R., and M.J. Samways. 2001. Width of grassland linkages for the conservation of butterflies in South African afforested areas. *Biological Conservation* 101:85–96.

Pulliam, H.R. 1988. Sources, sinks, and population regulation. *American Naturalist* 132:652–661.

Pulliam, H.R., and B.J. Danielson. 1991. Sources, sinks, and habitat selection: A landscape perspective on population dynamics. *American Naturalist* 137:50–66.

Pulliam, H.R., J.B. Dunning Jr., and J. Liu. 1992. Population dynamics in complex landscapes: A case study. *Ecological Applications* 2:165–177.

Pykälä, J., and R.K. Heikkinen. 2005. Complementarity-based algorithms for selecting sites to preserve grassland plant species. *Agriculture, Ecosystems and Environment* 106: 41–48.

Qi, Y., M. Henderson, M. Xu, J. Chen, P.J. Shi, C.Y. He, and G.W. Skinner. 2004. Evolving core-periphery interactions in a rapidly expanding urban landscape: The case of Beijing. *Landscape Ecology* 19:375–388.

Quinn, J.F., and S.P. Harrison. 1988. Effects of habitat fragmentation and isolation on species richness: Evidence from biogeographic patterns. *Oecologia* 75:132–140.

Quinn, J.F., and G.R. Robinson. 1987. The effects of experimental subdivision on flowering plant diversity in a California annual grassland. *Journal of Ecology* 75:837–856.

Quinn, J.F., C.L. Wolin, and M.L. Judge. 1989. An experimental analysis of patch size, habitat subdivision, and extinction in a marine intertidal snail. *Conservation Biology* 3:242–251.

Ramp, J.M. 2005. Restoration genetics and pollination of the rare vernal pool endemic *Lasthenia conjugens* (Asteraceae). PhD diss., University of Colorado.

Rand, T.A., and S.M. Louda. 2004. Exotic weed invasion increases the susceptibility of native plants to attack by a biocontrol herbivore. *Ecology* 85:1548–1554.

———. 2006. Spillover of agriculturally subsidized predators as a potential threat to native insect herbivores in fragmented landscapes. *Conservation Biology* 20:1720–1729.

Ranney, J.W., M.C. Bruner, and J.B. Levenson. 1981. The importance of edge in the structure and dynamics of forest islands. In Burgess and Sharpe, 68–95.

Rantalainen, M.-L., H. Fritze, J. Haimi, T. Pennanen, and H. Setälä. 2005. Species richness and food web structure of soil decomposer community as affected by the size of habitat fragment and habitat corridors. *Global Change Biology* 11:1614–1627.

Real, L.A., and J.H. Brown, eds. 1991. *Foundations of ecology: Classic papers with commentaries.* University of Chicago Press, Chicago, IL.

Real, L.A., and J.E. Childs. 2006. Spatial-temporal dynamics of rabies in ecological communities. In Collinge and Ray 2006, 168–185.

Redford, K.H. 1984. The termitaria of *Cornitermes cumulans* (Isopotera: Termitidae) and their role in determining a potential keystone species. *Biotropica* 16:112–119.

———. 1992. The empty forest. *BioScience* 42:412–422.

Redford, K.H., P. Coppolillo, E.W. Sanderson, G.A.B. da Fonseca, E. Dinerstein, C. Groves, G. Mace, S. Maginnis, R.A. Mittermeier, R.F. Noss, D. Olson, J.G. Robinson, A. Vedder, and M. Wright. 2003. Mapping the conservation landscape. *Conservation Biology* 17:116–131.

Redford, K.H., E.W. Sanderson, J.G. Robinson, and A. Vedder. 2000. *Landscape species and their conservation: Report from a WCS meeting, May 2000.* Wildlife Conservation Society, Bronx, NY. Quoted in Sanderson et al. 2002.

Rejmánková, E., J. Greico, N. Achee, P. Masuoka, K.O. Pope, D.R. Roberts, and R.M. Higashi. 2006. Freshwater community interactions and malaria. In Collinge and Ray 2006, 90–104.

Rejmánková, E., K.O. Pope, D.R. Roberts, M.G. Lege, R. Andre, J. Greico, and Y. Alonzo. 1998. Characterization and detection of *Anopheles vestitipennis* and *Anopheles punctimacula* (Diptera: Culicidae) larval habitats in Belize with field survey and SPOT satellite imagery. *Journal of Vector Ecology* 23:74–88.

Rejmánková, E., D.R. Roberts, S. Manguin, K.O. Pope, J. Komarek, and R.A. Post. 1996. *Anopheles albimanus* (Diptera: Culicidae) and cyanobacteria: An example of larval habitat selection. *Environmental Entomology* 25:1058–1067.

Ricketts, T.H. 2001. The matrix matters: Effective isolation in fragmented landscapes. *American Naturalist* 158:87–99.

Ricketts, T.H., G.C. Daily, P.R. Ehrlich, and J.P. Fay. 2001. Countryside biogeography of moths in a fragmented landscape: Biodiversity in native and agricultural habitats. *Conservation Biology* 15:378–388.

Ricketts, T.H., J. Regetz, I. Steffan-Dewenter, S.A. Cunningham, C. Kremen, A. Bogdanski, B. Gemmill-Herren, S.S. Greenleaf, A.M. Klein, M.M. Mayfield, L.A. Morandin, A. Ochieng, and B.F. Viana. 2008. Landscape effects on crop pollination services: Are there general patterns? *Ecology Letters* 11:499–515.

Ricketts, T.H., N.M. Williams, and M.M. Mayfield. 2006. Connectivity and ecosystem services: Crop pollination in agricultural landscapes. In Crooks and Sanjayan 2006, 255–289.

Ricklefs, R.E., and I.J. Lovette. 1999. The roles of island area per se and habitat diversity in the species-area relationships of four Lesser Antillean faunal groups. *Journal of Animal Ecology* 68:1142–1160.

Riebsame, W.E., ed. 1997. *Atlas of the new West.* W.W. Norton and Company, New York, NY.

Ries, L., D.M. Debinski, and M.L. Wieland. 2002. Conservation value of roadside prairie restoration to butterfly communities. *Conservation Biology* 15:401–411.

Ries, L., R.J. Fletcher, J. Battin, and T.D. Sisk. 2004. Ecological responses to habitat edges: Mechanisms, models, and variability explained. *Annual Review of Ecology and Systematics* 35:491–522.

Riley, S.P.D., J.P. Pollinger, R.M. Sauvajot, E.C. York, C. Bromley, T.K. Fuller, and R.K. Wayne. 2006. A southern California freeway is a physical and social barrier to gene flow in carnivores. *Molecular Ecology* 15:1733–1741.

Ripple, W.J., K.T. Hershey, and R.G. Anthony. 2000. Historical forest patterns of Oregon's central Coast Range. *Biological Conservation* 93:127–133.

Robichaud, I., M.-A. Villard, and C.S. Machtans. 2002. Effects of forest regeneration on songbird movements in a managed forest landscape of Alberta, Canada. *Landscape Ecology* 17:247–262.

Robinson, G.R., and S.N. Handel. 1993. Forest restoration on a closed landfill: Rapid addition of new species by bird dispersal. *Conservation Biology* 7:271–278.

———. 2000. Directing spatial patterns of recruitment during an experimental urban woodland reclamation. *Ecological Applications* 10:174–188.

Robinson, G.R., S.N. Handel, and J. Mattei. 2002. Experimental techniques for evaluating the success of restoration projects. *Korean Journal of Ecology* 25:1–7.

Robinson, G.R., S.N. Handel, and V.R. Schmalhofer. 1992. Survival, reproduction, and recruitment of woody plants after 14 years on a reforested landfill. *Environmental Management* 16:265–271.

Robinson, G.R., R.D. Holt, M.S. Gaines, S.P. Hamburg, M.L. Johnson, H.S. Fitch, and E.A. Martinko. 1992. Diverse and contrasting effects of habitat fragmentation. *Science* 257:524–526.

Robinson, G.R., and J.F. Quinn. 1988. Extinction, turnover and species diversity in an experimentally fragmented California annual grassland. *Oecologia* 76:71–82.

Robinson, W.D. 1999. Long-term changes in the avifauna of Barro Colorado Island, Panama, a tropical forest isolate. *Conservation Biology* 13:85–97.

Roland, J., and P.D. Taylor. 1997. Insect parasitoid species respond to forest structure at different spatial scales. *Nature* 386:710–713.

Romanuk, T.N., and J. Kolasa. 2004. Population variability is lower in diverse rock pools when the obscuring effects of local processes are removed. *Ecoscience* 11:455–462.

Root, K.V. 1998. Evaluating the effects of habitat quality, connectivity, and catastrophes on a threatened species. *Ecological Applications* 8:854–865.

Rosenberg, D.K., B.R. Noon, and E.C. Meslow. 1997. Biological corridors: Form, function, and efficacy. *BioScience* 47:677–687.

Rosenzweig, M.L. 1995. *Species diversity in space and time.* Cambridge University Press, Cambridge, UK.

Rouget, M., D.M. Richardson, R.M. Cowling, J.W. Lloyd, and A.T. Lombard. 2003. Current patterns of habitat transformation and future threats to biodiversity in terrestrial ecosystems of the Cape Floristic Region, South Africa. *Biological Conservation* 112:63–85.

Ruckelshaus, M., C. Hartway, and P.M. Kareiva. 1997. Assessing the data requirements of spatially explicit dispersal models. *Conservation Biology* 11:1298–1306.

Russell, D., and C. Harshbarger. 2003. *Groundwork for community-based conservation.* Alta-Mira Press, Lanham, MD.

Ryall, K.L., and L. Fahrig. 2006. Response of predators to loss and fragmentation of prey habitat: A review of theory. *Ecology* 87:1086–1093.

Salzer, J.S., I.B. Rwego, T.L. Goldberg, M.S. Kuhlenschmidt, and T.R. Gillespie. 2007. *Giardia* sp. and *Cryptosporidium* sp. infections in primates in fragmented and undisturbed forest in western Uganda. *Journal of Parasitology* 93:439–440.

Sanderson, E.W. 2006. How many animals do we want to save? The many ways of setting population target levels for conservation. *BioScience* 56:911–922.

Sanderson, E.W., K.H. Redford, A. Vedder, P.B. Coppolillo, and S.E. Ward. 2002. A conceptual model for conservation planning based on landscape species requirements. *Landscape and Urban Planning* 58:41–56.

Santelmann, M.V., D. White, K. Freemark, J.I. Nassauer, J.M. Eilers, K.B. Vaché, B.J. Danielson, R.C. Corry, M.E. Clark, S. Polasky, R.M. Cruse, J. Sifneos, H. Rustigian, C. Coiner, J. Wu, and D. Debinski. 2004. Assessing alternative futures for agriculture in Iowa, U.S.A. *Landscape Ecology* 19:357–374.

Santos, T., and J.L. Tellería. 1994. Influence of forest fragmentation on seed consumption and dispersal of Spanish juniper. *Biological Conservation* 70:129–134.

Saunders, D.A., G.W. Arnold, A.A. Burbridge, and A.J. Hopkins, eds. 1987. *Nature conservation: The role of remnants of native vegetation.* Surrey Beatty and Sons, Chipping Norton, Sydney, Australia.

Saunders, D.A., and R.J. Hobbs, eds. 1991. *Nature conservation 2: The role of corridors.* Surrey Beatty and Sons, Chipping Norton, Sydney, Australia.

Saunders, D.A., R.J. Hobbs, and P. Ehrlich, eds. 1993. *Nature conservation 3: Reconstruction of fragmented ecosystems.* Surrey Beatty and Sons, Chipping Norton, Sydney, Australia.

Saunders, D.A., R.J. Hobbs, and C.R. Margules. 1991. Biological consequences of ecosystem fragmentation: A review. *Conservation Biology* 5:18–32.

Schloegel, L.M., J.-M. Hero, L. Berger, R. Speare, K. McDonald, and P. Daszak. 2006. The decline of the sharp-snouted day frog (*Taudactylus acutirostris*): The first documented case of extinction by infection in a free-ranging wildlife species? *EcoHealth* 3:35–40.

Schmidt, K.A., and R.S. Ostfeld. 2001. Biodiversity and the dilution effect in disease ecology. *Ecology* 82:609–619.

Schmiegelow, F.K.A., C.S. Machtans, and S.J. Hannon. 1997. Are boreal birds resilient to fragmentation? An experimental study of short-term responses. *Ecology* 78:1914–1932.

Schrott, G.R., K.A. With, and A.W. King. 2005. Demographic limitations of the ability of habitat restoration to rescue declining populations. *Conservation Biology* 19:1181–1193.

Schultz, C.B. 1998. Dispersal behavior and its implications for reserve design in a rare Oregon butterfly. *Conservation Biology* 12:284–292.

———. 2001. Restoring resources for an endangered butterfly. *Journal of Applied Ecology* 38:1007–1019.

Schultz, C.B., and E.E. Crone. 2005. Patch size and connectivity thresholds for habitat restoration. *Conservation Biology* 19:887–896.

Schumaker, N.H. 1996. Using landscape indices to predict habitat connectivity. *Ecology* 77:1210–1225.

Schwartz, M.W., L.R. Iverson, and A.M. Prasad. 2001. Predicting the potential future distribution of four tree species in Ohio using current habitat availability and climatic forcing. *Ecosystems* 4:568–581.

Scott, J.M., B. Csuti, J.D. Jacobi, and J.E. Estes. 1987. Species richness: A geographic approach to protecting future biodiversity. *BioScience* 37:782–788.

Scott, J.M., F. Davis, B. Csuti, R.F. Noss, B. Butterfield, C. Groves, H. Anderson, S. Caicco, F. D'Erchia, T.C. Edwards Jr., J. Ulliman, and R.G. Wright. 1993. Gap analysis: A geographic approach to protection of biological diversity. *Wildlife Monographs* 123:3–41.

Seabloom, E.W., A.P. Dobson, and D.M. Stoms. 2002. Extinction rates under nonrandom patterns of habitat loss. *Proceedings of the National Academy of Sciences* 99:11229–11234.

Seabloom, E.W., and A. van der Valk. 2003. Plant diversity, composition, and invasion of restored and natural prairie pothole wetlands: Implications for restoration. *Wetlands* 23:1–12.

Shanker, K., and R. Sukumar. 1998. Community structure and demography of small-mammal populations in insular montane forests in southern India. *Oecologia* 116:243–251.

Shepherd, S., and D.M. Debinski. 2005. Evaluation of isolated and integrated prairie reconstructions as habitat for prairie butterflies. *Biological Conservation* 126:51–61.

Shochat, E., Z. Abramsky, and B. Pinshow. 2001. Breeding bird species diversity in the Negev: Effects of scrub fragmentation by planted forests. *Journal of Applied Ecology* 38:1135–1147.

Silva, M., L. Hartling, and S.B. Opps. 2005. Small mammals in agricultural landscapes of Prince Edward Island (Canada): Effects of habitat characteristics at three different spatial scales. *Biological Conservation* 126:556–568.

Simberloff, D.S. 1969. Experimental zoogeography of islands: A model for insular colonization. *Ecology* 50:296–314.

———. 1998. Flagships, umbrellas, and keystones: Is single-species management passé in the landscape era? *Biological Conservation* 83:247–257.

Simberloff, D.S., and J. Cox. 1987. Consequences and costs of conservation corridors. *Conservation Biology* 1:63–71.

Simberloff, D.S., J.A. Farr, J. Cox, and D.W. Mehlman. 1992. Movement corridors: Conservation bargains or poor investments? *Conservation Biology* 6:493–504.

Simberloff, D.S., and N. Gotelli. 1984. Effects of insularisation on plant species richness in the prairie-forest ecotone. *Biological Conservation* 29:27–46.

Simberloff, D.S., and E.O. Wilson. 1969. Experimental zoogeography of islands: The colonization of empty islands. *Ecology* 50:278–296.

———. 1970. Experimental zoogeography of islands: A two-year record of colonization. *Ecology* 51:934–937.

Sisk, T.D., N.M. Haddad, and P.R. Ehrlich. 1997. Bird assemblages in patchy woodlands: Modeling the effects of edge and matrix habitats. *Ecological Applications* 7:1170–1180.

Sjögren-Gulve, P. and C. Ray. 1996. Using logistic regression to model metapopulation dynamics: Large-scale forestry extirpates the pool frog. In McCullough 1996, 111–137.

Skelly, D.K., S.R. Bolden, M.P. Holland, L.K. Freidenburg, N.A. Freidenfelds, and T.R. Malcolm. 2006. Urbanization and disease in amphibians. In Collinge and Ray 2006, 153–167.

Smith, D.S. 1993. Greenway case studies. In Smith and Hellmund 1993, 161–208.

Smith, D.S., and P.C. Hellmund, eds. 1993. *Ecology of greenways.* University of Minnesota Press, Minneapolis, MN.

Smith, V.H., B.L. Foster, J.P. Grover, R.D. Holt, M.A. Leibold, and F. deNoyelles Jr. 2005. Phytoplankton species richness scales consistently from laboratory microcosms to the world's oceans. *Proceedings of the National Academy of Sciences of the USA* 102:4393–4396.

Snäll, T., J. Pennanen, L. Kivisto, and I. Hanski. 2005. Modelling epiphyte metapopulation dynamics in a dynamic forest landscape. *Oikos* 109:209–222.

Soranno, P.A., S.L. Hubler, S.R. Carpenter, and R.C. Lathrop. 1996. Phosphorus loads to surface waters: A simple model to account for spatial pattern of land use. *Ecological Applications* 6:865–878.

Soulé, M.E., ed. 1986. *Conservation biology: The science of scarcity and diversity.* Sinauer Associates, Sunderland, MA.

———, ed. 1987. *Viable populations for conservation.* Cambridge University Press, Cambridge, UK.

Soulé, M.E., D.T. Bolger, A.C. Alberts, J. Wright, M. Sorice, and S. Hill. 1988. Reconstructed dynamics of rapid extinctions of chaparral-requiring birds in urban habitat islands. *Conservation Biology* 2:75–92.

Soulé, M.E., J.A. Estes, J. Berger, and C. Martínez del Rio. 2003. Ecological effectiveness: Conservation goals for interactive species. *Conservation Biology* 17:1238–1250.

Soulé, M.E., J.A. Estes, B. Miller, and D.L. Honnold. 2005. Strongly interacting species: Conservation policy, management, and ethics. *BioScience* 55:168–176.

Soulé, M.E., and R.F. Noss. 1998. Rewilding and biodiversity: Complementary goals for continental conservation. *Wild Earth* 8:18–28.

Soulé, M.E., and J. Terborgh, eds. 1999. *Continental conservation: Design and management principles for long-term, regional conservation networks.* Island Press, Washington, DC.

Soulé, M.E., and B.A. Wilcox, eds. 1980. *Conservation biology: An evolutionary-ecological perspective.* Sinauer Associates, Sunderland, MA.

Southern Rockies Ecosystem Project. 2004. *The state of the southern Rockies ecoregion.* Colorado Mountain Club Press, Golden, CO.

Spies, T.A., W.J. Ripple, and G.A. Bradshaw. 1994. Dynamics and pattern of a managed coniferous forest landscape in Oregon. *Ecological Applications* 4:555–568.

Spirn, A.W. 1985. Urban nature and human design: Renewing the great tradition. *Journal of Planning Education and Research* 5:39–51.

Stanton, M.L., T.M. Palmer, T.P. Young, A. Evans, and M.L. Turner. 1999. Sterilization and canopy modification of a swollen thorn acacia tree by a plant-ant. *Nature* 401:578–581.

Steffan-Dewenter, I., U. Munzenberg, and T. Tscharntke. 2001. Pollination, seed set and seed predation on a landscape scale. *Proceedings of the Royal Society of London, B* 268:1685–1690.

Steinitz, C.F., C. Adams, L. Alexander, L. Eberhart, K. Hickey, B. Lyon, A. Mellinger, and A. Price. 1997. *An alternative future for the region of Camp Pendleton, California.* Harvard University Graduate School of Design, Cambridge, MA.

Steinitz, C.F., H. Arias, and A. Shearer. 2003. *Alternative futures for changing landscapes: The Upper San Pedro River Basin in Arizona and Sonora.* Island Press, Washington, DC.

Steinitz, C.F., R. Faris, M. Flaxman, K. Karish, A.D. Mellinger, T. Canfield, and L.A. Sucre. 2005. A delicate balance: Conservation and development scenarios for Panama's Coiba National Park. *Environment* 5:24–39.

Stith, B.M., J.W. Fitzpatrick, G.E. Woolfenden, and B. Pranty. 1996. Classification and conservation of metapopulations: A case study of the Florida scrub jay. In McCullough, 187–215.

Storfer, A., M.A. Murphy, J.S. Evans, C.S. Goldberg, S. Robinson, S.F. Spear, R. Dezzani, E. Delmelle, L. Vierling, and L.P. Waits. 2007. Putting the "landscape" in landscape genetics. *Heredity* 98:128–142.

Stouffer, P.C., and R.O. Bierregaard Jr., 1995. Use of Amazonian forest fragments by understory insectivorous birds. *Ecology* 76:2429–2445.

Stratford, J.A., and P.C. Stouffer. 1999. Local extinctions of terrestrial insectivorous birds in a fragmented landscape near Manaus, Brazil. *Conservation Biology* 13:1416–1423.

Styrsky, J.D., and M.D. Eubanks. 2007. Ecological consequences of interactions between ants and honeydew-producing insects. *Proceedings of the Royal Society of London, B* 274:151–164.

Suarez, A.V., D.T. Bolger, and T.J. Case. 1998. Effects of fragmentation and invasion on native ant communities in coastal southern California. *Ecology* 79:2041–2056.

Summerville, K.S., A.C. Bonte, and L.C. Fox. 2007. Short-term temporal effects on community structure of Lepidoptera in restored and remnant tallgrass prairies. *Restoration Ecology* 15:179–188.

Summerville, K.S., C.J. Conoan, and R.M Steichen. 2006. Species traits as predictors of Lepidopteran composition in restored and remnant tallgrass prairies. *Ecological Applications* 16:891–900.

Summerville, K.S., and T.O. Crist. 2004. Contrasting effects of habitat quantity and quality on moth communities in fragmented landscapes. *Ecography* 27:3–12.

Summerville, K.S., R.M. Steichen, and M.N. Lewis. 2005. Restoring Lepidopteran communities to oak savnnas: Contrasting influences of habitat quantity and quality. *Restoration Ecology* 13:120–128.

Suttcliffe, O.L., and C.D. Thomas. 1996. Open corridors appear to facilitate dispersal by ringlet butterflies (*Aphantopus hyperantus*) between woodland clearings. *Conservation Biology* 10:1359–1365.

Syphard, A.D., K.C. Clarke, and J. Franklin. 2005. Using a cellular automaton model to forecast the effects of urban growth on habitat pattern in Southern California. *Ecological Complexity* 2:185–203.

Tabarelli, M., J.M. Cardosa da Silva, and C. Gascon. 2004. Forest fragmentation, synergisms and the impoverishment of neotropical forests. *Biodiversity and Conservation* 13:1419–1425.

Tabor, G.M. 2002. Defining conservation medicine. In *Conservation medicine: Ecological health in practice*, ed. A.A. Aguirre, R.S. Ostfeld, G.M. Tabor, C. House, and M.C. Pearl, 8–16. Oxford University Press, Oxford, UK.

Talley, T.S., D. Wright, and M. Holyoak. 2006. Assistance with the 5-year review of the valley elderberry longhorn beetle (*Desmocerus californicus dimorphus*). Sacramento Office, U.S. Fish and Wildlife Service, Sacramento, CA.

Tallmon, D.A., E.S. Jules, N.J. Radke, and L.S. Mills. 2003. Of mice and men and *Trillium*: Cascading effects of forest fragmentation. *Ecological Applications* 13:1193–1203.

Tan, K.W. 2006. A greenway network for Singapore. *Landscape and Urban Planning* 76: 45–66.

Tanner, J.E. 2003. Patch shape and orientation influences on seagrass epifauna are mediated by dispersal abilities. *Oikos* 100:517–524.

Taylor, A.D. 1990. Metapopulations, dispersal, and predator-prey dynamics: An overview. *Ecology* 71:429–433.

Taylor, B., D. Skelly, L.K. Demarchis, M.D. Slade, D. Galusha, and P.M. Rabinowitz. 2005. Proximity to pollution sources and risk of amphibian limb malformation. *Environmental Health Perspectives* 113:1497–1501.

Taylor, P.D., L. Fahrig, K. Henein, and G. Merriam. 1993. Connectivity is a vital element of landscape structure. *Oikos* 68:571–573.

Temperton, V.M., R.J. Hobbs, T. Nuttle, and S. Halle, eds. 2004. *Assembly rules and restoration ecology: Bridging the gap between theory and practice*. Island Press, Washington, DC.

Terborgh, J. 1986. Keystone plant resources in the tropical forest. In Soulé, 330–344.

Terborgh J., L. Lopez, P. Nuñez, M. Rao, G. Shahabuddin, G. Orihuela, M. Riveros, R. Ascanio, G.H. Adler, T.D. Lambert, and L. Balbas. 2001. Ecological meltdown in predator-free forest fragments. *Science* 294:1923–1926.

Terborgh, J., and B. Winter. 1980. Some causes of extinction. In Soulé and Wilcox, 119–133.

Tewksbury, J.J., L. Garner, S. Garner, J.D. Lloyd, V. Saab, and T.E. Martin. 2006. Tests of landscape influence: Nest predation and brood parasitism in fragmented ecosystems. *Ecology* 87:759–768.

Tewksbury, J.J., D.J. Levey, N.M. Haddad, S. Sargent, J.L. Orrock, A. Weldon, B.J. Danielson, J. Brinkerhoff, E.I. Damschen, and P. Townsend. 2002. Corridors affect plants, animals, and their interactions in fragmented landscapes. *Proceedings of the National Academy of Science* 99:12923–12926.

Theobald, D.M. 2004. Placing exurban land-use change in a human modification framework. *Frontiers in Ecology and the Environment* 2:139–144.

———. 2006. Exploring the functional connectivity of landscapes using landscape networks. In Crooks and Sanjayan 2006, 416–444.

Theobald, D.M., J.R. Miller, and N.T. Hobbs. 1997. Estimating the cumulative effects of development on wildlife habitat. *Landscape and Urban Planning* 39:25–36.

Thomas, C.D., and I. Hanski. 1997. Butterfly metapopulations. In Hanski and Gilpin, 359–386.

Thomas, J.A., N.A.D. Bourn, R.T. Clarke, K.E. Stewart, D.J. Simcox, G.S. Pearman, R. Curtis, and B. Goodger. 2001. The quality and isolation of habitat patches both determine where butterflies persist in fragmented landscapes. *Proceedings of the Royal Society of London, B* 268:1791–1796.

Thorp, R.W. 1990. Vernal pool flowers and host-specific bees. In *Vernal pool plants: Their habitat and biology*, ed. D.H. Ikeda and R.A. Schlising, 109–122. Studies from the Herbarium 8. California State University, Chico, CA.

Thrall, P.H., and J.J. Burdon. 1999. The spatial scale of pathogen dispersal: Consequences for disease dynamics and persistence. *Evolutionary Ecology Research* 1:681–701.

Tian, G.J., Z.F. Yang, and Y. Xie. 2007. Detecting spatiotemporal dynamic landscape patterns using remote sensing and the lacunarity index: A case study of Haikou City, China. *Environment and Planning B: Planning and Design* 34:556–569.

Tilman, D. 1994. Competition and biodiversity in spatially structured habitats. *Ecology* 75:2–16.

Tilman, D., and P.M. Kareiva, eds. 1997. *Spatial ecology: The role of space in population dynamics and interspecific interactions*. Princeton University Press, Princeton, NJ.

Tilman, D., C.L. Lehman, and P.M. Kareiva. 1997. Population dynamics in spatial habitats. In Tilman and Kareiva, 3–20.

Tilman, D., R.M. May, C.L. Lehman, and M.A. Nowak. 1994. Habitat destruction and the extinction debt. *Nature* 371:65–66.

Tischendorf, L., D.J. Bender, and L. Fahrig. 2003. Evaluation of patch isolation metrics in mosaic landscapes for specialist vs. generalist dispersers. *Landscape Ecology* 18:41–50.

Tischendorf, L., and L. Fahrig. 2000. On the usage and measurement of landscape connectivity. *Oikos* 90:7–19.

———. 2001. Reply: On the use of connectivity measures in spatial ecology. *Oikos* 95:152–155.

Tischendorf, L., A.A. Grez, T. Zaviezo, and L. Fahrig. 2005. Mechanisms affecting population density in fragmented habitat. *Ecology and Society* 10(1):7. Also available at www.ecologyandsociety.org/vol10/iss1/art7/.

Tomlin, C.D. 1990. *Geographic information systems and cartographic modeling*. 1990. Prentice-Hall, Upper Saddle River, NJ.

Tscharntke, T., and R. Brandl. 2004. Plant-insect interactions in fragmented landscapes. *Annual Review of Ecology and Systematics* 49:405–430.

Tscharntke T., I. Steffan-Dewenter, A. Kruess, and C. Thies. 2002. Contribution of small habitat fragments to conservation of insect communities of grassland-cropland landscapes. *Ecological Applications* 12:354–363.

Turchi, G.M., P.L. Kennedy, D. Urban, and D. Hein. 1995. Bird species richness in relation to isolation of aspen habitats. *Wilson Bulletin* 107:463–474.

Turchin, P. 1998. *Quantitative analysis of movement: Measuring and modeling population redistribution in animals and plants*. Sinauer Associates, Sunderland, MA.

Turner, M.G., ed. 1987. *Landscape heterogeneity and disturbance*. Springer-Verlag, New York, NY.

———. 1989. Landscape ecology: The effect of pattern on process. *Annual Review of Ecology and Systematics* 20:171–197.

———. 2005. Landscape ecology in North America: Past, present, and future. *Ecology* 86:1967–1974.

Turner, M.G., G.J. Arthaud, R.T. Engstrom, S.J. Hejl, J. Liu, S. Loeb, and K. McKelvey. 1995. Usefulness of spatially explicit population models in land management. *Ecological Applications* 5:12–16.

Turner, M.G., and R.H. Gardner, eds. 1991. *Quantitative methods in landscape ecology: The analysis and interpretation of landscape heterogeneity.* Springer-Verlag, New York, NY.

Turner, M.G., R.H. Gardner, V.H. Dale, and R.V. O'Neill. 1989. Predicting the spread of disturbance across heterogeneous landscapes. *Oikos* 55:121–129.

Turner, M.G., and W.H. Romme. 1994. Landscape dynamics in crown fire ecosystems. *Landscape Ecology* 9:59–77.

Turner, M.G., W.H. Romme, and R.H. Gardner. 1994. Landscape disturbance models and the long-term dynamics of natural areas. *Natural Areas Journal* 14:3–11.

Turner, W.R., K. Brandon, T.M. Brooks, R. Costanza, G.A.B. da Fonseca, and R. Portela. 2007. Global conservation of biodiversity and ecosystem services. *BioScience* 57:868–873.

Turton, S.M., and H.J. Freiburger. 1997. Edge and aspect effects on the microclimate of a small tropical forest remnant on the Atherton Tableland, northeast Australia. In Laurance and Bierregaard, 45–54.

Urban, D.L., R.V. O'Neill, and H.H. Shugart Jr. 1987. Landscape ecology. *BioScience* 37:119–127.

U.S. Fish and Wildlife Service. 1984. Valley Elderberry Longhorn Beetle recovery plan. U.S. Fish and Wildlife Service, Portland, OR.

Van Buskirk, J., and R.S. Ostfeld. 1998. Controlling Lyme disease by modifying the density and species composition of tick hosts. *Ecological Applications* 5:1133–1140.

van Nouhuys, S. 2005. Effects of habitat fragmentation at different trophic levels in insect communities. *Annales Zoologici Fennici* 42:433–447.

Van Zandt, P.A., E. Collins, J.B. Losos, and J.M. Chase. 2005. Implications of food web interactions for restoration of Missouri Ozark glade habitats. *Restoration Ecology* 13: 312–317.

Veddeler, D., C.H. Schulze, I. Steffan-Dewenter, D. Buchori, and T. Tscharntke. 2005. The contribution of tropical secondary forest fragments to the conservation of fruit-feeding butterflies: Effects of isolation and age. *Biodiversity and Conservation* 14:3577–3592.

Vellend, M. 2003. Habitat loss inhibits recovery of plant diversity as forests regrow. *Ecology* 84:1158–1164.

———. 2004. Parallel effects of land-use history on species diversity and genetic diversity of forest herbs. *Ecology* 85:3043–3055.

———. 2005. Land-use history and plant performance in populations of *Trillium grandiflorum. Biological Conservation* 124:217–224.

Vellend, M., K. Verheyen, H. Jacquemyn, A. Kolb, H. van Calster, G. Peterken, and M. Hermy. 2006. Extinction debt of forest plants persists for more than a century following habitat fragmentation. *Ecology* 87:542–548.

Verboom, J., R. Alkemade, J. Klijn, M.J. Metzger, and R. Reijnen. 2007. Combining biodiversity modeling with political and economic development scenarios for 25 EU countries. *Ecological Economics* 62:267–276.

Verboom, B., and R. van Apeldoorn. 1990. Effects of habitat fragmentation on the red squirrel, *Sciurus vulgaris* L. *Landscape Ecology* 4:171–176.

Verheyen, K., I. Fastenaekels, M. Vellend, L. De Keersmaeker, and M. Hermy. 2006. Landscape factors and regional differences in recovery rates of herb layer richness in Flanders (Belgium). *Landscape Ecology* 21:1109–1118.

Verner, J., M.L. Morrison, and C.J. Ralph, eds. 1986. *Wildlife 2000: Modeling habitat relationships of terrestrial vertebrates.* University of Wisconsin Press, Madison, WI.

Vial, F., S. Cleaveland, G. Rasmussen, and D.T. Haydon. 2006. Development of vaccination strategies for the management of rabies in African wild dogs. *Biological Conservation* 131:180–192.

Villafuerte, R., J.A. Litvaitis, and D.F. Smith. 1997. Physiological responses by lagomorphs to resource limitations imposed by habitat fragmentation: Implications for condition-sensitive predation. *Canadian Journal of Zoology* 75:148–151.

Villard, M.A., M.K. Trzcinski, and G. Merriam. 1999. Fragmentation effects on forest birds: Relative influence of woodland cover and configuration on landscape occupancy. *Conservation Biology* 13:774–783.

Viña, A., S. Bearer, X. Chen, G. He, M. Linderman, L. An, H. Zhang, Z. Ouyang, and J. Liu. 2007. Temporal changes in giant panda habitat connectivity across boundaries of Wolong Nature Reserve, China. *Ecological Applications* 17:1019–1030.

Virolainen, K.M., T. Suomi, J. Suhonen, and M. Kuitunen. 1998. Conservation of vascular plants in single large and several small mires: Species richness, rarity and taxonomic diversity. *Journal of Applied Ecology* 35:700–707.

Voss, J.D., and L.L. Richardson. 2006. Nutrient enrichment enhances black band disease progression in corals. *Coral Reefs* 25:569–576.

Vuilleumier, S., and N. Perrin. 2006. Effects of cognitive abilities on metapopulation connectivity. *Oikos* 113:139–147.

Wales, B.A. 1972. Vegetation analysis of north and south edges in a mature oak-hickory forest. *Ecological Monographs* 42:451–471.

Wallin, D.O., F.J. Swanson, and B. Marks. 1994. Landscape pattern response to changes in pattern generation rules: Land-use legacies in forestry. *Ecological Applications* 4:569–580.

Walmsley, A. 1995. Greenways and the making of urban form. *Landscape and Urban Planning* 33:81–127.

Walters, J.R., H.A. Ford, and C.B. Cooper. 1999. The ecological basis of sensitivity of brown treecreepers to habitat fragmentation: A preliminary assessment. *Biological Conservation* 90:13–20.

Wang, B.C., V.L. Sork, M.T. Leong, and T.B. Smith. 2007. Hunting of mammals reduces seed removal and dispersal of the Afrotropical tree *Antrocaryon klaineanum* (Anacardiaceae). *Biotropica* 39:340–347.

Warner, R.R., and P.L. Chesson. 1985. Coexistence mediated by recruitment fluctuations: A field guide to the storage effect. *American Naturalist* 125:769–787.

Watson, J.E.M., R.J. Whittaker, and T.P. Dawson. 2004. Avifaunal responses to habitat fragmentation in the threatened littoral forests of south-eastern Madagascar. *Journal of Biogeography* 31:1791–1807.

Watt, A.S. 1947. Pattern and process in the plant community. *Journal of Ecology* 35:1–22.

Watts, C.H., and R.K. Didham. 2006a. Influences of habitat isolation on invertebrate colonization of *Sporadanthus ferrugineus* in a mined peat bog. *Restoration Ecology* 14:412–419.

———. 2006b. Rapid recovery of an insect-plant interaction following habitat loss and experimental wetland restoration. *Oecologia* 148:61–69.

Wayne, A.F., A. Cowling, D.B. Lindenmayer, C.G. Ward, C.V. Vellios, C.F. Donnelly, and M.C. Calvey. 2006. The abundance of a threatened arboreal marsupial in relation to anthropogenic disturbances at local and landscape scales in Mediterranean-type forests in south-western Australia. *Biological Conservation* 127:463–476.

Wear, D.N., M.G. Turner, and R.J. Naiman. 1998. Land cover along an urban-rural gradient: Implications for water quality. *Ecological Applications* 8:619–630.

Webb, N.R., and P.J. Hopkins. 1984. Invertebrate diversity on fragmented *Calluna* heathland. *Journal of Applied Ecology* 21:921–933.

Webb, N.R., and J.A. Thomas. 1994. Conserving insect habitats in heathland biotopes: A question of scale. In *Large-scale ecology and conservation biology*, ed. P.J. Edwards, R.M. May, and N.R. Webb, 129–151. Blackwell Scientific, Oxford, UK.

Webb, N.R., and A.H. Vermaat. 1990. Changes in vegetational diversity on remnant heathland fragments. *Biological Conservation* 53:253–264.

Wegner, J.F., and G. Merriam. 1979. Movements by birds and small mammals between a wood and adjoining farmland habitats. *Journal of Applied Ecology* 16:349–357.

Weiher, E., and P. Keddy, eds. 1999. *Ecological assembly rules: Perspectives, advances, retreats.* Cambridge University Press, Cambridge, UK.

Weldon, A. 2006. How corridors reduce Indigo Bunting nest success. *Conservation Biology* 20:1300–1305.

Weng, Q.H. 2002. Land use change analysis in the Zhujiang Delta of China using satellite remote sensing, GIS and stochastic modeling. *Journal of Environmental Management* 64:273–284.

Whitcomb, R.R., C.S. Robbins, J.F. Lynch, B.L. Whitcomb, M.K. Klimkiewicz, and D. Bystrak. 1981. Effects of forest fragmentation on avifauna of the eastern deciduous forest. In Burgess and Sharpe, 125–205.

White, D., P.G. Minotti, M.J. Barczak, J.C. Sifneos, K.E. Freemark, M.V. Santelmann, C.F. Steinitz, A.R. Kiester, and E.M. Preston. 1997. Assessing risks to biodiversity from future landscape change. *Conservation Biology* 11:349–360.

White, P.S. 1996. Spatial and biological scales in reintroduction. In *Restoring diversity: Strategies for reintroduction of endangered plants*, ed. D.A. Falk, C.I. Millar, and M. Olwell, 49–86. Island Press, Washington, DC.

Wiens, J.A. 1976. Population responses to patchy environments. *Annual Review of Ecology and Systematics* 7:81–120.

———. 1989. Spatial scaling in ecology. *Functional Ecology* 3:385–397.

———. 1996. Wildlife in patchy environments: Metapopulations, mosaics and management. In McCullough 1996, 53–84.

———. 1997. The emerging role of patchiness in conservation biology. In *The ecological basis of conservation: Heterogeneity, ecosystems, and biodiversity*, ed. S.T.A. Pickett, R.S. Ostfeld, M. Shachak, and G.E. Likens, 93–107. Chapman and Hall, New York, NY.

Wiens, J.A., C.S. Crawford, and J.R. Gosz. 1985. Boundary dynamics: A conceptual framework for studying landscape ecosystems. *Oikos* 45:421–427.

Wiens, J.A., M.R. Moss, M.G. Turner, and D. Mladenoff, eds. 2006. *Foundation papers in landscape ecology.* Columbia University Press, New York, NY.

Wiens, J.A., N.C. Stenseth, B. Van Horne, and R.A. Ims. 1993. Ecological mechanisms and landscape ecology. *Oikos* 66:369–380.

Wiersma, Y.F., and D.L. Urban. 2005. Beta diversity and nature reserve system design in the Yukon, Canada. *Conservation Biology* 19:1262–1272.

Wilcove, D.S. 1985. Nest predation in forest tracts and the decline of migratory songbirds. *Ecology* 66:1211–1214.

Wilcove, D.S., C.H. McLellan, and A.P. Dobson. 1986. Habitat fragmentation in the temperate zone. In Soulé 1986, 237–256.

Wilcove, D.S., D. Rothstein, J. Dubow, A. Phillips, and E. Losos. 1998. Quantifying threats to imperiled species in the United States. *BioScience* 48:607–615.

Wilcox, B.A. 1980. Insular ecology and conservation. In Soulé and Wilcox 1980, 95–117.

Wilcox, C., B.J. Cairns, and H.P. Possingham. 2006. The role of habitat disturbance and recovery in metapopulation persistence. *Ecology* 87:855–863.

Williams, J.C., C.S. ReVelle, and S.A. Levin. 2005. Spatial attributes and reserve design models: A review. *Environmental Modeling and Assessment* 10:163–181.

Williams, J.C., and S.A. Snyder. 2005. Restoring habitat corridors in fragmented landscapes using optimization and percolation models. *Environmental Modeling and Assessment* 10:239–250.

Williams, N.M., and C. Kremen. 2007. Floral resource distribution among habitats determines productivity of a solitary bee, *Osmia lignaria*, in a mosaic agricultural landscape. *Ecological Applications* 17:910–921.

Williams, N.S.G., M.J. McDonnell, and E.J. Seager. 2005. Factors influencing the loss of an endangered ecosystem in an urbanizing landscape: A case study of native grasslands from Melbourne, Australia. *Landscape and Urban Planning* 71:35–49.

Williamson, M. 1981. *Island populations.* Oxford University Press, New York, NY.

Willis, E.O. 1974. Populations and local extinctions of birds on Barro Colorado Island, Panama. *Ecological Monographs* 44:153–169.

Wilson, K.A., M.F. McBride, M. Bode, and H.P. Possingham. 2006. Prioritizing global conservation efforts. *Nature* 440:337–340.

Wilson, E.O. 1992. *The diversity of life.* Harvard University Press, Cambridge, MA.

Wilson, E.O., and E.O. Willis. 1975. Applied biogeography. In *Ecological structure of species communities*, ed. M.L. Cody and J.M. Diamond, 522–534. Harvard University Press, Cambridge, MA.

Winter, M., and J. Faaborg. 1999. Patterns of area-sensitivity in grassland-nesting birds. *Conservation Biology* 1:1424–1436.

Wintle, B.A., S.A. Bekessy, L.A. Venier, J.L. Pearce, and R.A. Chisholm. 2005. Utility of dynamic-landscape metapopulation models for sustainable forest management. *Conservation Biology* 19:1930–1943.

Wirsing, A.J., T.D. Steury, and D.L. Murray. 2002. A demographic analysis of a southern snowshoe hare population in a fragmented habitat: Evaluating the refugium model. *Canadian Journal of Zoology* 80:169–177.

With, K.A. 1994. Using fractal analysis to assess how species perceive landscape structure. *Landscape Ecology* 9:25–36.

———. 1997. The application of neutral landscape models in conservation biology. *Conservation Biology* 11:1069–1080.

———. 2002. The landscape ecology of invasive spread. *Conservation Biology* 16:1192–1203.

———. 2004. Assessing the risk of invasive spread in fragmented landscapes. *Risk Analysis* 24:803–815.

With, K.A., and T.O. Crist. 1995. Critical thresholds in species' responses to landscape structure. *Ecology* 76:2446–2459.

With, K.A., and A.W King. 1999. Extinction thresholds for species in fractal landscapes. *Conservation Biology* 13:314–326.

———. 2001. Analysis of landscape sources and sinks: The effect of spatial pattern on avian demography. *Biological Conservation* 100:75–88.

With, K.A., G.R. Schrott, and A.W. King. 2006. The implications of metalandscape connectivity for population viability in migratory songbirds. *Landscape Ecology* 21:157–167.

Wolf, A.T. 2001. Conservation of endemic plants in serpentine landscapes. *Biological Conservation* 100:35–44.

Wolf, A.T., and S.P. Harrison. 2001. Effects of habitat size and patch isolation on reproductive success of the serpentine morning glory. *Conservation Biology* 15:111–121.

Woodroffe, R. 2001. Assessing the risks of intervention: Immobilization, radio-collaring and vaccination of African wild dogs. *Oryx* 35:234–244.

Woodroffe, R., C.A. Donnelly, W.T. Johnston, F.J. Bourne, C.L. Cheeseman, R.S. Clifton-Hadley, D.R. Cox, G. Gettinby, R.G. Hewinson, A.M. Le Fevre, J.P. McInerney, and W.I. Morrison. 2005. Spatial association of *Mycobacterium bovis* infection in cattle and badgers (*Meles meles*). *Journal of Applied Ecology* 42:852–862.

Woodroffe, R., and J.R. Ginsberg. 1998. Edge effects and the extinction of populations inside protected areas. *Science* 280:2126–2128.

Wright, S.J. 1981. Intra-archipelago vertebrate distributions: The slope of the species-area relation. *American Naturalist* 118:726–748.

Xia, L., and O. Yang. 2007. Traffic disturbance to migration of Tibetan antelopes along Quinghai–Tibet railway and highway. Society for Conservation Biology, 21st Annual Meeting, Port Elizabeth, South Africa, July 1–5. Oral presentation abstracts, 60.

Yahner, R.H. 1988. Changes in wildlife communities near edges. *Conservation Biology* 2:333–339.

Young, A., and N. Mitchell. 1994. Microclimate and vegetation edge effects in a fragmented podocarp-broadleaf forest in New Zealand. *Biological Conservation* 67:63–72.

Young, T.P., J.M. Chase, and R.T. Huddleston. 2001. Community succession and assembly: Comparing, contrasting, and combining paradigms in the context of ecological restoration. *Ecological Restoration* 19:5–18.

Young, T.P., D.A. Petersen, and J.J. Clary. 2005. The ecology of restoration: Historical links, emerging issues, and unexplored realms. *Ecology Letters* 8:662–673.

Yu, X.J., and C.N. Ng. 2007. Spatial and temporal dynamics of urban sprawl along two urban-rural transects: A case study of Guangzhou, China. *Landscape and Urban Planning* 79:96–109.

Zartman, C.E. 2003. Habitat fragmentation impacts on epiphyllous bryophyte communities in central Amazonia. *Ecology* 84:948–954.

Zhu, H., Z.F. Xu, H. Wang, and B.G. Li. 2004. Tropical rain forest fragmentation and its ecological and species diversity changes in southern Yunnan. *Biodiversity and Conservation* 13:1355–1372.

Zollner, P.A., and S.L. Lima. 2005. Behavioral tradeoffs when dispersing across a patchy landscape. *Oikos* 108:219–230.

Zube, E.H. 1986. The advance of ecology. *Landscape Architecture* 76:58–67.

Zube, E.H. 1995. Greenways and the US national park system. *Landscape and Urban Planning* 33:17–25.

Index